Student's Aid to Gross Anatomy

By the Same Author

Clinical Anatomy for Medical Students
Atlas of Clinical Anatomy
Gross Anatomy Dissector
Clinical Neuroanatomy for Medical Students
Clinical Embryology for Medical Students
An Atlas of Normal Radiographic Anatomy
(with Alvin C. Wyman)
Functional and Clinical Histology for Medical Students

Student's Aid to Gross Anatomy

Richard S. Snell, M.D., Ph.D.
Professor and Chairman
Department of Anatomy
George Washington School of Medicine
and Health Sciences
Washington, D.C.

ACC APPLETON-CENTURY-CROFTS/Norwalk, Connecticut

0-8385-8687-2

86 87 88 89 90 / 10 9 8 7 6 5 4 3 2 1

Prentice-Hall of Australia, Pty. Ltd., Sydney
Prentice-Hall Canada, Inc.
Prentice-Hall Hispanoamericana, S.A., Mexico
Prentice-Hall of India Private Limited, New Delhi
Prentice-Hall International (UK) Limited, London
Prentice-Hall of Japan, Inc., Tokyo
Prentice-Hall of Southeast Asia (Pte.) Ltd., Singapore
Whitehall Books Ltd., Wellington, New Zealand
Editora Prentice-Hall do Brasil Ltda., Rio de Janeiro

Library of Congress Cataloging-in-Publication Data
Snell, Richard S.
Student's aid to gross anatomy.

Includes index.
1. Anatomy, Human. I. Title. [DNLM: 1. Anatomy—outlines. QS 18 S671s]
QM23.2.S56 1986 611 86-8044
ISBN 0-8385-8687-2

Design: Lynn Luchetti
Cover: M. Chandler Martylewski

PRINTED IN THE UNITED STATES OF AMERICA

For the student who is reviewing for examinations

Contents

Preface

This book has been written as an abstract of the larger textbooks which cover human gross anatomy system by system. It is designed to meet the needs of medical and dental students who are reviewing for examinations and need a handy source of essential material.

Scattered throughout the book are simple line drawings that are keyed to the text. This book attempts to omit the nonessential and includes information that is particularly pertinent to clinicians. The terminology, with few exceptions, is an anglicized version of that used in the internationally recognized *Nomina Anatomica.*

I thank the many students, clinical colleagues and friends who have made valuable suggestions regarding the content of this book. I wish to express my sincere thanks to Ira Grunther, B.S., for his excellent artwork. Special appreciation is due to Barbara Chambers and Betty Hodge for their skill and patience in typing the manuscript.

Finally to the staff of Appleton-Century-Crofts go my gratitude and appreciation for their assistance throughout the preparation of this book.

R.S.S.

Student's Aid to Gross Anatomy

1

Bones

The skeleton is arranged in two main divisions, the axial skeleton and the appendicular skeleton (*Fig. 1–1*).

The **axial skeleton** is made up of the bones that form the main axis of the support of the body, namely, the skull, the hyoid bone, the vertebral column, the sternum, and the ribs, and the **appendicular skeleton** is made up of bones of the upper and lower limbs that are attached as appendages to the axial skeleton. The appendicular skeleton includes the girdles, which connect the bones of the limbs to the axial skeleton. As you are reading the description of each bone it would be helpful to have a set of bones to study and if possible to have access to an articulated skeleton.

AXIAL SKELETON

Skull

The skull rests on the superior end of the vertebral column and is composed of a number of separate bones.

The bones of the skull may be divided into those of the **cranium** or brain case and those of the face. The vault is the upper part of the cranium, and the **base of the skull** is the lowest part of the cranium. The skull bones are made up of **external** and **internal**

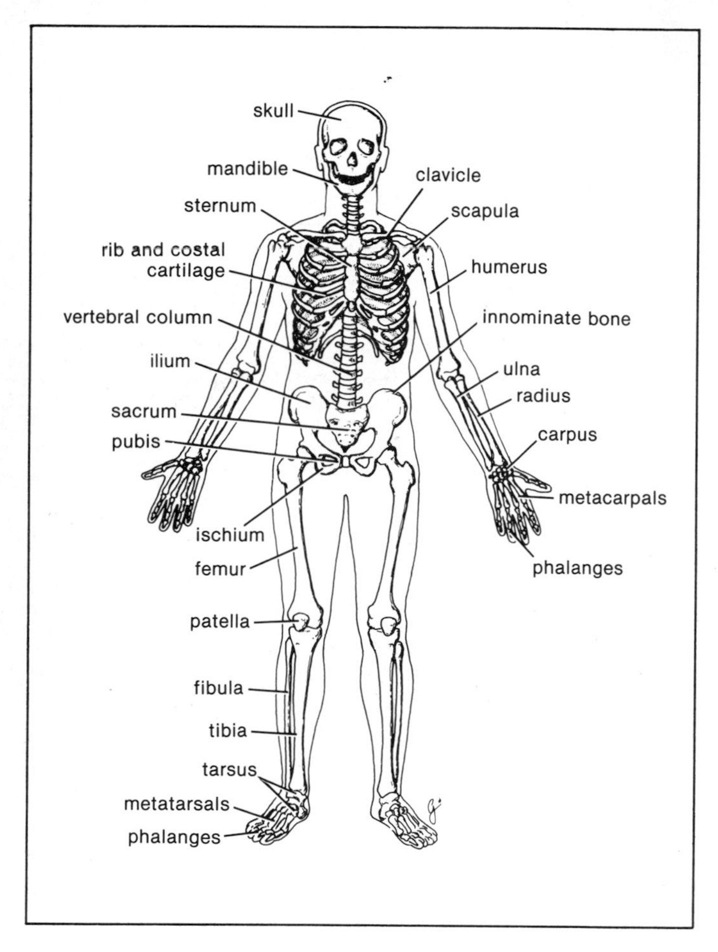

Figure 1–1. Anterior aspect of skeleton.

tables of compact bone, separated by a layer of spongy bone, called the **diploë**. The bones are covered on the outer and inner surfaces with periosteum.

The bones of the skull, with the exception of the mandible, are united at immobile joints called **sutures**. A suture, which means a seam or stitch, possesses a small amount of connective tissue between the bones; the connective tissue is known as the **sutural ligament**. Three important skull sutures should be identified. The **coronal suture,** which lies between the frontal bones and the parietal bones, the **lambdoid suture** between the parietal bones and the occipital bone, and the **sagittal suture,** which lies between the two parietal bones (*Fig. 1–2*).

The mandible articulates with the remainder of the skull at the mobile temporomandibular joint (*Fig. 1–2*).

Fontanelles. In the early embryo the vault of the skull consists of a fibrous membrane. As development proceeds, islands of bone are formed in the membrane that slowly grow toward one another. At birth areas of membrane still remain between the bones and these soft areas are known as the **fontanelles**. At childbirth the membranous parts of the skull allow it to be compressed and permit the skull bones to override one another. This mechanism greatly reduces the size of the baby's head and facilitates the passage of the head through the female genital tract.

The two most important fontanelles are called the anterior and posterior fontanelles. The **anterior fontanelle** is diamond-shaped and lies between the two halves of the frontal bone and the two parietal bones. The anterior fontanelle is replaced by bone and is closed by 18 months of age. The **posterior fontanelle** is triangular in shape and lies between the two parietal bones and the occipital bone. The posterior fontanelle is usually closed by the end of the first year.

Anterior View of the Skull. The frontal bone or forehead bone curves downward to make the upper margins of the orbits (*Fig. 1–2*). The **superciliary arches** can be seen on either side.

The **orbital margins** are bounded by the frontal bone superi-

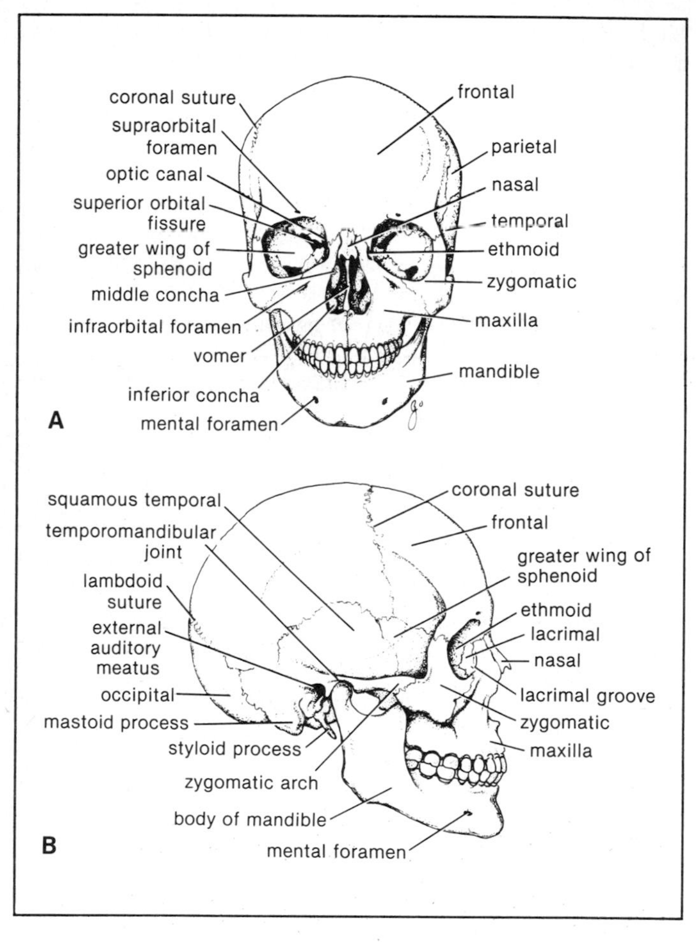

Figure 1–2. **A.** Bones of the anterior aspect of the skull. **B.** Bones of the lateral aspect of the skull.

orly, the zygomatic bone laterally, the maxilla inferiorly, and the processes of the maxilla and the frontal bone medially.

Within the **frontal bone**, just above the orbital margins, are two hollow spaces lined with mucous membrane, called the **frontal air sinuses**. These communicate with the nasal cavity and serve as voice resonators.

The two **nasal bones** form the bridge of the nose. Their lower borders, with the maxillae, make the **anterior nasal aperture**. The nasal cavity is divided into two by the bony nasal septum, which is largely formed by the **vomer**. The **superior and middle conchae** are shelves of bone that jut into the nasal cavity from the **ethmoid** on each side; the **inferior conchae** are separate bones.

The two **maxillae** form the upper jaw, the anterior part of the hard palate, part of the lateral walls of the nasal cavities, and part of the floors of the orbital cavities. The two bones meet in the midline at the **intermaxillary suture** and form the lower margin of the nasal aperture. Below the orbit the maxilla is perforated by the **infraorbital foramen**. The alveolar process projects downward and together with the opposite side forms the **alveolar arch**, which carries the upper teeth. Within each maxilla is a large pyramid-shaped cavity lined with mucous membrane, called the **maxillary sinus**. This communicates with the nasal cavity and serves as a voice resonator.

The **zygomatic bone** forms the prominence of the cheek and part of the lateral wall and floor of the orbital cavity. Medially, it articulates with the maxilla, and laterally, it articulates with the zygomatic process of the temporal bone to form the **zygomatic arch**.

Paranasal Sinuses. The paranasal sinuses are cavities within certain bones of the skull that communicate with the nasal cavity. The sinuses are lined with mucous membrane. The function of the cavities is to lighten the skull and provide the voice with resonators. The latter function can be understood when one realizes that the columns of air within the larynx, the pharynx, and the nasal cavities are in free communication with one another. Persons whose sinuses become filled with infected secretions, as in sinusitis following a severe cold, have an altered quality to their voices that are less resonant.

The paranasal sinuses are the **frontal sinuses** in the frontal bone, the **maxillary sinus** in each of the maxillary bones, the **ethmoid sinuses** in the ethmoid bones, and the **sphenoid sinuses** in the sphenoid bone. For further details concerning the sinuses, *see page 348.*

Lateral View of the Skull. The **frontal bone** forms the anterior part of the side of the skull and articulates with the parietal bone at the **coronal suture** (*Fig. 1–2*).

The **parietal bones** form the sides and roof of the cranium and articulate with each other in the midline at the **sagittal suture.** They articulate with the occipital bone behind, at the **lambdoid suture.**

The skull is completed from the side by the squamous part of the **occipital bone**; parts of the **temporal bone**, namely, the **squamous, tympanic, mastoid process, styloid process,** and **zygomatic process**; and the **greater wing of the sphenoid**. Note the position of the **external auditory meatus.** The ramus and the body of the mandible lie inferiorly.

Note that the thinnest part of the lateral wall of the skull is where the anteroinferior corner of the parietal bone articulates with the greater wing of the sphenoid; this point is referred to as the **pterion.**

Clinically, the pterion is a very important area, because it overlies the anterior division of the **middle meningeal artery** and **vein** (*see p. 184*).

Inferior View of the Skull. If the mandible is discarded, the anterior part of this aspect of the skull is seen to be formed by the **hard palate**. The **palatal processes of the maxillae** and the **horizontal plates of the palatine bones** can be identified.

Above the posterior edge of the hard palate are the **choanae** (posterior nasal apertures). These are separated from each other by the posterior margin of the **vomer** and are bounded laterally by the **medial pterygoid plates** of the sphenoid bone. The inferior end of the medial pterygoid plate is prolonged as a curved spike of bone, the **pterygoid hamulus**.

The following bony structures should be identified. The **lateral pterygoid plate**, the **infratemporal fossa**, the **tuberosity of the maxilla**, the **petrous part of the temporal bone**, the **styloid process**, the **mastoid process**, the **occipital condyles**, and the **external occipital protruberance**, and the **superior nuchal line**. Identify also the **mandibular fossa** of the **temporal bone** and the **articular tubercle**, which form the upper articular surfaces for the temporomandibular joint.

The following fissures or foramina should also be located because they allow passage of the cranial nerves and other important structures from the skull: the **foramen ovale**, the **foramen spinosum**, the **carotid canal**, the **jugular foramen**, the **foramen lacerum**, the **stylomastoid foramen**, and the **hypoglossal canal** (*Fig. 1–3*). In addition note the important **foramen magnum** and the opening of the bony part of the **auditory tube**.

Superior View of the Base of the Skull. The interior of the base of the skull is conveniently divided up into three cranial fossae or depressions: anterior, middle, and posterior (*Fig. 1–3*). The anterior cranial fossa is separated from the middle cranial fossa by the lesser wing of the sphenoid, and the middle cranial fossa is separated from the posterior cranial fossa by the petrous portion of the temporal bone.

The **anterior cranial fossa** lodges the frontal lobes of the cerebral hemispheres, the lateral parts of the **middle cranial fossa** lodge the temporal lobes of the cerebral hemipheres, and the very deep **posterior cranial fossa** lodges parts of the hindbrain, namely, the cerebellum, the pons, and the medulla oblongata.

Identify the **sphenoid bone**, which occupies the central position in the cranial floor. The sphenoid resembles a bat having a centrally placed **body** with **greater and lesser wings** that are outstretched on each side. The sphenoid bone stabilizes the center of the skull by being attached by sutures to the frontal, parietal, occipital, and ethmoid bones. The sphenoid bone also forms part of the inferior and lateral walls of each orbital cavity. The body

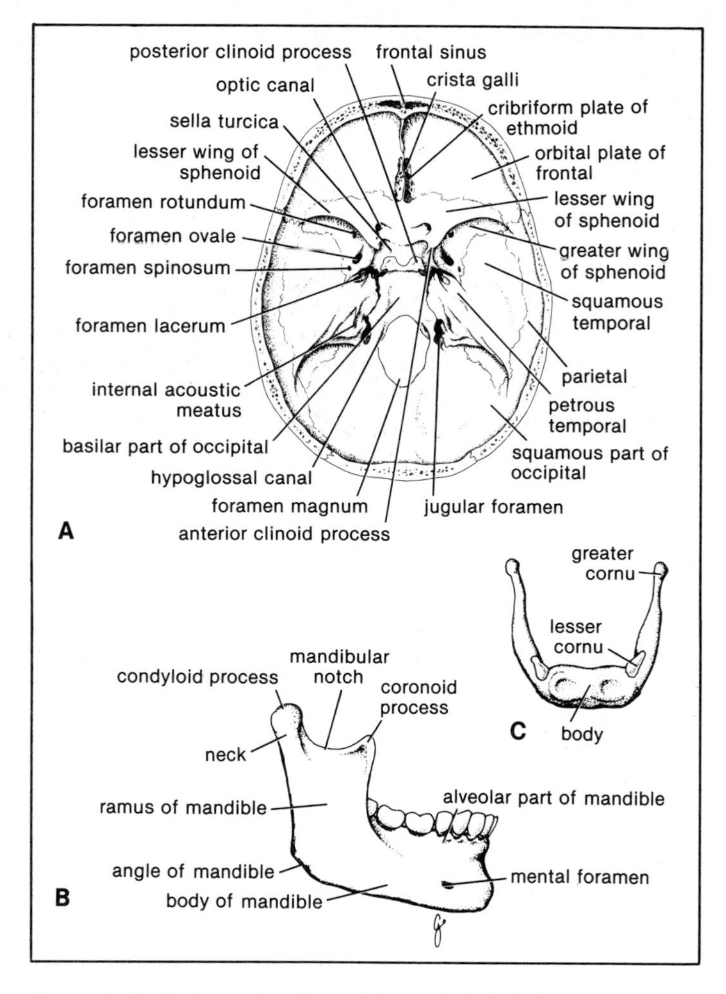

Figure 1–3. **A.** Internal surface of base of skull. **B.** Lateral aspect of right side of mandible. **C.** Anterior aspect of hyoid bone.

of the sphenoid contains the **sphenoid air sinuses**, which are lined with mucous membrane and communicate with the nasal cavity; they serve as voice resonators.

The following foraminae should be identified (*Fig. 1–3*): The **perforations of the cribriform plate of the ethmoid** that transmit the olfactory nerves. The **optic canal** that transmits the optic nerve and the ophthalmic artery. The slit-like **superior orbital fissure** that transmits the oculomotor, trochlear, branches of the ophthalmic division of the trigeminal, and the abducent nerves. The **foramen rotundum** that transmits the maxillary division of the trigeminal nerve. The **foramen ovale** that transmits the mandibular division of the trigeminal nerve. The **foramen spinosum** that transmits the middle meningeal artery. The large irregular **foramen lacerum** that allows the passage of the internal carotid artery from the carotid canal into the cranial cavity.

In the posterior cranial fossa there is the **foramen magnum** that transmits the medulla oblongata. Here the medulla becomes continuous with the spinal cord. The foramen also allows passage of the spinal roots of the accessory nerves and the two vertebral arteries.

The **hypoglossal canal** transmits the hypoglossal nerve and the **jugular foramen** transmits the glossopharyngeal, the vagus, and the accessory nerves. It is here that the sigmoid venous sinus leaves the skull to become the internal jugular vein.

The **internal acoustic meatus** pierces the posterior surface of the petrous part of the temporal bone and transmits the vestibulocochlear nerve and the facial nerve.

Mandible

The mandible or lower jaw is the largest and strongest bone of the face and it articulates with the skull at the **temporomandibular joint**. The mandible consists of a horseshoe-shaped **body** and a pair of **rami** (*Fig. 1–3*). The body of the mandible meets the ramus on each side at the **angle of the mandible**.

The **body of the mandible**, on its external surface in the midline, has a faint ridge, the **symphysis menti** indicating the line of fusion of the two halves during development. The **mental foramen** can be seen below the second premolar tooth; it transmits one of the terminal branches of the trigeminal nerve, the inferior alveolar nerve.

On the medial surface of the body of the mandible can be seen the **submandibular fossa,** which lodges the superficial part of the submandibular salivary gland; this lies below the posterior part of the **mylohyoid line.** The **sublingual fossa,** for the **sublingual gland,** lies above the anterior part of the mylohyoid line.

The upper border of the body of the mandible is called the **alveolar** part; in the adult it contains 16 sockets for the roots of the lower teeth.

The **ramus of the mandible** is vertically placed and has an anterior **coronoid process** and **posterior condyloid process,** or **head,** and the two processes are separated by the **mandibular notch.** Below the condyloid process is a short **neck** (*Fig. 1–3*). On the medial surface of the ramus is the **mandibular foramen,** for the inferior alveolar nerve. The foramen leads into the **mandibular canal,** which opens on the lateral surface of the body of the mandible at the **mental foramen.**

The condyloid process or head of the mandible articulates with the temporal bone at the temporomandibular joint.

Hyoid Bone

The hyoid bone is a single bone found in the midline of the neck below the mandible and above the larynx. It does not articulate with any other bones. The hyoid bone is U-shaped and consists of a body and two greater and two lesser cornua (*Fig. 1–3*). The lesser cornu on each side is attached to the styloid process of the temporal bone of the skull by the **stylohyoid ligament.** The body and the greater cornua are attached to the thyroid cartilage of the larynx by the **thyrohyoid membrane.** The hyoid bone is very mobile and moves upward and downward with the larynx when swallowing occurs. The hyoid bone forms a firm bony base for the tongue and is suspended in position by muscles that connect it to the mandible, to the styloid process of the temporal bone, to the thyroid cartilage, to the sternum, and to the scapula.

Vertebral Column

The vertebral column is the central pillar of the body (*Fig. 1–1*). The head is balanced on the upper end of the pillar and the ribs, the thoracic and abdominal viscera are suspended from the front. The vertebral column serves to protect the spinal cord and supports the weight of the head and the trunk, which it transmits to the hip bones and the lower limbs. It is a flexible structure because it is segmented and made up of irregular bones called **vertebrae**, their joints, and pads of fibrocartilage called **intervertebral discs**. The intervertebral discs form one-quarter of the length of the column.

The vertebrae are arranged in the following groups (*Fig. 1–4*):

- Cervical (7)
- Thoracic (12)
- Lumbar (5)
- Sacral (5 fused to form the sacrum)
- Coccygeal (4, the lower 3 are commonly fused)

Between the vertebrae are spaces called **intervertebral foramina**. These allow the spinal nerves, which leave the spinal cord, to be distributed to the different parts of the body.

The vertebral column when viewed from the side shows four curves (*Fig. 1–4*). In the cervical region the column curves forward producing a posterior concavity; in the thoracic region the column curves backward producing a posterior convexity; and in the lumbar region the column curves forward again producing a posterior concavity. Finally in the sacral region the column curves backward producing a posterior convexity.

During the later months of pregnancy, with the increase in size and weight of the fetus, women tend to increase the posterior lumbar concavity in an attempt to preserve their center of gravity.

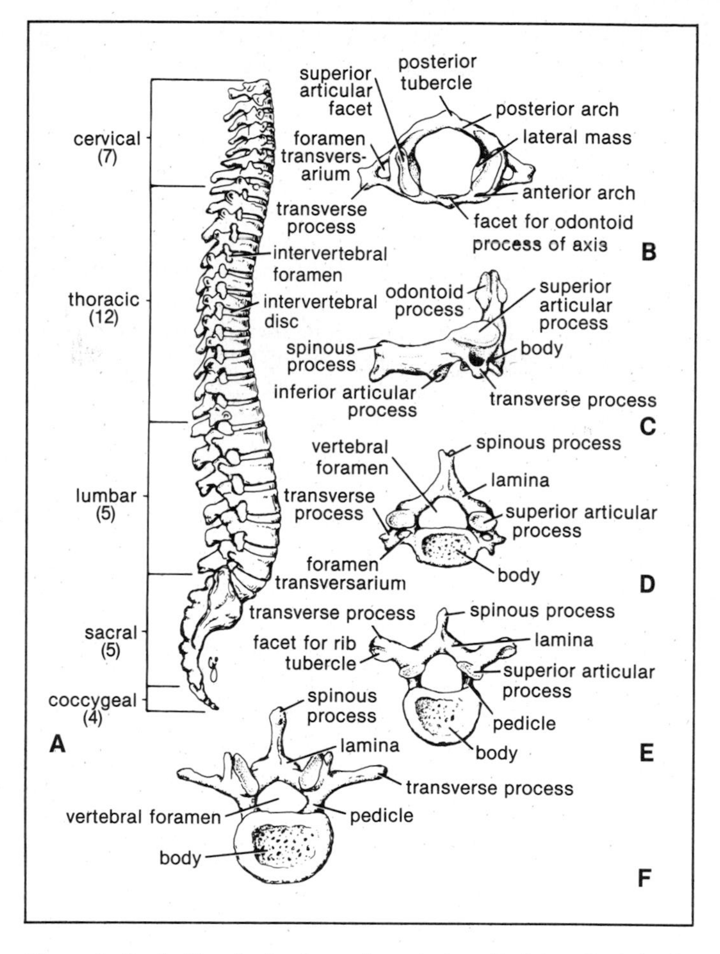

Figure 1–4. **A.** Vertebral column, lateral view. **B.** Atlas. **C.** Axis. **D.** Typical cervical vertebra (fourth). **E.** Typical thoracic vertebra (sixth). **F.** Typical lumbar vertebra (third).

In old age the intervertebral discs atrophy, resulting in a loss of height and the vertebral column tends to show a continuous posterior convexity and the individual has a bent forward appearance (kyphosis).

General Characteristics of a Vertebra. Although vertebrae show regional differences, they all possess a common pattern (*Fig. 1-4*).

A **typical vertebra** consists of a rounded **body** anteriorly and a **vertebral arch** posteriorly. These enclose a space called the **vertebral foramen**, through which run the spinal cord and its coverings. The vertebral arch consists of a pair of cylindrical **pedicles**, which form the sides of the arch, and a pair of flattened **laminae**, which complete the arch posteriorly. The vertebral arch gives rise to seven processes, one spinous, two transverse, and four articular (*Fig. 1-4*).

The **spinous process**, or **spine** is directed posteriorly from the junction of the two laminae. The **transverse processes** are directed laterally from the junction of the laminae and the pedicles. Both the spinous and transverse processes serve as levers and receive attachments of muscles and ligaments.

The **articular processes** are vertically arranged and consist of two superior and two inferior processes. They arise from the junction of the laminae and the pedicles, and their articular surfaces are covered with hyaline cartilage. The two superior articular processes of one vertebral arch articulate with the two inferior articular processes of the arch above, forming two synovial joints.

The pedicles are notched on their upper and lower borders, forming the **superior and inferior vertebral notches**. The superior notch of one vertebra and the inferior notch of an adjacent vertebra together form an **intervertebral foramen**. These foramina in an articulated skeleton serve to transmit the spinal nerves and the blood vessels.

Cervical Part of the Vertebral Column. The cervical part of the vertebral column shows a forward convexity and is made up of seven vertebrae. A **typical cervical vertebra** has the following characteristics (*Fig. 1–4*). Each transverse process possesses a **foramen transversarium** for the passage of the vertebral artery and veins. The spines are small and bifid. The body is small and broader from side to side than from front to back; there are small synovial joints on each side. The vertebral foramen is large and triangular in shape. The superior articular processes have small, flat articular facets that face backward and upward; the inferior articular processes have facets that face downward and forward.

The first, second, and seventh cervical vertebrae are atypical. The **first cervical vertebra**, or **atlas**, has no body and no spinous process (*Fig. 1–4*). It is merely a ring of bone consisting of anterior and posterior arches and a lateral mass on each side. Each lateral mass has articular surfaces on its upper and lower aspects. The bone articulates above with the occipital condyles, forming the **atlantooccipital joints**, where nodding movements of the head take place. Below, the bone articulates with the axis, forming the **atlantoaxial joints**, where rotation movements of the head take place.

The **second cervical vertebra**, or **axis**, has a peglike **odontoid process**, which projects from the superior surface of the body and represents the body of the atlas that has fused with the axis (*Fig. 1–4*).

The **seventh cervical vertebra**, or **vertebra prominens**, is so named because it has the longest spinous process. The spinous process is not bifid. The **transverse process** is large, but the **foramen transversarium** is small and does not transmit the vertebral artery.

Thoracic Part of the Vertebral Column. The thoracic part of the vertebral column is concave forward and is made up of 12 vertebrae, together with their intervertebral discs. Thoracic vertebrae have the following characteristics (*Fig. 1–4*):

1. The body is medium-sized and heart-shaped.
2. The vertebral foramen is relatively small and circular.
3. The spines are long and inclined downward.

4. Costal facets are present on the sides of the bodies where the heads of the ribs articulate, and on the transverse processes for articulation with the tubercles of the ribs. (T11 and T12 have no facets on the transverse processes.)
5. The superior articular processes bear facets that face backward and laterally, whereas the facets on the inferior articular processes face forward and medially. The inferior articular processes of the twelfth vertebra face laterally, as do those of the lumbar vertebrae.

Lumbar Part of the Vertebral Column. The lumbar part of the vertebral column shows a forward convexity and is made up of five vertebrae, together with their intervertebral discs.

A typical lumbar vertebra has the following characteristics (*Fig. 1–4*). The body of each lumbar vertebra is massive and kidney-shaped. The pedicles are strong and directed backward. The laminae are thick, and the vertebral foramina are triangular in shape. The transverse processes are long and slender. The spinous process is short, flat, and quadrangular in shape and projects directly backward. The articular surfaces of the superior articular processes face medially, and those of the inferior articular processes face laterally.

The lumbar vertebrae have no facets for articulation with ribs and no foramina in the transverse processes. The fifth lumbar vertebra articulates with the base of the sacrum at the **lumbosacral joint.**

The intervertebral discs in the lumbar region are thicker than in other regions of the vertebral column. They are wedge-shaped and are responsible for the normal lordosis (forward convexity) found in the lumbar region.

Sacral Part of the Vertebral Column. The sacral part of the vertebral column shows a forward concavity (*Fig. 1–4*) and is made up of five rudimentary vertebrae that are fused together to form a single wedge-shaped bone, the **sacrum**. The upper border, or base, of the bone articulates with the fifth lumbar vertebra. The narrow inferior border articulates with the coccyx. Laterally, the sacrum articulates with the two innominate, or hip bones, to form the sacroiliac joints (*Fig. 1–1*). The anterior and upper margin of the first sacral vertebra bulges forward as the posterior margin of the pelvic inlet and is known as the **sacral promontory**. This is an important obstetrical landmark used when measuring the size of the pelvis.

The vertebral foramina are present and form the **sacral canal**. The laminae of the fifth sacral vertebra, and sometimes those of the fourth also, fail to meet in the midline, forming the **sacral hiatus**. The **sacral canal** contains part of the cauda equina, filum terminale, and meninges down as far as the lower border of the second sacral vertebra. The lower part of the sacral canal contains the lower sacral and coccygeal nerve roots, the filum terminale, and fibrofatty material.

The anterior and posterior surfaces of the sacrum have four foramina on each side for the passage of the anterior and posterior rami of the upper four sacral spinal nerves.

Coccyx. The coccyx consists of four vertebrae fused together to form a single small triangular bone, which articulates at its base with the lower end of the sacrum (*Fig. 1–4*). The first coccygeal vertebra is commonly not fused, or is incompletely fused with the second vertebra.

Thoracic Bones

The thorax or chest has in its walls a bony and cartilaginous cage formed by the sternum, the costal cartilages, the ribs, and the bodies of the thoracic vertebrae (*Fig. 1–1*). The thoracic cage is cone-shaped having a narrow inlet, superiorly, and a broad outlet, inferiorly. The cage is flattened from front to back. The function of the cage is to participate in the movements of respiration and to protect the underlying thoracic viscera, especially the heart and the lungs, and the upper abdominal viscera, namely, the liver, the spleen, and the stomach.

Sternum. The sternum or breast bone is a flat narrow bone that lies in the midline of the anterior chest wall and may be divided into three parts: (1) manubrium sterni; (2) body of the sternum; and (3) xiphoid process.

The **manubrium** is the triangular upper part of the sternum, and it articulates with the clavicles and the first and upper part of the second costal cartilages on each side (*Fig. 1–1*). The manubrium has a depression on its superior border called the **suprasternal (jugular) notch**.

The **body of the sternum** articulates above with the manubrium by means of a fibrocartilaginous joint, the **manubriosternal joint**. Below, it articulates with the xiphoid process at the **xiphisternal joint**. On each side are notches for articulation with the lower part of the second costal cartilage and the third to the seventh costal cartilages (*Fig. 1–1*).

The **xiphoid process** (*Fig. 1–1*) is the lowest and smallest part of the sternum. It is a plate of hyaline cartilage that becomes ossified at its proximal end in adult life. The xiphoid process has no ribs or costal cartilages attached to it. However, it does provide attachment for some muscles of the anterior abdominal wall.

The sternal angle **(angle of Louis)**, formed by the articulation of the manubrium with the body of the sternum, can be recognized by the presence of a transverse ridge on the anterior aspect of the sternum. The transverse ridge lies at the level of the second costal cartilage, the point from which all costal cartilages and ribs are counted in clinical practice. The sternal angle lies opposite to the intervertebral disc between the fourth and fifth thoracic vertebrae. The xiphisternal joint lies opposite the body of the ninth thoracic vertebra.

Costal Cartilages. Costal cartilages are bars of hyaline cartilage connecting the upper seven ribs to the lateral edge of the sternum, and the eighth, ninth, and tenth ribs to the cartilage immediately above. The cartilages of the eleventh and twelfth ribs end in the abdominal musculature (*Fig. 1–1*).

Ribs. There are 12 pairs of ribs, the majority of which encircle the trunk from the vertebral column behind to the region of the sternum in front. The ribs are attached posteriorly to the thoracic vertebrae (*Fig. 1–5*). The upper seven pairs are attached anteriorly to the sternum by their costal cartilages. The eighth, ninth, and tenth pairs of ribs are attached anteriorly to each other and to the seventh rib by means of their costal cartilages. The eleventh and twelfth pairs have no anterior attachment and are referred to as **floating ribs**.

A **typical rib** is a long, twisted flat bone having a rounded, smooth superior border and a sharp, thin inferior border (*Fig. 1–5*). The inferior border overhangs and forms the **costal groove**, which accommodates the intercostal vessels and nerves.

A rib has a **head, neck, tubercle, angle,** and **shaft**. The **head** has two facets for articulation with the numerically corresponding vertebral body and that of the vertebra immediately above. The **neck** is a constricted portion situated between the head and the tubercle. The **tubercle** has a facet for articulation with the transverse process of the numerically corresponding vertebra. The **angle** is where the rib bends sharply forward. The **shaft** of each rib is attached to the corresponding costal cartilage.

The first rib is **atypical**. It is important because of its close relationship to the lower nerves of the brachial plexus and the main vessels to the arm, namely, the subclavian artery and vein that cross the rib. This rib is flattened from above downward. It has a tubercle on the inner border known as the **scalene tubercle**, for the insertion of the scalenus anterior muscle.

APPENDICULAR SKELETON

Bones of the Upper Limb

The bones of the upper limbs consist of the shoulder girdle, the arm, the forearm, and the hand.

Bones of the Shoulder Girdle

The bones of the shoulder girdle consist of the clavicle and the scapula, which articulate with one another at the acromioclavicular joint (*Fig. 1–1*). The shoulder girdle has no articulation with

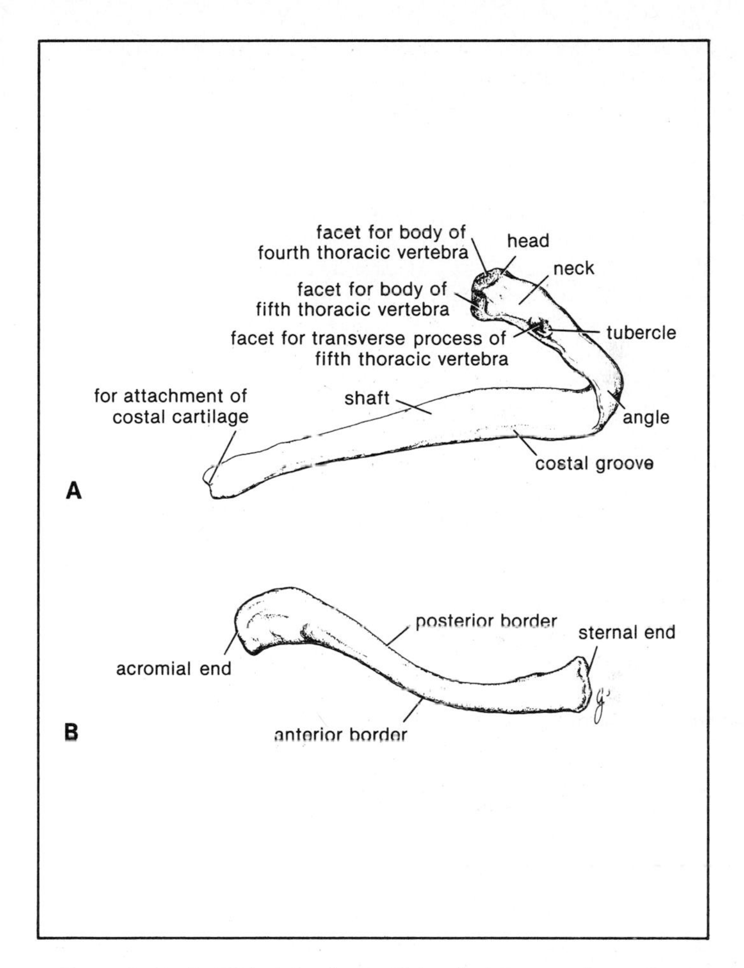

Figure 1–5. **A.** Fifth right rib. **B.** Right clavicle, superior surface.

the vertebral column but articulates with the sternum at the sternoclavicular joints.

Clavicle

The clavicle or collar bone is a long slender S-shaped bone (*Fig. 1–5*) that lies horizontally and articulates with the sternum and first costal cartilage medially, and with the acromion process of the scapula laterally. The clavicle acts as a strut, which holds the upper limb away from the trunk. It also serves to transmit forces from the upper limb to the axial skeleton, and it provides attachment for muscles. The clavicle lies just beneath the skin throughout its length, its medial two-thirds being convex forward, and its lateral third, concave forward.

Scapula

The scapula or shoulder blade is a flat triangular bone (*Fig. 1–6*), that lies on the posterior thoracic wall between the second and seventh ribs. On its posterior surface the **spine of the scapula** projects backward. The lateral end of the spine is free and forms the **acromion**, which articulates with the clavicle. The superolateral angle of the scapula forms the pear-shaped **glenoid cavity**, or **fossa**, which articulates with the head of the humerus at the shoulder joint. The **coracoid process** projects upward and forward above the glenoid cavity and provides attachment for muscles and ligaments. Medial to the base of the coracoid process is the **suprascapular notch**. The anterior surface of the scapula is concave and forms the shallow **subscapular fossa**. The posterior surface of the scapula is divided by the spine into the **supraspinous fossa** above and an **infraspinous fossa** below (*Fig. 1–6*).

Bones of the Arm

Humerus

The humerus is the longest and largest bone of the upper limbs. It articulates with the scapula at the shoulder joint and the radius and the ulna at the elbow joint. The upper end of the humerus

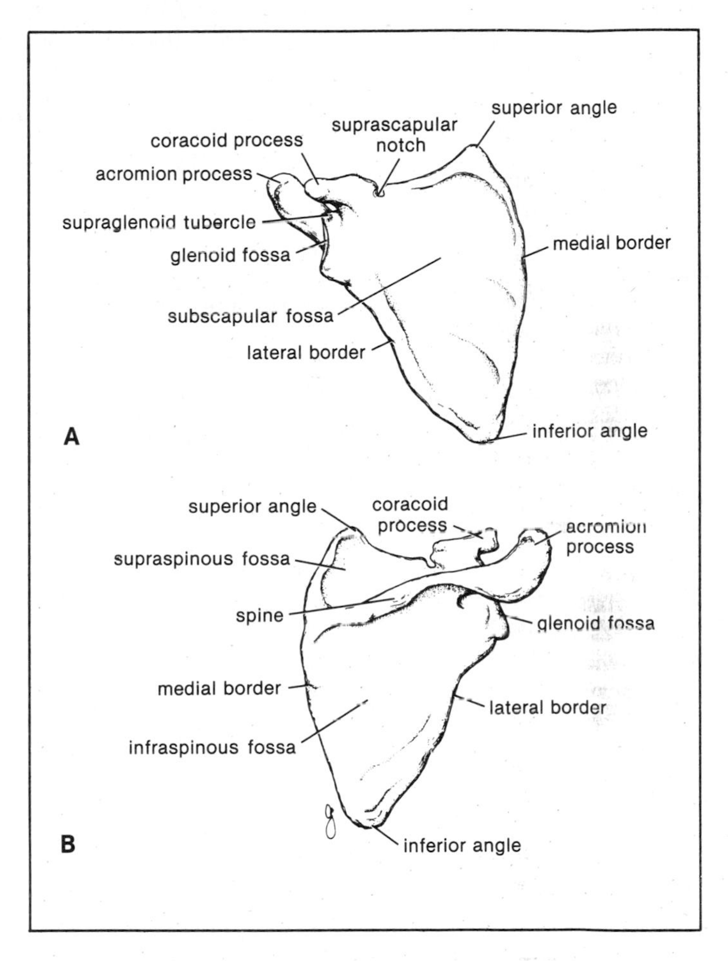

Figure 1–6. Right scapula. **A.** Anterior view. **B.** Posterior view.

has a **head** (*Fig. 1–7*), which forms about one-third of a sphere and articulates with the glenoid cavity of the scapula. Immediately below the head is the **anatomical neck**. Below the neck are the **greater and lesser tuberosities**, separated from each other by the **bicipital groove**. Where the upper end of the humerus joins the shaft, there is a narrow region that is frequently fractured and is called the **surgical neck**. About halfway down the lateral aspect of the shaft, there is a roughened elevation, called the **deltoid tuberosity** that receives the insertion of the deltoid muscle. Behind and below the tuberosity is a **spiral groove**, which accommodates the radial nerve.

The lower end of the humerus possesses the **medial** and **lateral epicondyles** for the attachment of muscles and ligaments, the rounded **capitulum** for articulation with the head of the radius, and the pulley-shaped **trochlea** for articulation with the trochlear notch of the ulna (*Fig. 1–7*). Above the capitulum is the **radial fossa**, which receives the head of the radius when the elbow is flexed. Above the trochlea, anteriorly, is the **coronoid fossa**, which during the same movement receives the coronoid process of the ulna. Above the trochlea, posteriorly, is the **olecranon fossa**, which receives the olecranon process of the ulna when the elbow joint is extended.

Bones of the Forearm

The bones of the forearm are the radius and the ulna.

Radius

The radius is the lateral bone (thumbside) of the forearm. Its upper end articulates with the humerus at the elbow joint and with the ulna at the superior radioulnar joint. Its lower end articulates with the scaphoid and lunate bones of the carpus at the wrist joint, and with the ulna at the inferior radioulnar joint.

At the **upper end** of the radius is the small **circular head** (*Fig. 1–7*). The upper surface of the head is concave and articulates with the convex capitulum of the humerus. The circumference of the head articulates with the radial notch of the ulna. Below the

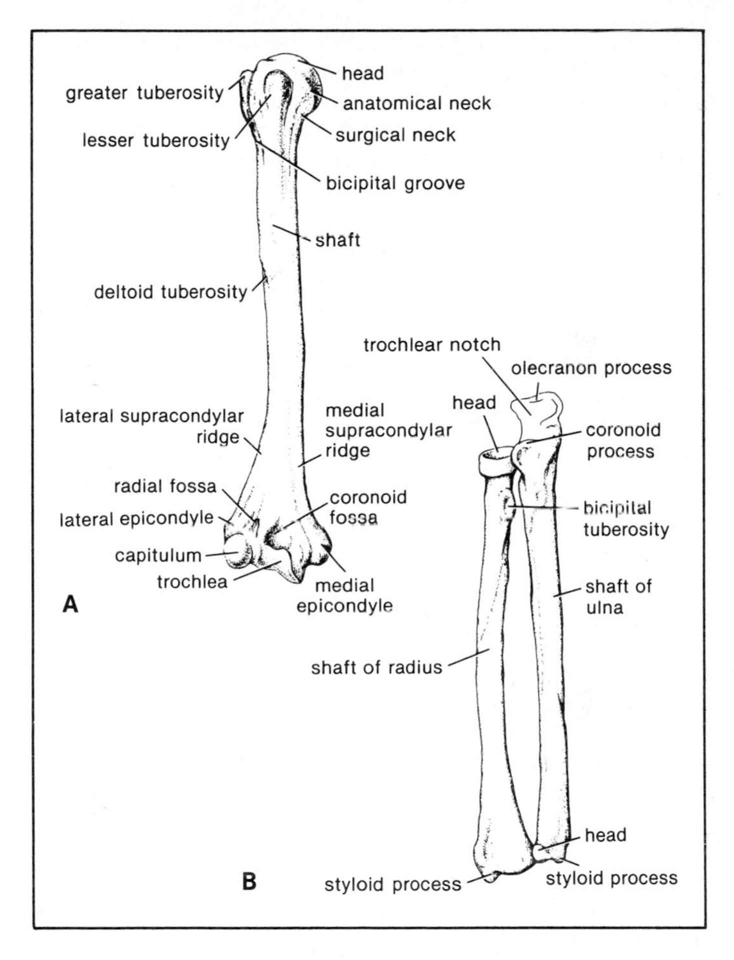

Figure 1–7. **A.** Right humerus, anterior view. **B.** Right radius and ulna, anterior view.

head, the bone is constricted to form the **neck**. Below the neck is the **bicipital tuberosity** for the insertion of the biceps brachii muscle.

The **shaft** of the radius, as compared to that of the ulna, is wider below than above (*Fig. 1–7*). It has a sharp **interosseous border** medially for the attachment of the interosseous membrane that serves to bind the radius and the ulna bones together. The shaft of the radius has an **oblique line**, or ridge, on its anterior surface, which extends downward and laterally from the bicipital tuberosity to the **pronator tubercle** on its lateral surface. The pronator tubercle is for the insertion of the pronator teres muscle.

At the **lower end** of the radius is the **styloid process**; this projects distally from its lateral margin (*Fig. 1–7*). On the medial surface is the **ulnar notch**, which articulates with the round head of the ulna. The inferior articular surface articulates with the scaphoid bone laterally and the lunate bone medially. On the posterior aspect of the lower end is a small tubercle, the **dorsal tubercle**, which is grooved on its medial aspect by the tendon of the extensor pollicis longus.

Ulna

The ulna is the medial bone (little finger side) and it is longer than the radius of the forearm (*Fig. 1–7*). The upper end of the ulna articulates with the humerus at the elbow joint, and with the head of the radius at the superior radioulnar joint. Its lower end articulates with the radius at the inferior radioulnar joint, but it is excluded from the wrist joint by an articular disc (*see p. 63*).

The **upper end** of the ulna is large and is known as the **olecranon process**. This forms the prominence of the elbow. It has a notch on its anterior surface, the **trochlear notch**, which articulates with the trochlea of the humerus. Below the trochlear notch is the triangular **coronoid process**, which has on its lateral surface the **radial notch** for articulation with the head of the radius.

The **shaft** of the ulna tapers from above down. It has three surfaces and three borders. The **lateral**, or **interosseous border**, is sharp and gives attachment to the interosseous membrane. The

posterior border is rounded and subcutaneous and can be easily palpated throughout its length. The **anterior border** is also rounded, but it is covered by muscles. Below the radial notch is a depression, the **supinator fossa**, which gives clearance for the movement of the bicipital tuberosity of the radius. The posterior border of the fossa is sharp and is known as the **supinator crest**; it gives origin to the supinator muscle.

At the **lower end** of the ulna is the small rounded **head**, which has projecting from its medial aspect the **styloid process** (*Fig. 1–7*).

Bones of the Hand

The bones of the hand are the carpal bones, the metacarpal bones, and the phalanges.

Carpal Bones. There are eight small carpal bones in the region of the wrist that are strongly united to one another by ligaments. The bones are made up of two rows of four (*Fig. 1–8*). The **proximal row** consists of (from lateral to medial) the **scaphoid** (navicular), **lunate, triquetral**, and **pisiform** bones. The **distal row** consists of (from lateral to medial) the **trapezium, trapezoid, capitate,** and **hamate** bones. Together, the bones of the carpus present on their anterior surface a concavity, to the lateral and medial edges of which is attached a strong membranous band, the **flexor retinaculum**, which forms a bridge. The bridge and the bones form a tunnel, known as the **carpal tunnel** for the passage of the median nerve and the long flexor tendons of the fingers.

Metacarpals and Phalanges. There are five metacarpal bones in the hand, each of which has a proximal **base**, a **shaft**, and a distal **head** (*Fig. 1–8*). The metacarpal bones are numbered from one to five starting with the metacarpal of the thumb. The first metacarpal bone of the thumb is the shortest and most mobile. The bases of the metacarpal bones articulate with the distal row of the carpal bones; the heads, which form the knuckles, articulate with the proximal phalanges (*Fig. 1–8*).

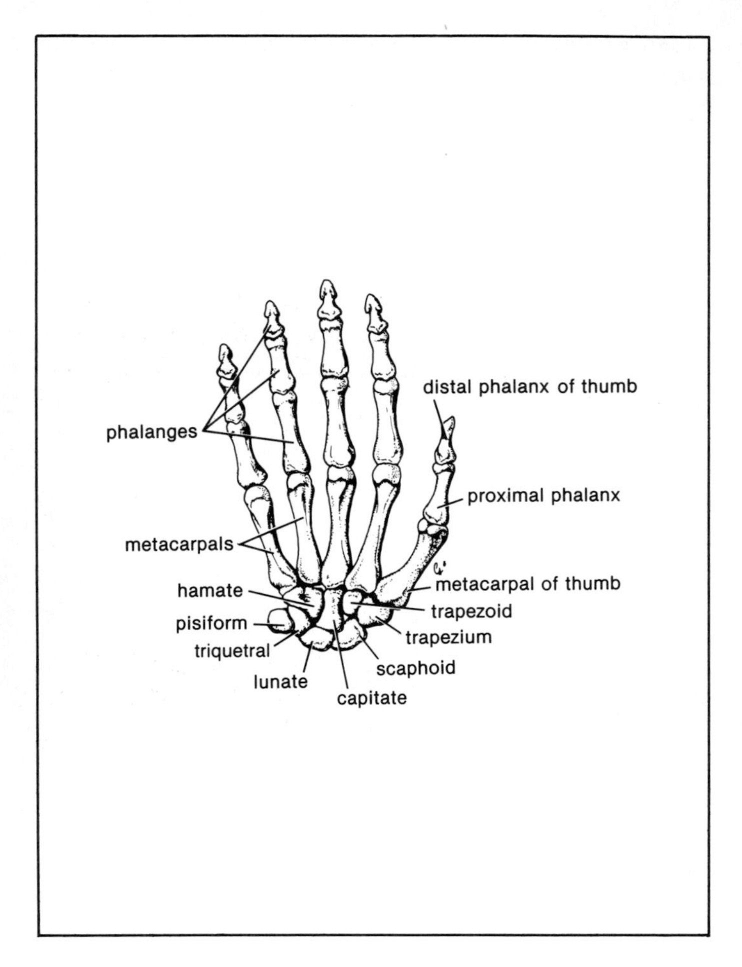

Figure 1–8. Bones of right hand, anterior view.

There are three phalanges for each of the fingers, but only two for the thumb (*Fig. 1–8*). Each phalanx has a proximal **base**, a **shaft**, and a distal **head**.

Bones of the Lower Limb

The bones of the lower limb consist of the pelvic girdle, the thigh, the leg, and the foot.

Bones of the Pelvic Girdle

The pelvic girdle is composed of four bones: the two innominate bones, or hip bones, the sacrum, and the coccyx (*Fig. 1–1*). The pelvic girdle provides a strong and stable connection between the trunk and the lower limbs; it has to be strong because of the large weight it carries. It is interesting to compare the pelvic girdle with the shoulder girdle because the latter is relatively unstable but possesses a great deal more mobility than the pelvic girdle.

The two innominate bones articulate with each other anteriorly at the **symphysis pubis** and posteriorly with the sacrum at the **sacroiliac joints**. The pelvic girdle with its joints form a strong basin-shaped structure called the **pelvis**.

The pelvis is divided into two parts by the **pelvic brim**, which is formed by the **sacral promontory** behind, the **iliopectineal lines** laterally, and the **symphysis pubis** anteriorly. Above the brim is the **false pelvis**, or **greater pelvis**, which forms part of the abdominal cavity. Below the brim is the **true pelvis**, or **lesser pelvis**.

True Pelvis. The true pelvis is a bowl-shaped structure that contains and protects the lower parts of the intestinal and urinary tracts and the internal organs of reproduction. The true pelvis has an inlet, an outlet, and a cavity. The **pelvic inlet** is bounded posteriorly by the sacral promontory, laterally by the iliopectineal lines, and anteriorly by the symphysis pubis. The **pelvic outlet** is bounded posteriorly by the coccyx, laterally by the ischial tuberosities, and anteriorly by the pubic arch. The **pelvic cavity** lies between the inlet and the outlet. It is a short, curved canal, with a shallow anterior wall and a much deeper posterior wall.

Innominate, or Hip Bone. Each innominate bone in the child consists of the superior ilium, the posterior and inferior ischium, and the anterior and inferior pubis (*Fig. 1–9*). At puberty these three bones fuse together to form one large, irregular bone. The hip bones articulate with the sacrum at the sacroiliac joints and form the anterolateral walls of the pelvis; they also articulate with one another anteriorly at the symphysis pubis.

On the outer surface of the innominate bone is a deep depression, the **acetabulum**, which articulates with the hemispherical head of the femur (*Fig. 1–9*). Behind the acetabulum is a large notch, the **greater sciatic notch**, which is separated from the **lesser sciatic notch** by the **spine of the ischium**. The inferior margin of the acetabulum is deficient and is marked by the **acetabular notch**. The articular surface of the acetabulum is limited to a horseshoe-shaped area and is covered with hyaline cartilage. The floor of the acetabulum is nonarticular and is called the **acetabular fossa**.

The **ilium**, which is the upper flattened part of the bone, possesses the **iliac crest** (*Fig. 1–9*). The **iliac crest** runs between the **anterior and posterior superior iliac spines**. Below these spines are the corresponding **inferior iliac spines**. On the inner surface of the ilium is the large **auricular surface** for articulation with the sacrum. The **iliopectineal line** runs downward and forward around the inner surface of the ilium and serves to divide the false from the true pelvis.

The **ischium** is the inferior and posterior part of the innominate bone and possesses an **ischial spine** and an **ischial tuberosity** (*Fig. 1–9*).

The **pubis** is the anterior part of the innominate bone and has a **body** and **superior and inferior pubic rami**. The body of the pubis bears the **pubic crest** and the **pubic tubercle** and articulates with the pubic bone of the opposite side at the **symphysis pubis**.

In the lower part of the innominate bone is a large opening, the **obturator foramen**, which is bounded by the parts of the ischium and pubis (*Fig. 1–9*).

Sex Differences of the Pelvis. The sex differences of the pelvis are easily recognized. They exist because the female pelvis is

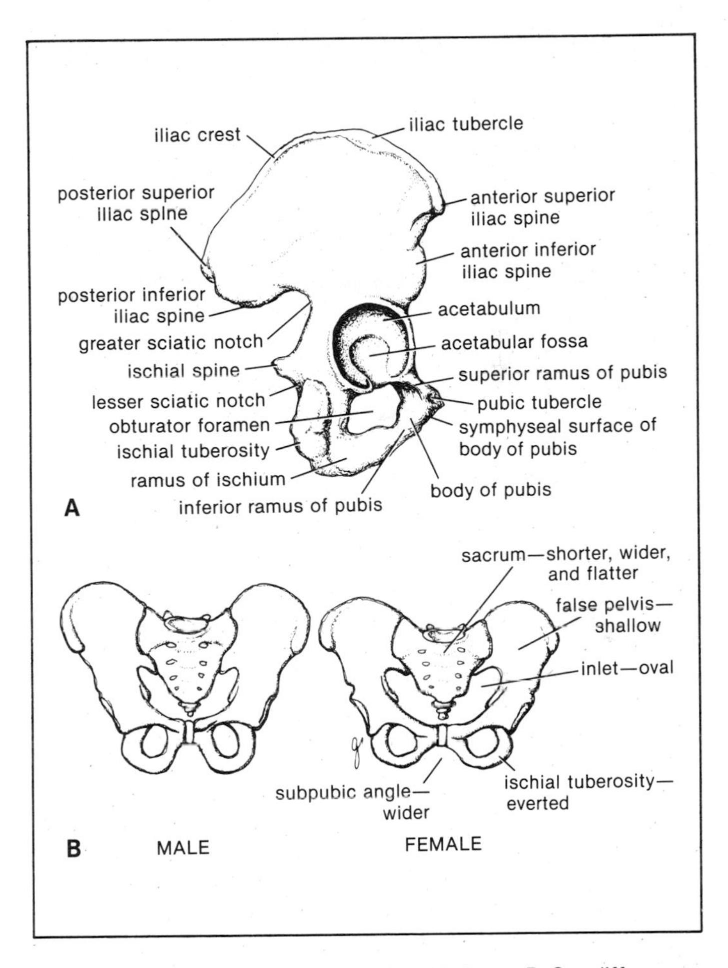

Figure 1–9. **A.** External surface of right hip bone. **B.** Sex differences of the pelvis.

broader than that of the male to allow easier passage of the fetal head and because female bones are more slender than those of the male.

1. The false pelvis is shallow in the female and deep in the male.
2. The pelvic inlet is transversely oval in the female but is heart-shaped in the male. This is due to the indentation produced by the promontory of the sacrum in the male.
3. The pelvic cavity is roomier in the female than in the male, and the distance between the inlet and the outlet is much shorter.
4. The pelvic outlet is larger in the female than the male. The ischial tuberosities are everted in the female and turned in in the male.
5. The sacrum is shorter, wider, and flatter in the female than in the male.
6. The subpubic angle, or pubic arch, is more rounded and wider in the female than in the male (*Fig. 1–9*).

Bones of the Thigh

The bones of the thigh consist of the femur and the patella or kneecap.

Femur

The femur is the longest bone in the body. It articulates above with the acetabulum to form the hip joint and below it articulates with the tibia and the patella to form the knee joint. The upper end of the femur has a head, neck, and greater and lesser trochanters (*Fig. 1–10*). The **head** forms about two-thirds of a sphere and articulates with the acetabulum of the hip bone to form the hip joint. In the center of the head there is a small depression, called the **fovea capitis**, for the attachment of the ligament of the head. Part of the blood supply to the head of the femur from the obturator artery is conveyed along this ligament and enters the bone at the fovea.

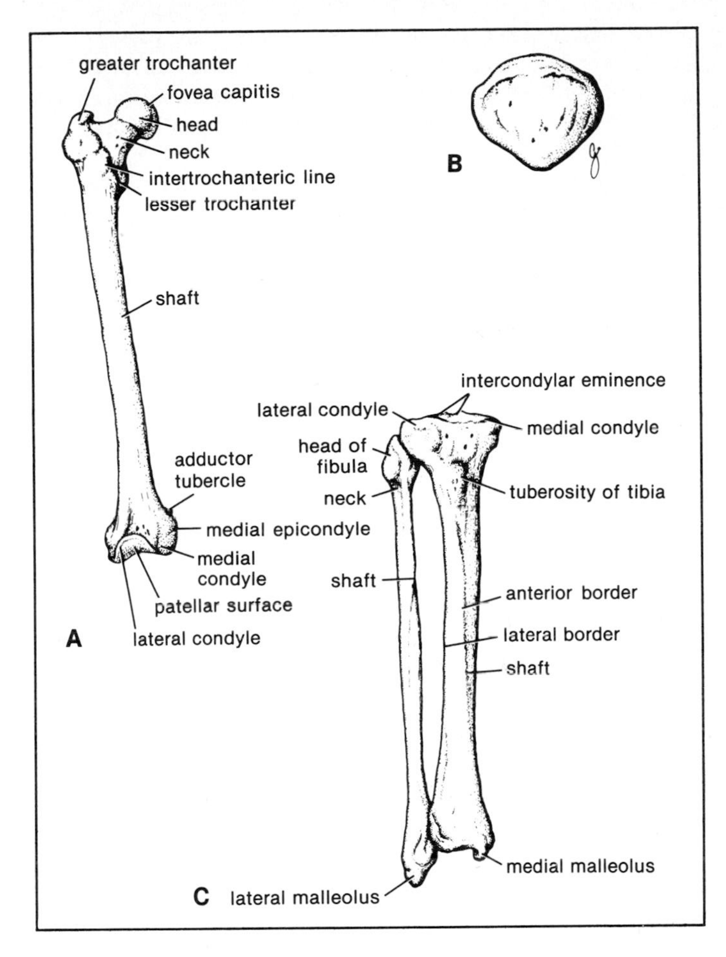

Figure 1–10. A. Right femur, anterior view. **B.** Right patella, anterior view. **C.** Right tibia and fibula, anterior view.

The **neck**, which connects the head to the shaft, passes downward, backward, and laterally makes an angle of about 125 degrees (slightly less in the female) with the long axis of the shaft. The size of this angle is important because it can be altered by disease.

The **greater** and **lesser trochanters** are large eminences situated at the junction of the neck and the shaft. Connecting the two trochanters are the **intertrochanteric line**, anteriorly, and a prominent **intertrochanteric crest**, posteriorly, on which is the **quadrate tubercle**.

The **shaft** of the femur shows a general forward convexity. It is smooth and rounded on its anterior surface, but has posteriorly a ridge, the **linea aspera**. The margins of the linea aspera diverge above and below. The medial margin continues below as the **medial supracondylar ridge** to the **adductor tubercle** on the medial condyle. The lateral margin becomes continuous below with the **lateral supracondylar ridge**. On the posterior surface of the shaft below the greater trochanter is the **gluteal tuberosity** for the insertion of the gluteus maximus muscle. The tuberosity is continuous below with the linea aspera. The shaft becomes broader toward its distal end and forms a flat triangular area on its posterior surface, called the **popliteal surface**.

The lower end of the femur has **lateral** and **medial condyles**, separated posteriorly by the **intercondylar notch**. The anterior surfaces of the condyles are joined by an articular surface for the patella. The two condyles take part in the formation of the knee joint. Above the condyles are the **medial** and **lateral epicondyles** (*Fig. 1–10*). The **adductor tubercle** is continuous with the medial epicondyle.

Patella

The patella (*Fig. 1–10*) is a sesamoid bone (i.e., a bone that develops within a tendon) lying within the quadriceps tendon. It is triangular in shape, and its apex lies inferiorly; the apex is connected to the tuberosity of the tibia by the ligamentum patellae.

The posterior surface articulates with the condyles of the femur. It is situated in an exposed position in front of the knee joint.

The upper, lateral, and medial margins of the patella give attachment to the different parts of the quadriceps femoris muscle. It is prevented from being displaced laterally during the action of the quadriceps muscle by the lower horizontal fibers of the vastus medialis and by the large size of the lateral condyle of the femur.

Bones of the Leg

The bones of the leg are the tibia and the fibula.

Tibia

The tibia or shin bone is the large medial bone of the leg (*Fig. 1–10*). It supports the greater part of the weight on the leg. The tibia articulates with the condyles of the femur and the head of the fibula above, and with the talus and the distal end of the fibula below. It has an expanded upper end, a smaller lower end, and a shaft.

At the upper end are the **lateral and medial condyles**, which articulate with the lateral and medial condyles of the femur, the **lateral and medial semilunar cartilages** intervening. Separating the upper articular surfaces of the tibial condyles is the **intercondylar eminence**. The lateral condyle possesses on its lateral aspect an **oval articular facet for the head of the fibula**.

The **shaft of the tibia** is triangular in cross section, presenting three borders and three surfaces. Its anterior and medial borders, with the medial surface between them, are subcutaneous. The anterior border is prominent and forms the shin. At the junction of the anterior border with the upper end of the tibia is the **tuberosity**, which receives the attachment of the ligamentum patellae. The anterior border becomes rounded below, where it becomes continuous with the medial malleolus. The lateral or interosseous border gives attachment to the **interosseous membrane**, which binds the tibia and the fibula together. The posterior surface of the shaft shows an oblique line, the **soleal line** for the attachment of the soleus muscle.

The lower end of the tibia is slightly expanded and on its inferior aspect shows a saddle-shaped articular surface for the talus. The lower end is prolonged downward and medially to form the **medial malleolus.** The lateral surface of the medial malleolus articulates with the talus. The lower end of the tibia shows a wide, rough depression on its lateral surface for articulation with the fibula.

Fibula

The fibula is the slender needle-shaped lateral bone of the leg (*Fig. 1–10*). It takes no part in the articulation at the knee joint, but below, it forms the lateral malleolus of the ankle joint. It has an expanded upper end, a shaft, and a lower end.

The **upper end, or head**, has a **styloid process** and possesses an **articular surface** for articulation with the lateral condyle of the tibia.

The **shaft of the fibula** is long and slender, and its shape is subject to considerable variation. The medial or **interosseous border** gives attachment to the interosseous membrane.

The **lower end of the fibula** forms the **lateral malleolus**, which is subcutaneous. On the medial surface of the lateral malleolus is an **articular facet** for articulation with the lateral aspect of the talus.

Bones of the Foot

The bones of the foot are the **tarsal bones**, the **metatarsal bones,** and the **phalanges.**

Tarsal Bones. The tarsal bones are the calcaneum, the talus, the navicular, the cuboid, and the three cuneiform bones.

Calcaneum. The calcaneum is the largest bone of the foot (*Fig. 1–11*). It articulates above with the talus and in front with the cuboid bone. The posterior surface forms the prominence of the

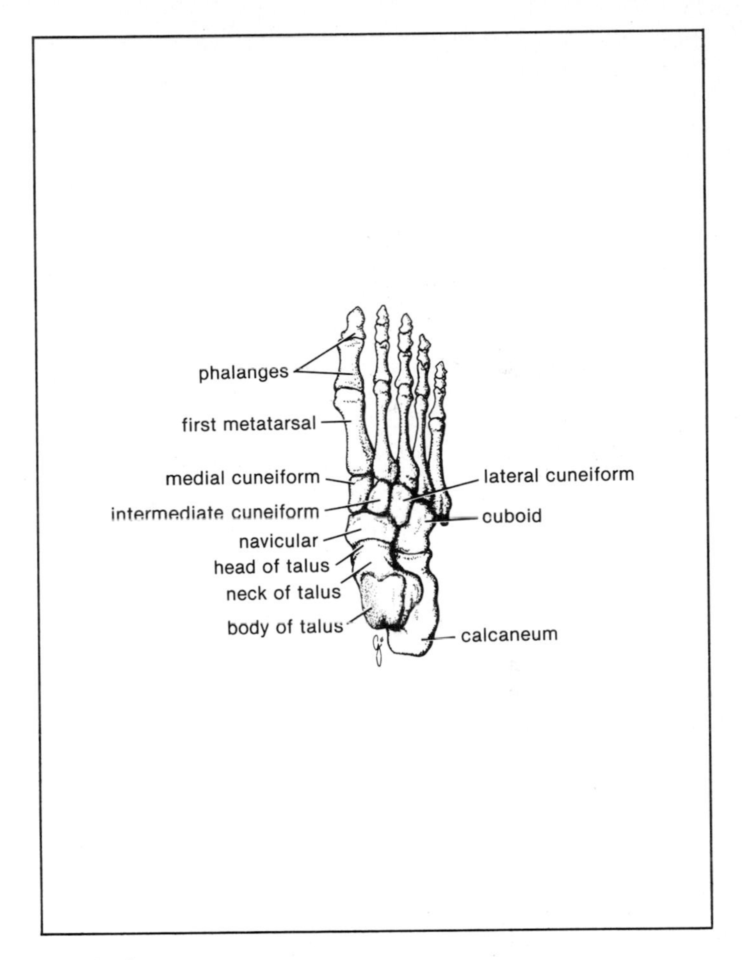

Figure 1-11. Dorsal aspect of bones of right foot.

heel. The medial surface possesses a large shelf-like ridge, termed the **sustentaculum tali**, which assists in the support of the talus.

Talus. The talus articulates above at the ankle joint with the tibia and the fibula, below with the calcaneum, and in front with the navicular bone. It possesses a **head**, a **neck**, and a **body** (*Fig. 1–11*). Numerous important ligaments are attached to the talus, but no muscles are attached to this bone.

The remaining tarsal bones should be identified and the following features noted.

Navicular Bone. The **tuberosity** of the navicular bone can be seen and felt on the medial border of the foot (2.5 cm) (1 inch) in front of and below the medial malleolus; it gives attachment to the main part of the tibialis posterior tendon (*Fig. 1–11*).

Cuboid Bone. There is a deep groove on its inferior aspect, which lodges the tendon of the peroneus longus muscle.

Cuneiform Bones. These are small, wedge-shaped bones, that articulate proximally with the navicular bone and distally with the first three metatarsal bones (*Fig. 1–11*). Their wedge shape contributes greatly to the formation and maintenance of the transverse arch of the foot.

Metatarsal Bones and Phalanges. The metatarsal bones and the phalanges resemble the metacarpals and the phalanges of the hand, and each possesses a distal **head**, a **shaft**, and a proximal **base**. The metatarsals are numbered from the medial to the lateral side. The big toe like the thumb possesses only two phalanges. The other four toes each have three phalanges.

The **fifth metatarsal** has a prominent **tubercle** on its base, which can be easily palpated along the lateral border of the foot. The tubercle gives attachment to the peroneus brevis tendon.

2

Joints

GENERAL DESCRIPTION

The term joint or articulation is used to describe the site where two or more bones of the skeleton come together. At most joints, the bones are held together by flexible connective tissues that permit the muscles to act on the bones and therefore bring about movements of the different parts of the body. Joints may be classified according to their mobility into immovable, slightly movable, or freely movable joints. It is much more satisfactory to classify joints according to their structure, namely: fibrous, cartilaginous, and synovial.

Fibrous Joints

Fibrous joints have no joint cavity and the bones are held together by dense fibrous tissue (*Fig. 2–1*). In the sutures of the vault of the skull, the flat bones are held together by dense sutural ligaments and there is no movement. The **sagittal suture**, which unites the two parietal bones, the **coronal suture**, which unites the frontal bone to the two parietal bones, and the **lambdoid suture**, which unites the two parietal bones to the occipital bone, are good examples.

In some fibrous joints, the length of the fibrous tissue uniting the bones is longer than the sutural ligaments and a small amount of movement is possible. The inferior tibiofibular joints are examples of this type of fibrous joint.

A further type of fibrous joint exists where a cone-shaped peg fits into a socket. An example of this type is seen with the roots of the teeth fitting into the alveolar processes of the maxillae and mandible. The fibrous tissue joining the tooth to the bone socket is called the **peridontal** membrane.

Cartilaginous Joints

Cartilaginous joints have no joint cavity. They may be divided into two types, primary and secondary. A **primary cartilaginous joint** is one in which the bones are united by a plate or bar of hyaline cartilage. Thus, the union between the **epiphysis** and the **diaphysis** of a growing bone and that between the first rib and the manubrium sterni are examples of such a joint. No movement is possible. In a growing bone the plate of hyaline cartilage, the **epiphyseal cartilage**, that exists between the epiphysis and the diaphysis is eventually replaced by bone as growth ceases and the cartilaginous joint is thus only temporary.

In a **secondary cartilaginous joint**, the bones are united by a plate of fibrocartilage; the articular surfaces of the bones are covered by a thin layer of hyaline cartilage. Examples are the intervertebral joints between the bodies of adjacent vertebrae and the symphysis pubis, which is a joint between the pubic bones of the pelvis (*Fig. 2–1*). A small amount of movement is possible at these joints, and in the case of the symphysis pubis, the degree of movement may be increased during the later stages of pregnancy as the result of the action of hormones on the ligaments of the joint.

Synovial Joints

The articular surfaces of the bones are covered by a thin layer of hyaline cartilage separated by a joint cavity (*Fig. 2–1*). This arrangement permits a great degree of freedom of movement. The cavity of the joint is lined by **synovial membrane**, which extends from the margins of one articular surface to those of the other. The synovial membrane is protected on the outside by a tough fibrous membrane referred to as the **capsule** of the joint. The articular surfaces are covered and are lubricated by a thin layer of viscous fluid called **synovial fluid**. The synovial fluid is produced by the synovial membrane and contains hyaluronic acid and also phagocytic cells that remove debris derived from the wear and tear in the joint. In certain synovial joints, for example in the knee joint, discs or wedges of fibrocartilage are interposed between the articular surfaces of the bones. These are referred to as **articular discs**.

The degree of movement in a synovial joint is limited by the shape of the bones participating in the joint, the coming together of adjacent anatomical structures (for example, the thigh against the anterior abdominal wall on flexing the hip joint), and the presence of fibrous **ligaments** uniting the bones. Most ligaments lie outside the joint capsule, and are known as **extracapsular ligaments**. Examples of extracapsular ligaments are the strong iliofemoral, pubofemoral, and ischiofemoral ligaments of the hip joint. **Intracapsular ligaments** lie within the capsule but they are excluded from the joint cavity by coverings of synovial membrane. Examples of intracapsular ligaments are the cruciate ligaments found in the knee joint (*Fig. 2–9*).

Types of Synovial Joints

Hinge Joints. A hinge joint resembles the hinge of a door, in which movement is permitted in only one plane. The elbow and knee joints are two good examples (*Figs. 2–4 and 2–9*).

Gliding Joints. A gliding joint is one where the flat articular surfaces slide one upon the other. The acromioclavicular and sternoclavicular joints are examples of this type of joint (*Fig. 2–3*).

Pivot Joints. A pivot joint is one where one bone is permitted to rotate on another. For example, the head of the radius rotates on the ulna in the superior radioulnar joint and the atlas rotates on the axis in the upper part of the vertebral colulmn.

Ball-and-Socket Joints. A ball-and-socket joint is one in which a hemispherical-shaped end of a bone fits into a cup-shaped cavity of another bone (*Fig. 2–1*). The hip joint and the shoulder joint are two good examples. A considerable amount of movement is permitted at this type of joint.

Biaxial Joints (Condyloid Joints). Biaxial joints, as their name implies, permit movements in two planes only but do not allow rotation. The wrist or radiocarpal joint (*Fig. 2–5*) and the carpometacarpal joint of the thumb are biaxial joints.

Movements of Synovial Joints. The movements that are possible at synovial joints may be divided into four principal types (1) gliding movements, (2) angular movements, (3) rotation movements, and (4) circumduction.

Gliding Movements. Gliding movements are a very simple type of movement where the flat articular surfaces merely slide over one another without angular or rotatory motion. This type of movement occurs between the carpal and tarsal bones and the acromioclavicular and sternoclavicular joints mentioned above.

Angular Movements. Angular movements involve an increase or a decrease in the size of the angle between the bones. The type of angular movement is named according to the direction in which the movement takes place. **Flexion** is a movement in a sagittal plane and usually results in a decrease in the angle between the anterior surfaces of the bones, as in flexion of the elbow joint. The knee joint is an exception to this definition in that the angle between the posterior surfaces of the bones is decreased in flexion. **Extension** means a straightening of the joint in the sagittal plane and usually takes place in a posterior direction. **Abduction** is the movement of a bone away from the midline of the body in the coronal plane. **Adduction** is the movement of a bone toward the body in the coronal plane.

Rotation Movements. Rotation is the movement of the bone around its long axis. Examples of this type of movement are seen in the rotation of the radius on the ulna at the superior and inferior radioulnar joints; also the atlas vertebra with the skull rotating around the axis vertebra.

Circumduction. Circumduction is the term used to describe a combination in sequence of angular movements and rotation movements; it includes flexion, extension, abduction, and adduction. A good example of this type of movement is seen at the shoulder joint.

The special types of movements that include **protraction** and **retraction** of the mandible at the temporomandibular joints, **pronation** and **supination** of the forearm bones at the radioulnar joints, and **inversion** and **eversion** of the foot at the subtalar and transverse tarsal joints are described later with the particular joints.

Joint Stability

The strength or stability of a joint depends on three main factors: (1) the shape, size, and arrangement of the articular surfaces; (2) the ligaments; and (3) the tone of the muscles around the joint.

Articular Surfaces. The ball-and-socket arrangement of the hip joint (*Fig. 2-1*) where the hemispherical head of the femur fits into the socket-shaped acetabulum of the hip bone is a good example of how bone shape plays an important role in joint stability. On the other hand the small flat articular surfaces of the acromion and the clavicle contribute nothing to the strength of the acromioclavicular joint (*Fig. 2-3*).

Ligaments. Dense fibrous ligaments often strongly connect the bones of a joint together such as the strong collateral ligaments found in the elbow and wrist joints. Unfortunately, if the stress on the ligaments is continued for an excessively long time, the fibrous ligaments will permanently stretch. Examples of this are seen in the bones of the foot where excessive stress may lead to the falling of the arches of the feet.

Elastic ligaments not only support the joints but assist in the return of the bones to their original position after movement. The ligamentum flavum uniting the laminae of adjacent vertebrae assists in the return of the flexed vertebral column to the anatomical position.

Muscle Tone. In most joints, muscle tone, i.e., the tension of muscles, is a major factor contributing to the strength of a joint. For example, the tone of the short muscles around the shoulder joint and the muscles acting on the knee joint greatly strengthen these joints.

Nerve Supply of Joints

The capsule and the ligaments receive an abundant sensory nerve supply. Overstretching of the capsule and the ligaments produces reflex contraction of muscles around the joint, which tends to protect the joint; excessive stretching produces pain. The motor nerve supply to a muscle acting on a joint usually gives a branch to supply the joint on which it is acting.

JOINTS OF THE SKULL

Sutures of the Skull

In the vault of the skull the flat bones are joined together by fibrous joints called sutures (*Fig. 2–1*). The periosteum covering the outer surface of the bones becomes continuous with the endosteum covering the inner surface at the suture, forming the **sutural ligament**. These dense ligaments do not permit any movement between the bones. Examples are the **sagittal suture**, the **coronal suture**, and the **lambdoid suture** (*see Fig. 1–2*).

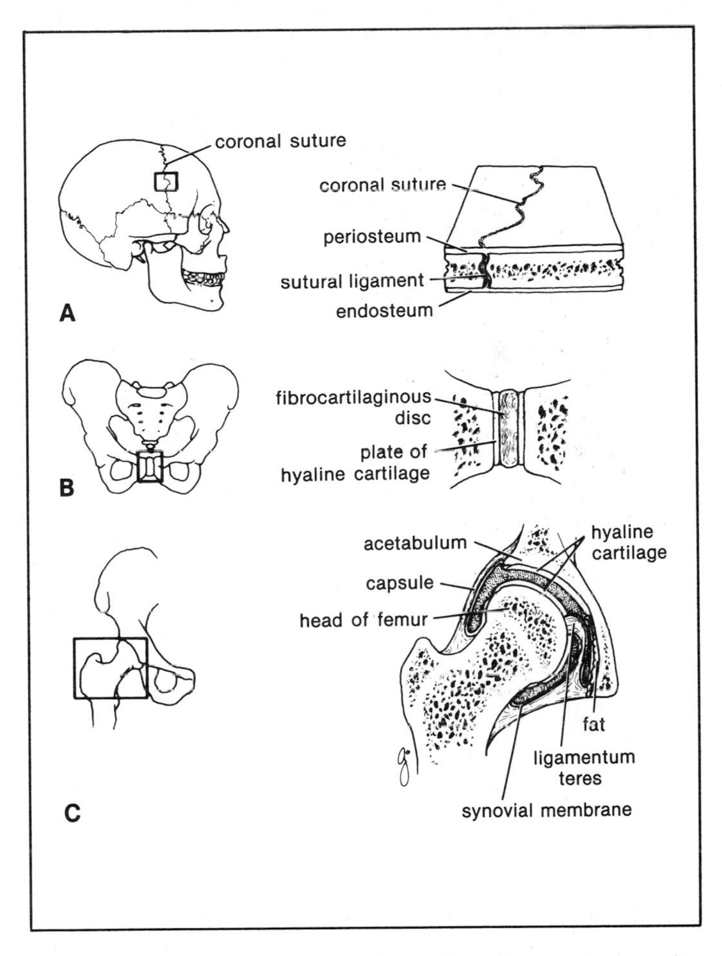

Figure 2–1. **A.** Coronal suture, a fibrous joint. **B.** Symphysis pubis, a cartilaginous joint. **C.** Hip joint, a synovial joint.

Temporomandibular Joint

Articulation. Above, the articular tubercle and the mandibular fossa of the temporal bone; below, the head of the mandible; the surfaces are covered with fibrocartilage (*Fig. 2–2*).

Type. Synovial. The fibrocartilaginous disc divides the joint into upper and lower cavities.

Capsule. Encloses the joint.

Ligaments. The **lateral temporomandibular ligament** is attached above to the articular tubercle at the root of the zygomatic arch and below to the neck of the mandible. The fibers extend downward and backward. The ligament limits the posterior movement of the mandible. The **sphenomandibular ligament** is attached above to the spine of the sphenoid bone and below to the lingula of the mandibular foramen. Its function is not known. The **stylomandibular ligament** extends from the styloid process to the angle of the mandible. Its function is unknown. The **articular disc** is an oval disc of fibrocartilage (*Fig. 2–2*) that is attached in front to the tendon of the lateral pterygoid muscle and posteriorly by fibrous bands to the head of the mandible. Its circumference is attached to the capsule. The disc often shows evidence of degeneration. The disc permits gliding movement in the upper part of the joint and hinge movement in the lower part of the joint.

Synovial Membrane. This lines the capsule.

Nerve Supply. Auriculotemporal nerve and masseteric nerve, branches of the mandibular division of the trigeminal nerve.

Movements. The head of the mandible and the articular disc can be **protruded** forward by the lateral pterygoid muscle and **retracted** backward by the posterior fibers of the temporalis muscle. The **mouth is opened** by the head of the mandible rotating on the undersurface of the articular disc around a horizontal axis.

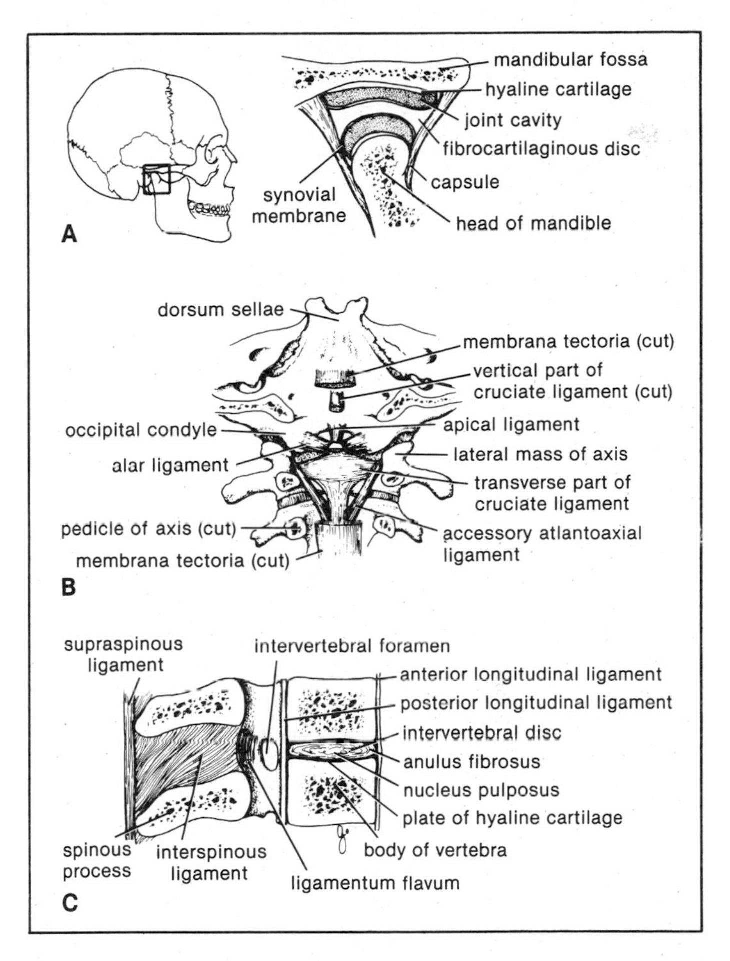

Figure 2–2. **A.** Temporomandibular joint, sagittal section. **B.** Atlantooccipital joints, posterior view. **C.** Joints between lumbar vertebrae, sagittal section.

The digastrics, the geniohyoids, and the mylohyoids depress the mandible. At the same time the lateral pterygoid muscle pulls forward the neck of the mandible and the articular disc. The mandible is **elevated** by the temporalis, the masseter, and the medial pterygoids. The head of the mandible is retracted by the posterior fibers of the temporalis and the articular disc is pulled backward by the fibroelastic tissue, which connects the disc to the temporal bone posteriorly. **Lateral chewing** movements are brought about by alternately protruding and retracting the mandible on each side.

JOINTS OF THE VERTEBRAL COLUMN

Atlantooccipital Joints

The atlantooccipital joints (*Fig. 2–2*) are synovial joints that are formed between the occipital condyles, which are found on either side of the foramen magnum of the skull, above, and the facets on the superior surfaces of the lateral masses of the atlas, below. They are enclosed by a capsule.

Ligaments. These are the **anterior atlantooccipital membrane** that connects the anterior arch of the atlas to the anterior margin of the foramen magnum and the **posterior atlantooccipital membrane** that connects the posterior arch of the atlas to the posterior margin of the foramen magnum.

Movements. Flexion, extension, and lateral flexion. No rotation.

The joints are strengthened by the ligaments that connect the skull to the axis and by the tone of the muscles that surround the joints.

Atlantoaxial Joints

The atlantoaxial joints (*Fig. 2–2*) are three synovial joints, one of which is between the odontoid process and the anterior arch of the atlas, whereas the other two are between the lateral masses of the bones. The joints are enclosed by capsules.

Ligaments. The **apical ligament** is a median-placed structure connecting the apex of the odontoid process to the anterior margin of the foramen magnum.

The **alar ligaments** lie one on each side of the apical ligament and connect the odontoid process to the medial sides of the occipital condyles.

The **cruciate ligament** consists of a transverse part and a vertical part. The transverse part is attached on each side to the inner aspect of the lateral mass of the atlas and binds down the odontoid process to the anterior arch of the atlas. The vertical part runs from the posterior surface of the body of the axis to the anterior margin of the foramen magnum.

The **membrana tectoria** is an upward continuation of the posterior longitudinal ligament. It is attached above to the occipital bone just within the foramen magnum. It covers the posterior surface of the odontoid process and the apical, alar, and cruciate ligaments.

Movements. The atlantoaxial joints permit a wide range of rotation of the atlas and thus of the head on the axis.

Joints of the Vertebral Column Below the Axis

With the exception of the first two cervical vertebrae, the remainder of the mobile vertebrae articulate with each other by means of cartilaginous joints between their bodies and by synovial joints between their articular processes (*Fig. 2–2*).

Joints Between Two Vertebral Bodies

The upper and lower surfaces of the bodies of adjacent vertebrae are covered by thin plates of hyaline cartilage. Sandwiched between the plates of hyaline cartilage is an intervertebral disc of fibrocartilage (*Fig. 2–2*). The collagen fibers of the disc strongly unite the bodies of the two vertebrae.

In the lower cervical region, small synovial joints are present at the sides of the intervertebral disc between the upper and lower surfaces of the bodies of the vertebrae.

Ligaments. The **anterior and posterior longitudinal ligaments** run as continuous bands down the anterior and posterior surfaces of the vertebral column from the skull to the sacrum (*Fig. 2–2*). These ligaments hold the vertebrae firmly together but at the same time permit a small amount of movement to take place between the vertebrae.

Intervertebral Discs

The intervertebral discs are responsible for one-quarter of the length of the vertebral column (*see Figs. 1–1 and 1–4*). They are thickest in the cervical and lumbar regions, where the movements of the vertebral column are greatest. Each disc consists of a peripheral part, the anulus fibrosus, and a central part, the nucleus pulposus (*Fig. 2–2*).

The **anulus fibrosus** is composed of fibrocartilage, in which the collagen fibers are arranged in concentric layers or sheets. The collagen bundles pass obliquely between adjacent vertebral bodies, and their inclination is reversed in alternate sheets. The more peripheral fibers are strongly attached to the anterior and posterior longitudinal ligaments of the vertebral column.

The **nucleus pulposus** in the child is an ovid mass of gelatinous material containing a large amount of water, a small number of collagen fibers, and a few cartilage cells. It is normally under pressure and situated slightly nearer to the posterior than to the anterior margin of the disc.

The upper and lower surfaces of the bodies of adjacent vertebrae that abut onto the disc are covered with thin plates of hyaline cartilage. The semifluid nature of the nucleus pulposus allows it to change shape and permits one vertebra to rock forward or backward on another, as in flexion and extension of the vertebral column. With advancing age the water content of the nucleus pulposus diminishes and is replaced by fibrocartilage. In old age, the discs are thin and less elastic, and it is no longer possible to distinguish the nucleus from the anulus.

No discs are found between the first two cervical vertebrae or in the sacrum or coccyx.

A sudden increase in the compression load on the vertebral column, as when carrying a heavy load or when suddenly flexing the vertebral column, may cause the anulus fibrosus to rupture. This allows the nucleus pulposus to herniate and protrude into the vertebral canal where it may press upon the spinal nerve roots, the spinal nerve, or even the spinal cord.

Joints Between Two Vertebral Arches

The joints between two vertebral arches consist of synovial joints between the superior and inferior articular processes of adjacent vertebrae. The articular facets are covered with hyaline cartilage, and the joints are surrounded by a capsular ligament.

Ligaments. The **supraspinous ligament** runs between adjacent spines (*Fig. 2–2*). The **interspinous ligament** connects adjacent spines. The **ligamentum flavum** connects adjacent laminae. In the cervical region, the supraspinous and interspinous ligaments are greatly thickened to form the **ligamentum nuchae.** The latter extends from the spine of the seventh cervical vertebra to the external occipital protuberance of the skull, its anterior border being strongly attached to the cervical spines in between.

Nerve Supply of Vertebral Joints. The joints are supplied by branches of the corresponding spinal nerves.

CURVES OF THE VERTEBRAL COLUMN

The vertebral column is a flexible structure and exhibits the following curves.

Curves in the Sagittal Plane

In the developing fetus, the vertebral column has one continuous anterior concavity. As development proceeds, the **lumbosacral**

angle appears. After birth, when the child becomes able to raise the head and keep it balanced on the upper end of the vertebral column, the cervical part of the vertebral column becomes concave posteriorly. The development of these **secondary curves** is largely due to modification in the shape of the intervertebral discs.

In the adult in the standing position (*see Fig. 1–4*), the vertebral column therefore exhibits in the sagittal plane the following regional curves: cervical–posterior concavity, thoracic–posterior convexity, lumbar–posterior concavity, and sacral–posterior convexity. During the later months of pregnancy, with the increase in size and weight of the fetus, women tend to increase the posterior lumbar concavity in an attempt to preserve their center of gravity. In old age, the intervertebral discs atrophy, resulting in a loss of height and a gradual return of the vertebral column to a continuous anterior concavity.

Curves in the Coronal Plane

In late childhood it is quite common to find the development of minor lateral curves in the thoracic region of the vertebral column. This is normal and is usually due to the predominant use of one of the upper limbs. For example, right-handed persons will often have a slight right-sided thoracic convexity. Slight compensatory curves are always present above and below such a curvature.

Movements of the Vertebral Column

The following movements are possible: flexion, extension, lateral flexion, rotation, and circumduction.

Flexion is a forward movement, and **extension** is a backward movement. They are both extensive in the cervical and lumbar regions, but restricted in the thoracic region due to the presence of the ribs and costal cartilages, which are connected directly or indirectly to the sternum.

Lateral flexion is the bending of the body to one or the other side. It is extensive in the cervical and lumbar regions, but restricted in the thoracic region due to the presence of the ribs.

Rotation is a twisting of the vertebral column. This is most extensive in the lumbar region.

Circumduction is a combination of all these movements.

The type and range of movements possible in each region of the column is largely dependent on the thickness of the intervertebral discs and the shape and direction of the articular processes. In the thoracic region, as was noted, the ribs, the costal cartilages, and the sternum severely restrict the range of movement.

The vertebral column is moved by numerous muscles, many of which are attached directly to the vertebrae, whereas others, such as the sternocleidomastoid and the abdominal wall muscles, are attached to the skull or to the ribs or fasciae.

JOINTS OF THE RIBS

A typical rib articulates by its head and by its tubercle with the vertebral column posteriorly. Anteriorly, the first seven ribs with their costal cartilages articulate with the sternum (*see Fig. 1–1*); the eighth, ninth, and tenth ribs and their costal cartilages articulate with the costal cartilages above. The eleventh and twelfth ribs are free floating ribs anteriorly.

Joints of the Heads of the Ribs

The first and the lower three ribs have a simple synovial joint with its corresponding vertebral body. From the second to the ninth ribs, the head articulates by means of a synovial joint with facets on the corresponding vertebral body and that of the vertebra above. Within these latter joints there is a strong **intra-articular ligament** that connects the head to the intervertebral disc.

Joints of the Tubercles of the Ribs

The medial facet of the tubercle articulates by means of a synovial joint with the transverse process of the vertebra to which it corresponds numerically. (It is absent on the eleventh and twelfth ribs.)

Costochondral Joints

These are primary cartilaginous joints and no movements are possible.

JOINTS OF THE COSTAL CARTILAGES WITH THE STERNUM

The first costal cartilages articulate with the manubrium by means of primary cartilaginous joints permitting no movement (*see Fig. 1-1*). The second costal cartilages articulate with the manubrium and body of the sternum by a synovial joint. The third to the seventh costal cartilages articulate with the lateral border of the body of the sternum by synovial joints. (The sixth, seventh, eighth, ninth, and tenth costal cartilages articulate with one another along their borders by small synovial joints. The cartilages of the eleventh and twelfth ribs are embedded in the abdominal musculature.)

Movements of the Ribs and Costal Cartilages

The first ribs and their costal cartilages are fixed to the manubrium and are immobile. The raising and lowering of the ribs during respiration are accompanied by gliding movements in both the joints of the head and the tubercle permitting the neck of each rib to rotate around its own axis.

JOINTS OF THE STERNUM

Manubriosternal Joint. The manubriosternal joint is a secondary cartilaginous joint between the manubrium and the body of the sternum. The bony surfaces are covered with hyaline cartilage and joined by a disc of fibrocartilage. A small amount of angular movement is possible during respiration.

Xiphisternal Joint. The xiphisternal joint is a secondary cartilaginous joint between the xiphoid process (cartilage) and the body of the sternum. The xiphoid process usually fuses with the body of the sternum during middle age.

JOINTS OF THE UPPER EXTREMITY

Sternoclavicular Joint

Articulation. This occurs between the medial end of the clavicle, the manubrium sterni, and the first costal cartilage (*Fig. 2-3*).

Type. It is a synovial gliding joint.

Capsule. This encloses the joint.

Ligaments. The capsule is reinforced above, in front of, and behind the joint by strong ligaments, the **interclavicular** and the **sternoclavicular ligaments** (*Fig. 2-3*).

The joint is divided by the **articular disc** into two compartments. The disc is attached by its circumference to the capsule, to the superior margin of the articular surface of the clavicle, and to the first costal cartilage below. The disc serves as a strong intra-articular ligament.

Accessory Ligament. The **costoclavicular ligament** is a strong ligament, which runs from the junction of the first rib with the first costal cartilage to the inferior surface of the clavicle (*Fig. 2-3*).

Synovial Membrane. This lines the capsule of each compartment of the joint.

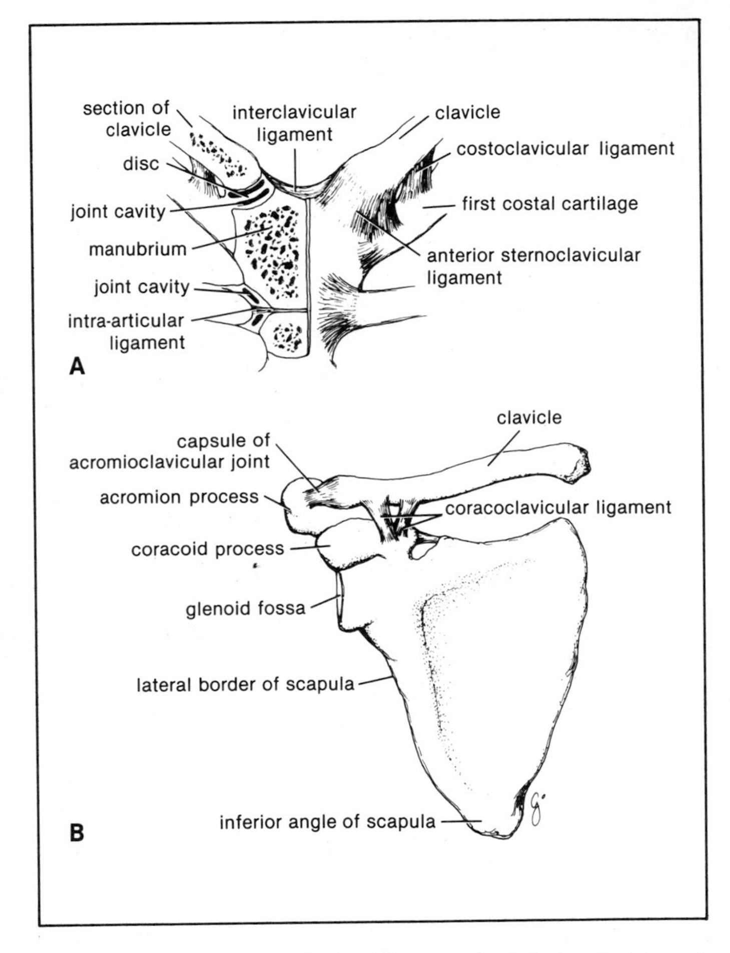

Figure 2–3. A. Sternoclavicular joints; on the left the clavicle and sternum have been cut and the joint opened. **B.** Acromioclavicular joint.

Nerve Supply. The supraclavicular nerve and the nerve to subclavius.

Movements. Anterior and posterior movements of the clavicle takes place in the medial compartment. Elevation and depression of the clavicle takes place in the lateral compartment. All these movements are associated with movements of the scapula.

Muscles Producing Movements. The anterior movement of the clavicle is produced by the serratus anterior muscle. The posterior movement is produced by the trapezius and rhomboid muscles. Elevation of the clavicle is produced by the trapezius, sternocleidomastoid, levator scapulae, and rhomboid muscles. Depression of the clavicle is produced by the pectoralis minor and the subclavius muscles.

Acromioclavicular Joint

Articulation. This occurs between the acromion process of the scapula and the lateral end of the clavicle (*Fig. 2–3*).

Type. This is a synovial gliding joint.

Capsule. This surrounds the joint.

Ligaments. The **acromioclavicular ligament** strengthens the capsule superiorly. An incomplete wedge-shaped fibrocartilaginous disc projects into the joint cavity from above. It is attached to the capsule superiorly.

Accessory Ligament. The very strong **coracoclavicular ligament** connects the coracoid process to the inferior surface of the clavicle (*Fig. 2–3*). It is largely responsible for suspending the weight of the scapula and the upper limb from the clavicle.

Synovial Membrane. This lines the capsule.

Nerve Supply. The suprascapular nerve.

Movements. A gliding movement when the scapula rotates, or when the clavicle is elevated or depressed.

Shoulder Joint

Articulation. This occurs between the rounded head of the humerus and the shallow, pear-shape glenoid cavity of the scapula. The glenoid cavity is deepened by the presence of a fibrocartilaginous rim called the **glenoid labrum** (*Fig. 2–4*).

Type. Synovial ball-and-socket joint.

Capsule. This surrounds the joint and is thin and lax, allowing a wide range of movement. It is strengthened by the tendons of the short muscles around the joint.

Ligaments. The **glenohumeral ligaments** are three weak bands of fibrous tissue that strengthen the interior of the anterior part of the capsule. The **transverse humeral ligament** strengthens the capsule and bridges the gap between the greater and lesser tuberosities of the humerus. It holds the tendon of the long head of the biceps muscle in place. The **coracohumeral ligament** strengthens the capsule above and extends from the root of the coracoid process to the greater tuberosity of the humerus.

Accessory Ligament. The **coracoacromial ligament** extends between the coracoid process and the acromion. It protects the superior aspect of the joint.

Synovial Membrane. This lines the capsule. It surrounds the tendon of the biceps and also protrudes forward through the anterior wall of the capsule to form a bursa, which lies beneath the subscapularis muscle.

Nerve Supply. The axillary and suprascapular nerves.

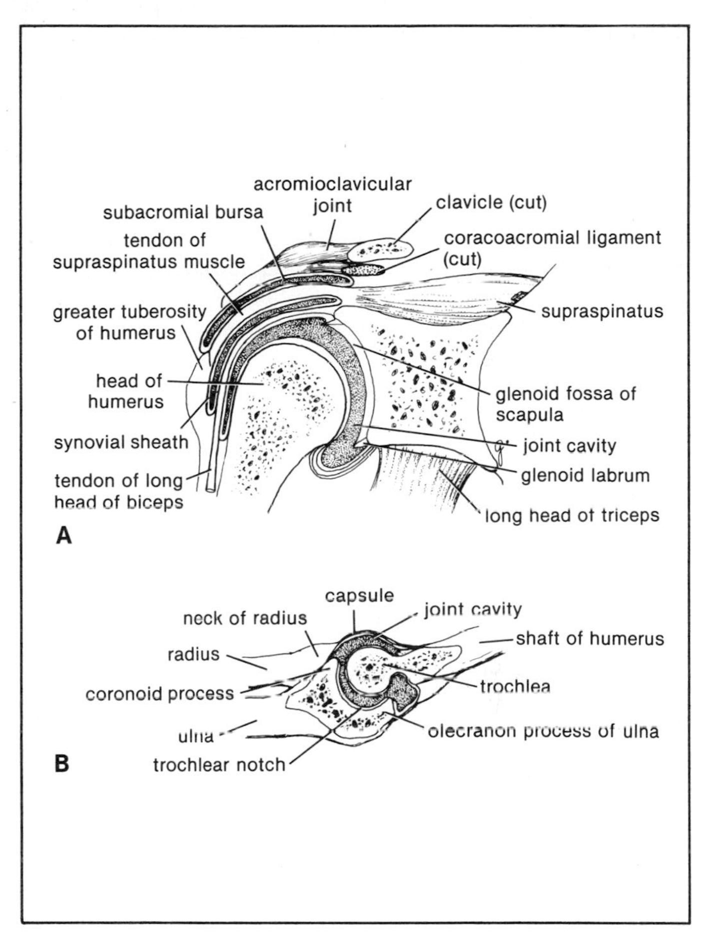

Figure 2–4. A. Right shoulder joint, coronal section. **B.** Right elbow joint, sagittal section.

Movements. The shoulder joint has a wide range of movement, and the stability of the joint has been sacrificed to permit this. (Compare with the hip joint, which is stable, but limited in its movements.) The strength of the joint depends on the tone of the short muscles that cross in front, above, and behind the joint, namely, the subscapularis, supraspinatus, infraspinatus, and teres minor. The tendons of insertion of these muscles are fused to the capsule of the joint and together form the important **rotator cuff.**

When the joint is abducted, the lower surface of the head of the humerus is supported by the long head of the triceps, which bows downward because of its length and gives little actual support to the humerus. In addition, the inferior part of the capsule is the weakest area.

The following movements are possible and the muscles producing the movements are:

Flexion. Anterior fibers of deltoid, pectoralis major, biceps, and coracobrachialis.

Extension. Posterior fibers of deltoid, latissimus dorsi, teres major.

Abduction. Middle fibers of deltoid, assisted by supraspinatus.

Adduction. Pectoralis major, latissimus dorsi, teres major, teres minor.

Lateral Rotation. Infraspinatus, teres minor, posterior fibers of deltoid.

Medial Rotation. Subscapularis, latissimus dorsi, teres major, anterior fibers of deltoid.

Circumduction. A combination of these movements.

Important Relations. *Anteriorly.* The axillary vessels and the brachial plexus. *Inferiorly.* The axillary nerve and the posterior circumflex humeral vessels as they lie in the quadrilateral space.

Elbow Joint

Articulation. This occurs between the trochlea and the capitulum of the humerus and the trochlear notch of the ulna and the head of the radius (*Fig. 2–4*).

Type. It is a synovial hinge joint.

Capsule. This encloses the joint.

Ligaments. The **lateral collateral ligament** (*Fig. 2–5*) is triangular in shape and is attached by its apex to the lateral epicondyle of the humerus, and by its base to the superior margin of the anular ligament and to the ulna. The **medial collateral ligament** is also triangular in shape and consists of three bands: (1) the anterior band, which passes from the medial epicondyle of the humerus to the medial margin of the coronoid process of the ulna, (2) the posterior band, which connects the medial epicondyle of the humerus to the olecranon, and (3) the transverse or oblique band, which passes between the ulnar attachments of the two preceeding bands (*Fig. 2–5*).

Synovial Membrane. This lines the capsule and is continuous below with the synovial membrane of the superior radioulnar joint.

Nerve Supply. Branches from the median, ulnar, musculocutaneous, and radial nerves.

Movements and the Muscles Producing the Movements

Flexion. Brachialis, biceps, brachioradialis, pronator teres.

Extension. Triceps, anconeus.

It should be noted that the extended forearm lies at an angle to the upper arm. This angle, which opens laterally, is called the **carrying angle** and is about 170 degrees in the male and 167 de-

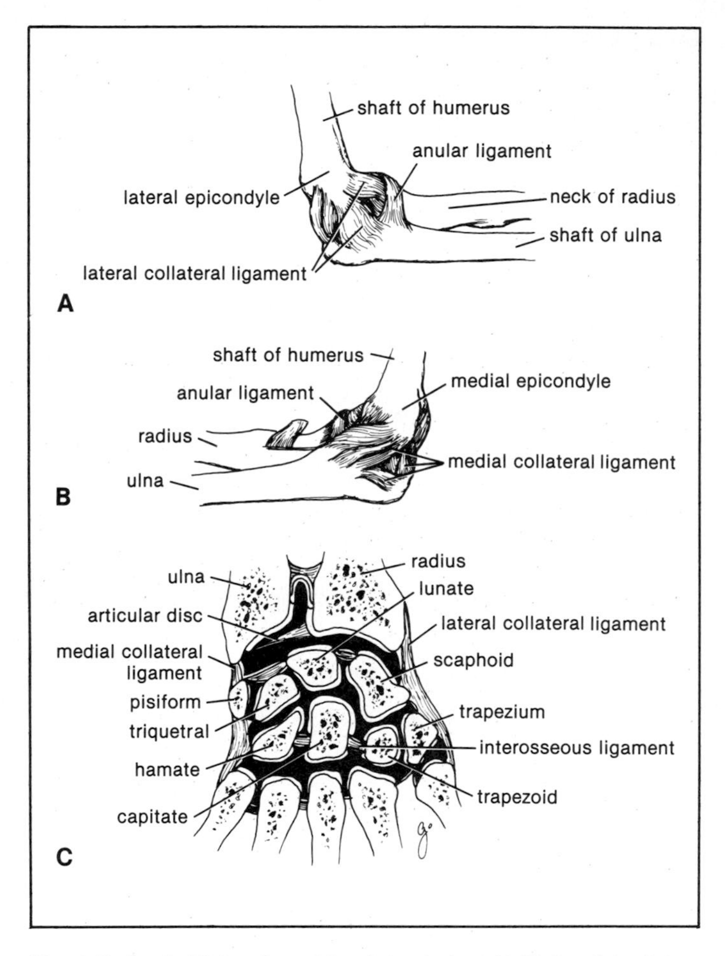

Figure 2–5. **A.** Right elbow joint, lateral view. **B.** Right elbow joint, medial view. **C.** Right wrist and carpal joints, coronal section.

grees in the female. The angle disappears when the elbow joint is fully flexed.

Important Relations. Anterior to the joint are the median nerve and the brachial artery. On the medial side of the joint is the ulnar nerve, which passes behind the medial epicondyle of the humerus.

Superior Radioulnar Joint

Articulation. The circumference of the head of the radius and the anular ligament and the radial notch of the ulna.

Type. Synovial pivot joint.

Capsule. The capsule encloses the joint.

Ligaments. The **anular ligament** forms a collar around the head of the radius and is attached to the margins of the radial notch of the ulna. The **quadrate ligament** is the name given to the lower part of the capsule that lies between the radius and the ulna.

Synovial Membrane. This lines the capsule and is continuous above with that of the elbow joint.

Nerve Supply. Branches of the median, ulnar, musculocutaneous, and radial nerves.

Movements. Pronation and supination of the forearm (see below).

Middle Radioulnar "Joint"

The shafts of the radius and the ulna are connected by the oblique cord and the interosseous membrane.

The **oblique cord** is an inconstant fibrous band that extends from the radius, below the tuberosity, to the coronoid process of the ulna. Its function is unknown.

The **interosseous membrane** unites the interosseous borders of the radius and the ulna. Its fibers run downward and medially, so that a force applied to the lower end of the radius is transmitted to the ulna. Its fibers are taut when the forearm is in the midprone position, thus the forearm bones are most stable when the hand assumes the position of function. The membrane provides attachment for muscles.

Inferior Radioulnar Joint

Articulation. The rounded head of the ulna and the ulnar notch on the radius.

Type. Synovial pivot joint.

Ligaments. Weak **anterior** and **posterior ligaments** strengthen the capsule.

Articular Disc. This fibrocartilaginous disc is triangular in shape. It is attached by its apex to the base of the styloid process of the ulna and by its base to the lower border of the ulnar notch of the radius. It binds together the distal ends of the radius and the ulna and shuts off the inferior radioulnar joint from the wrist joint (*Fig. 2–5*).

Synovial Membrane. This lines the capsule and forms a small pouch that projects upward anterior to the lower end of the interosseous membrane.

Nerve Supply. Anterior interosseous branch of the median and the deep branch of the radial.

Movements. The movements of pronation and supination of the forearm are rotary movements around a vertical axis at the superior and inferior radioulnar joints. The movement of pronation results in the hand rotating medially in such a manner that the palm comes to face posteriorly, and the thumb lies on the medial side. The movement of supination is a reversal of this process, so that the hand returns to the anatomical position, and the palm faces anteriorly. The following muscles produce these movements:

Pronation. Pronator teres, pronator quadratus.

Supination. Biceps, supinator.

Supination is the most powerful of the two movements because of the strength of the biceps muscle. This fact is made use of in the design of screw threads and the spiral corkscrews.

Wrist Joint (Radiocarpal Joint)

Articulation. Between the lower end of the radius and the triangular cartilaginous articular disc above, and the scaphoid, lunate, and triquetral bones below (*Fig. 2–5*). The proximal articular surface forms an ovoid concave surface, which is adapted to the distal ovoid convex surface.

Type. Synovial condyloid joint.

Capsule. The capsule encloses the joint.

Ligaments. **Anterior and posterior ligaments** strengthen the capsule. The **medial ligament** connects the styloid process of the ulna to the triquetral bone. The **lateral ligament** connects the styloid process of the radius to the scaphoid bone.

Synovial Membrane. This lines the capsule.

Nerve Supply. Anterior interosseous nerve from the median and the deep branches of the radial and ulnar nerves.

Movements. Flexion, extension, abduction, adduction, and circumduction. Rotation is not possible because the articular surfaces are ovoid in shape. The lack of rotation is compensated for by the movements of pronation and supination of the forearm. The following muscles produce the movements of the wrist joint:

Flexion. Flexor carpi radialis, flexor carpi ulnaris, palmaris longus, flexor digitorum superficialis, flexor digitorum profundus, and flexor pollicis longus.

Extension. Extensor carpi radialis longus, extensor carpi radialis brevis, extensor carpi ulnaris, extensor digitorum, extensor indicis, extensor digiti minimi, and extensor pollicis longus.

Abduction. Flexor carpi radialis, extensor carpi radialis longus and brevis, abductor pollicis longus, and extensor pollicis longus and brevis.

Adduction. Flexor and extensor carpi ulnaris.

Important Relations. *Anteriorly.* The median and ulnar nerves. *Laterally.* The radial artery.

Intercarpal Joints

Articulation. Between bones of the proximal row of the carpus, i.e., the scaphoid, lunate, triquetral, and pisiform; between bones of the distal row of the carpus, i.e., the trapezium, trapezoid, capitate, and hamate; and finally between the proximal and distal rows of carpal bones (*Fig. 2–5*).

Type. Synovial gliding joints.

Capsule. Encloses each joint.

Ligaments. The bones are connected by strong **anterior, posterior**, and **interosseous** ligaments (*Fig. 2–5*).

Synovial Membrane. Lines the capsules. The joint cavity extends not only between the proximal and distal rows of carpal bones, but also proximally between the individual bones forming the proximal row and distally between the bones of the distal row.

Nerve Supply. Anterior interosseous branch of the median and the deep branches of the radial and ulnar nerves.

Movements. Small amount of gliding movement.

Carpometacarpal and Intermetacarpal Joints

These are synovial gliding joints having **anterior, posterior**, and **interosseous ligaments**. They have a common joint cavity (*Fig. 2–5*).

Carpometacarpal Joint of the Thumb

Articulation. Trapezium and the saddle-shaped base of the first metacarpal bone.

Type. Synovial saddle joint (biaxial joint).

Capsule. Encloses the joint.

Synovial Membrane. Lines capsule and forms separate joint cavity.

Movements. The following muscles produce the movements:

Flexion. Flexor pollicus brevis and longus, opponens pollicis.

Extension. Extensor pollicis longus and brevis.

Abduction. Abductor pollicis longus and brevis.

Adduction. Adductor pollicis.

Rotation (Opposition). The thumb is rotated medially by the opponens pollicis.

Metacarpophalangeal Joints

Articulation. Convex heads of metacarpal bones and concave bases on the proximal phalanges.

Type. Synovial condyloid joints.

Capsule. Encloses the joint.

Ligaments. The **palmar ligaments** are strong and contain some fibrocartilage. The **collateral ligaments** are cordlike bands that join the head of the metacarpal bone to the base of the phalanx. The collateral ligaments are taut when the joint is in flexion and lax when the joint is in extension. The fingers are thus locked in flexion but can be abducted and adducted in extension.

Synovial Membrane. Lines the capsule.

Movements. The following muscles produce the movements:

Flexion. Lumbricals and interossei assisted by flexor digitorum superficialis and profundus.

Extension. Extensor digitorum, extensor indicis, and extensor digiti minimi.

Abduction. (Movement away from the midline of the third finger.) Dorsal interossei.

Adduction. (Movement toward the midline of the third finger.) Palmar interossei.

In the case of the metacarpal phalangeal joint of the thumb, **flexion** is performed by the flexor pollicis longus and brevis, **extension** by the extensor pollicis longus and brevis. The movements of abduction and adduction are performed at the carpometacarpal joint.

Interphalangeal Joints

Interphalangeal joints are synovial hinge joints that have a structure similar to that of the metacarpophalangeal joints.

JOINTS OF THE PELVIS

Sacroiliac Joint

Articulation. The auricular surfaces of the sacrum and the ilium (*Fig. 2-6*). The articular surfaces are irregular and interlock with one another.

Type. Synovial gliding joint.

Capsule. Encloses the joint.

Ligaments. **Anterior sacroiliac ligament** is a thickening of the capsule. The **interosseous sacroiliac ligament** is very strong. It is situated above and behind the joint cavity and binds the bones together. The **posterior sacroiliac ligament** overlies the interosseous ligament (*Fig. 2-6*).

Accessory Ligaments. The **sacrotuberous ligament** connects the back of the sacrum to the ischial tuberosity (*Fig. 2-6*). The **sacrospinous ligament** lies anterior to the sacrotuberous ligament and connects the sacrum to the ischial spine. These two ligaments prevent the lower end of the sacrum from rotating backward as the weight of the vertebral column is transmitted downward through the sacroiliac joints. The **iliolumbar ligament** connects the tip of the transverse process of the fifth lumbar vertebra to the iliac crest. This acts as a suspensory ligament of the vertebral column (*Fig. 2-6*).

Synovial Membrane. Lines the capsule.

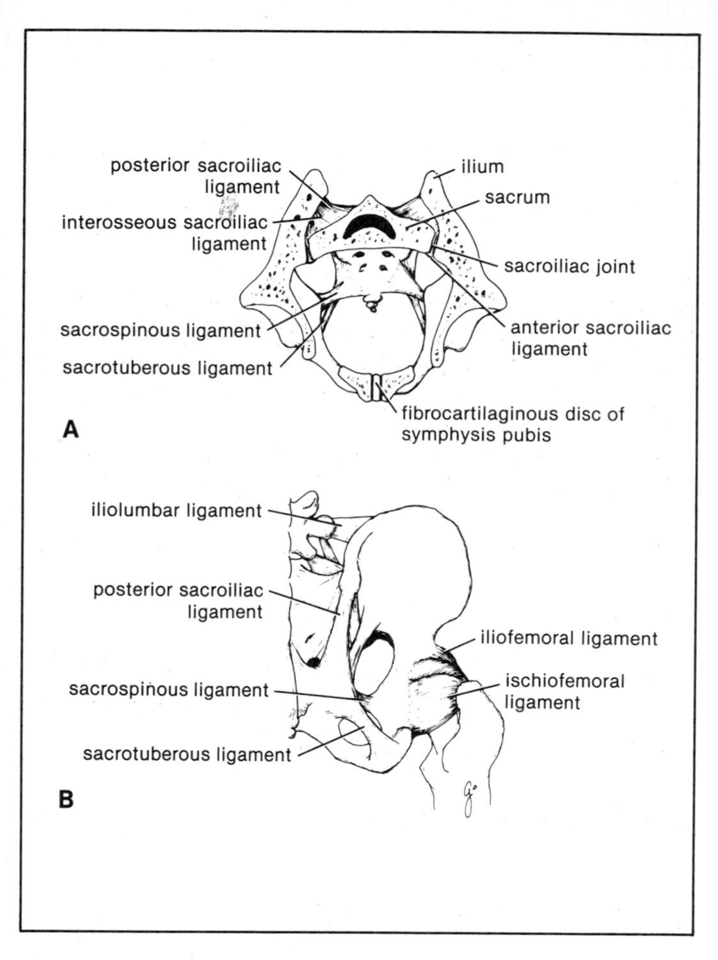

Figure 2–6. A. Sacroiliac joints and symphysis pubis, coronal section. **B.** Sacroiliac joints, posterior view.

Nerve Supply. Branches from the sacral spinal nerves.

Movements. A small amount of movement is possible. During pregnancy the ligaments undergo softening in response to hormones thus increasing the mobility and increasing the potential size of the pelvis during childbirth. Obliteration of the joint cavity occurs in both sexes after middle age.

Symphysis Pubis

Articulation. Apposed surfaces of the bodies of the pubic bones (*Fig. 2–6*).

Type. Secondary cartilaginous joint.

Ligaments. **Superior and inferior (arcuate) ligaments** extend from one pubic bone to the other. The **interpubic disc** is composed of fibrocartilage and connects together the two pubic bones. The articular surfaces are covered by a thin layer of hyaline cartilage. The disc may have a small cavity in the midline.

Movements. Almost no movement is possible. During pregnancy the ligaments undergo softening in response to hormones thus increasing the mobility and the potential size of the pelvis during childbirth.

JOINTS OF THE LOWER EXTREMITY

Hip Joint

Articulation. Between the hemispherical head of the femur and the cup-shaped acetabulum of the hip bone (*Fig. 2–7*). The articular surface of the acetabulum is horseshoe-shaped and is deficient inferiorly at the **acetabular notch**. The cavity of the acetabulum is deepened by the presence of a fibrocartilaginous rim, called the **acetabular labrum**. The labrum bridges across the

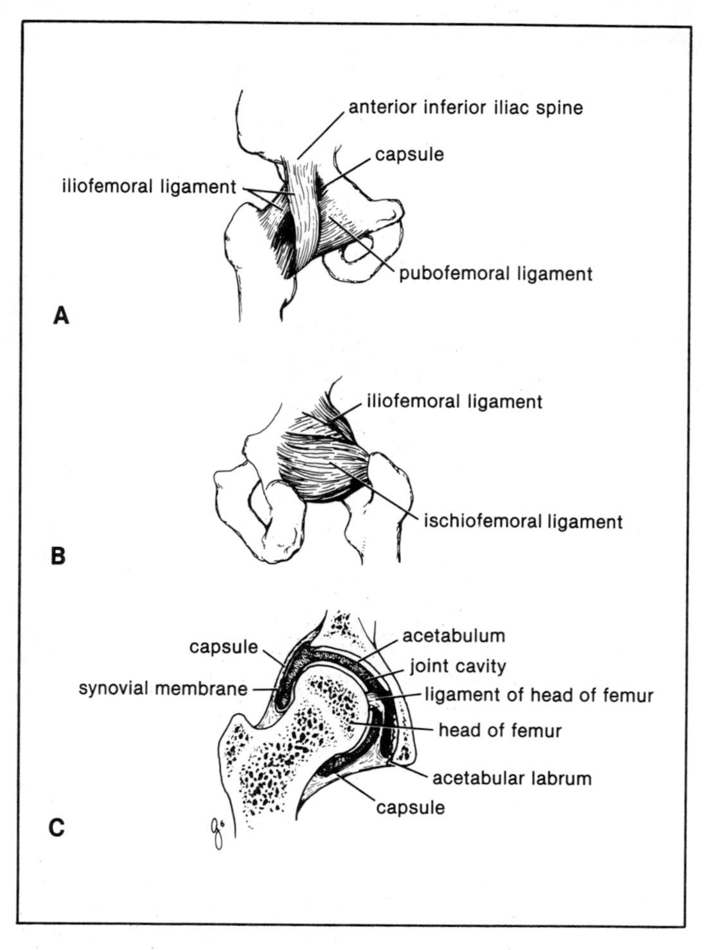

Figure 2–7. Right hip joint. **A.** Anterior view. **B.** Posterior view. **C.** Coronal section.

acetabular notch and is here called the **transverse acetabular ligament**. The articular surfaces are covered with hyaline cartilage.

Type. Synovial ball-and-socket joint.

Capsule. This encloses the joint and is attached medially to the acetabular labrum. Laterally it is attached to the intertrochanteric line of the femur in front and halfway along the posterior aspect of the neck of the bone behind.

Ligaments. The **iliofemoral ligament** is strong and shaped like an inverted Y. Its base is attached to the anterior inferior iliac spine above and the two limbs of the Y are attached to the upper and lower parts of the intertrochanteric line of the femur. The **pubofemoral ligament** is triangular in shape. The base is attached to the superior ramus of the pubis, and the apex is attached below to the lower part of the intertrochanteric line. The **ischiofemoral ligament** is spiral in shape and attached to the body of the ischium and laterally to the greater trochanter. The **ligament of the head of the femur** is flat and triangular in shape. It is attached by its apex to the fovea capitis of the femur and by its base to the transverse acetabular ligament. This ligament lies within the joint and is ensheathed by synovial membrane (*Fig. 2–7*).

Synovial Membrane. This lines the capsule and is attached to the margins of the articular surfaces (*Fig. 2–7*). It covers the portion of the neck of the femur that lies within the joint capsule. It ensheathes the ligament of the head of the femur and covers the acetabular fossa. It frequently communicates with the **psoas bursa** beneath the psoas tendon.

Nerve Supply. Femoral, obturator, and sciatic nerves and the nerve to the quadratus femoris.

Movements. The hip joint has a wide range of movement, but less so than the shoulder joint. The strength of the joint depends largely on the shape of the bones taking part in the articulation and on the very strong ligaments. When the knee is flexed, flexion is limited by the anterior surface of the thigh coming into contact with the anterior abdominal wall. When the knee joint is extended, flexion of the hip joint is limited by the tension of the hamstring group of muscles on the back of the thigh. Extension, which is the movement of the flexed thigh backward to the anatomical position, is limited by the tension of the iliofemoral ligament. Abduction is limited by the tension of the pubofemoral ligament, and adduction is limited by contact with the opposite limb and by the tension in the ligament of the head of the femur. Lateral rotation is limited by the tension in the iliofemoral and pubofemoral ligaments, and medial rotation is limited by the ischiofemoral ligament. The following movements take place:

Flexion. Iliopsoas, rectus femoris, sartorius, adductor muscles.

Extension. (A posterior movement of the flexed thigh.) Gluteus maximus, hamstring muscles.

Abduction. Gluteus medius and minimus, sartorius, tensor fasciae latae, piriformis.

Adduction. Adductor longus and brevis, adductor fibers of adductor magnus, pectineus, gracilis.

Lateral Rotation. Piriformis, obturator internus and externus, superior and inferior gemelli, quadratus femoris, gluteus maximus.

Medial Rotation. Anterior fibers of gluteus medius and minimus, tensor fasciae latae.

Circumduction. A combination of the above movements.

Important Relations. *Anteriorly.* Femoral vessels and nerve. *Posteriorly.* Sciatic nerve.

Knee Joint

Articulation. Above are the rounded condyles of the femur; below are the condyles of the tibia and their semilunar cartilages (*Fig. 2–8*); in front is the articulation between the lower end of the femur and the patella. The articular surfaces are covered with hyaline cartilage.

Type. The joint between the femur and the tibia is a synovial joint of the hinge variety, but some degree of rotatory movement is possible. The joint between the patella and the femur is a synovial gliding joint.

Capsule. This encloses the joint. On the front of the joint, however, the capsule is absent, permitting the synovial membrane to pouch upward beneath the quadriceps tendon, forming the **suprapatellar bursa** (*Fig. 2–8*).

Ligaments. These may be divided into those that lie outside the capsule and those that lie within the capsule.

Extracapsular Ligaments. The **ligamentum patellae** is attached above to the lower border of the patella and below to the tubercle of the tibia. It is a continuation of the tendon of the quadriceps femoris muscle. The **lateral collateral ligament** is cordlike and is attached above to the lateral condyle of the femur and below to the head of the fibula. The tendon of the popliteus muscle separates the ligament from the lateral semilunar cartilage (*Fig. 2–8*). The **medial collateral ligament** is a flat band that is attached above to the medial condyle of the femur and below to the medial surface of the shaft of the tibia. It is strongly attached to the medial semilunar cartilage (*Fig. 2–9*). The **oblique popliteal ligament** is a tendinous expansion derived from the semimembranosus muscle. It strengthens the posterior part of the capsule.

Intracapsular Ligaments. These are the **cruciate ligaments** that are two very strong ligaments that cross each other within the

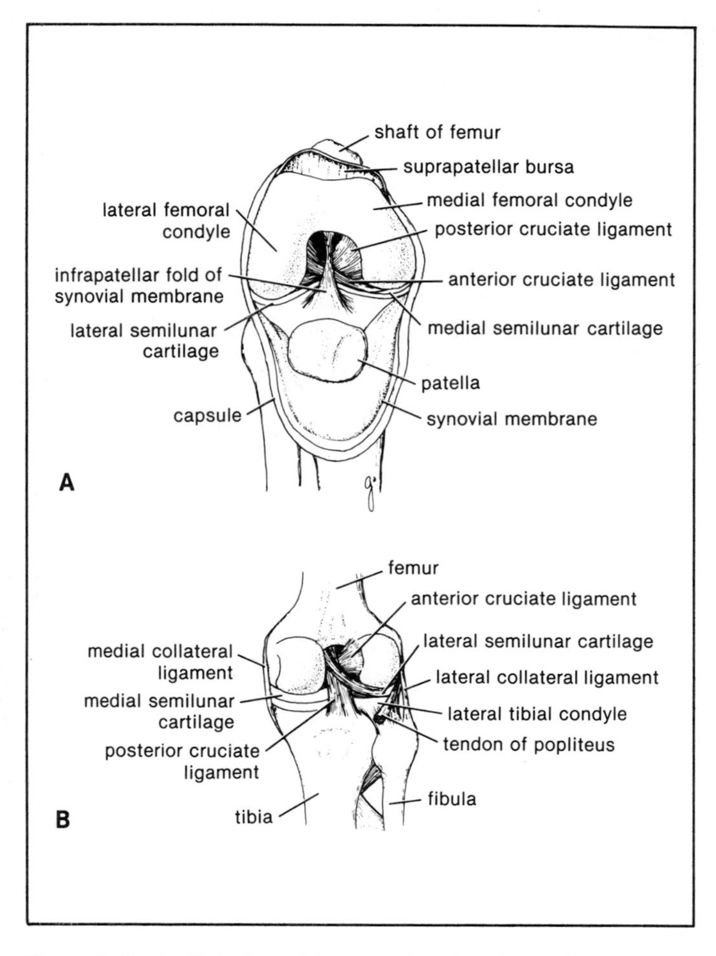

Figure 2–8. **A.** Right knee joint, anterior view, internal aspect; the capsule has been cut and the patella has been turned downward. **B.** Right knee joint, posterior view, internal aspect; the capsule and the synovial membrane have been removed.

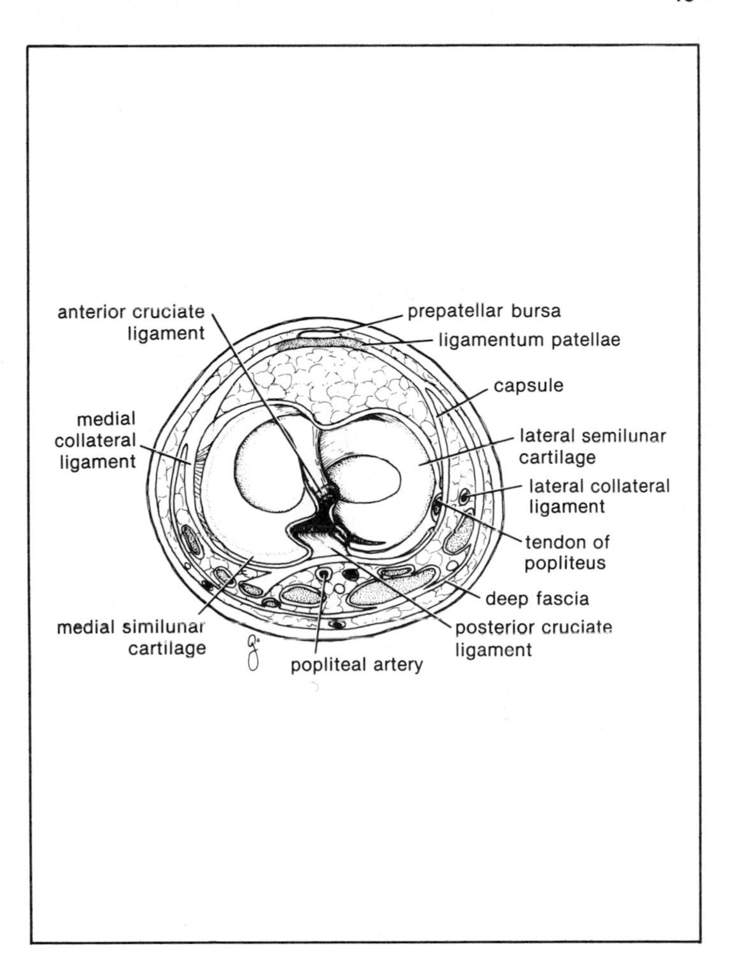

Figure 2–9. Cross section of knee joint, showing position of ligaments and semilunar cartilages.

joint cavity (*Figs. 2–8 and 2–9*). They are named anterior and posterior, according to their tibial attachments. The **anterior cruciate ligament** is attached to the anterior intercondylar area of the tibia and passes upward, backward, and laterally, to be attached to the posterior part of the medial surface of the lateral femoral condyle. The **posterior cruciate ligament** is attached to the posterior intercondylar area of the tibia and passes upward, forward, and medially, to be attached to the anterior part of the lateral surface of the medial femoral condyle.

Semilunar Cartilages. The semilunar cartilages are C-shaped sheets of fibrocartilage. The peripheral border of each is thick and attached to the capsule, and the inner border is thin and concave and forms a free edge (*Fig. 2–9*). The upper surfaces are in contact with the femoral condyles. The lower surfaces are in contact with the tibial condyles. Their function is to deepen the articular surfaces of the tibial condyles to receive the convex femoral condyles. Each cartilage is attached to the upper surface of the tibia by anterior and posterior horns. Because the medial cartilage is also attached to the medial collateral ligament it is relatively immobile.

Synovial Membrane. The synovial membrane lines the capsule and is attached to the margins of the articular surfaces. On the front of the joint it forms a pouch, which extends up beneath the quadriceps femoris muscle for three finger-breadths above the patella, forming the **suprapatellar bursa**.

At the back of the joint the synovial membrane is prolonged downward on the deep surface of the tendon of the popliteus, forming the **popliteal bursa**. The synovial membrane is reflected forward from the posterior part of the capsule around the front of the cruciate ligaments. As a result the cruciate ligaments lie behind the synovial cavity.

In the anterior part of the joint the synovial membrane is reflected backward from the ligamentum patellae to form the **infrapatellar fold**; the free borders of the fold are called the **alar folds**.

Bursae Related to the Knee Joint. There are numerous bursae related to the knee joint. They are found wherever skin, muscle, or tendon rubs against bone. The following bursae always communicate with the joint cavity. The **suprapatellar bursa** that lies beneath the quadriceps muscle. The **popliteal bursa** that surrounds the tendon of the popliteus muscle as it leaves the joint.

The **semimembranosus bursa**, which lies between the tendon of this muscle and the medial condyle of the tibia, may communicate with the joint cavity.

Nerve Supply of the Knee Joint. Femoral, obturator, common peroneal, and tibial nerves.

Movements of the Knee Joint. These are flexion, extension, and rotation. As the knee joint assumes the position of full extension, medial rotation of the femur results in a twisting and tightening of all the major ligaments of the joint, and the knee becomes a mechanically rigid structure; the semilunar cartilages are compressed like rubber cushions between the femoral and tibial condyles. The extended knee is said to be in the locked position.

Before flexion of the knee joint can occur, it is essential that the major ligaments be untwisted and slackened to permit movements between the joint surfaces. This unlocking or untwisting process is accomplished by the popliteus muscle, which laterally rotates the femur on the tibia. Once again the semilunar cartilages have to adapt their shape to the changing contour of the femoral condyles.

When the knee joint is flexed to a right angle, a considerable range of rotation is possible. In the flexed position, the tibia can also be moved passively forward and backward on the femur; this is possible because the major ligaments, especially the cruciate ligaments, are slack in this position. The following muscles produce the movements of the knee joint:

Flexion. Biceps femoris, semitendinosus, semimembranosus.

Extension. Quadriceps femoris.

Medial Rotation. Sartorius, gracilis, semitendinosus.

Lateral Rotation. Biceps femoris.

Superior Tibiofibular Joint

Articulation. Oval facet on lateral condyle of tibia and oval facet on head of fibula.

Type. Synovial gliding joint.

Capsule. Encloses the joint.

Ligaments. **Anterior and posterior ligaments** that strengthen the capsule.

Synovial Membrane. Lines the capsule.

Nerve Supply. Common peroneal nerve and nerve to popliteus.

Movements. Small amount of gliding movement during dorsiflexion of ankle joint. This is produced by the wider anterior part of the articular surface of the talus forcing the malleoli to separate during dorsiflexion of the ankle.

Middle Tibiofibular "Joint"

This "joint" is formed by the interosseous membrane connecting the interosseous borders of the tibia and the fibula. The majority of the fibers of the membrane are directed downward and laterally. The anterior tibial vessels pass forward in the interval above the membrane and the perforating branch of the peroneal artery pierces the membrane below. The membrane provides attachment for muscles on its anterior and posterior surfaces.

Inferior Tibiofibular Joint

Articulation. Fibular notch at the lower end of the tibia and the medial side of the lower end of the fibula.

Type. Fibrous joint.

Capsule. None.

Ligaments. The **interosseus ligament** is a strong band of fibrous tissue that binds the two bones together. The **anterior and posterior ligaments** are flat bands of fibrous tissue connecting the two bones in front and behind the interosseous ligament. The **inferior transverse ligament** runs from the medial surface of the upper part of the lateral malleolus to the posterior border of the lower end of the tibia.

Nerve Supply. Deep peroneal and tibial nerves.

Movements. A small amount of movement occurs during movements at the ankle joint. See movements at superior tibiofibular joint.

Important Relations. *Posteriorly.* The popliteal vessels and the tibial and common peroneal nerves.

Ankle Joint

Articulation. Between the lower end of the tibia, the two malleoli, and the body of the talus (*Fig. 2-10*). The **inferior transverse tibiofibular ligament** deepens the socket into which the body of the talus fits snugly.

Type. Synovial hinge joint.

Capsule. This encloses the joint.

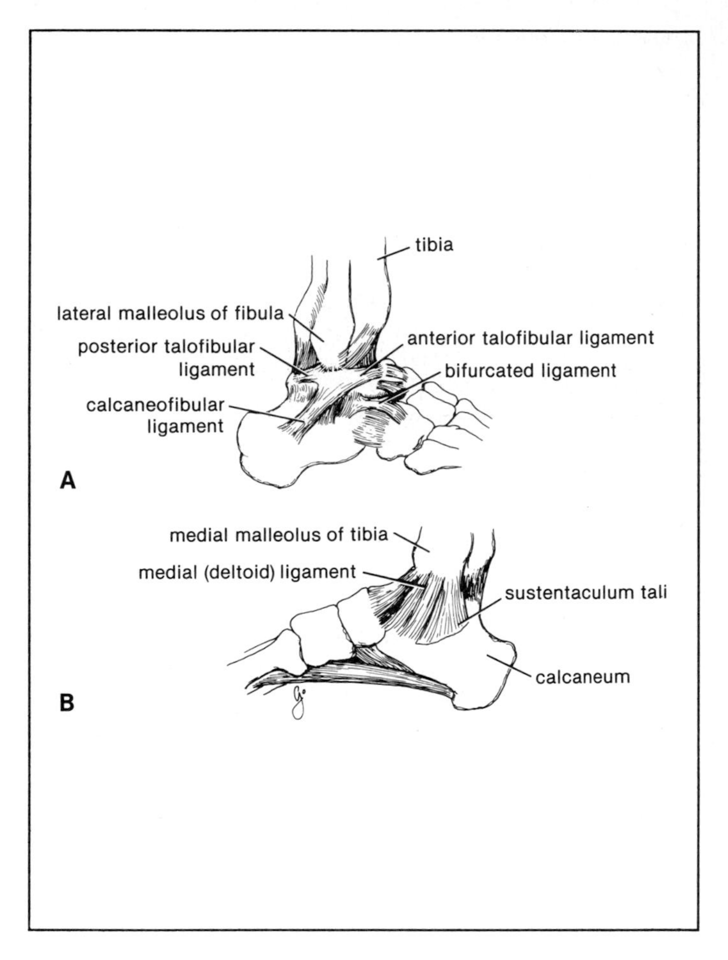

Figure 2–10. Right ankle joint. **A.** Lateral aspect. **B.** Medial aspect.

Ligaments. The **medial, or deltoid ligament** is very strong and is attached by its apex to the tip of the medial malleolus (*Fig. 2–10*). The deep fibers are attached below to the medial surface of the body of the talus, the superficial fibers are attached to the medial side of the talus, the sustentaculum tali, the plantar calcaneonavicular ligament, and the tuberosity of the navicular bone.

The **lateral ligament** is weaker than the medial ligament and consists of three distinct bands: (1) The **anterior talofibular ligament** runs from the lateral malleolus to the lateral surface of the talus, (2) the **calcaneofibular ligament** runs from the lateral malleolus to the lateral surface of the calcaneum, (3) the **posterior talofibular ligament** runs from the lateral malleolus to the posterior tubercle of the talus (*Fig. 2–10*).

Synovial Membrane. This lines the capsule.

Nerve Supply. Deep peroneal and tibial nerves.

Movements. These are dorsiflexion (toes pointing upward) and plantar flexion (toes pointing downward). The movements of inversion and eversion take place at the tarsal joints and **not at the ankle joint**. The following muscles perform the movements:

Dorsiflexion. Tibialis anterior, extensor hallucis longus, extensor digitorum longus, peroneus tertius.

Plantar Flexion. Gastrocnemius, soleus, plantaris, peroneus longus, peroneus brevis, tibialis posterior, flexor digitorum longus, flexor hallucis longus.

Important Relations. *Anteriorly.* The anterior tibial vessels and the deep peroneal nerve. *Posteriorly.* The tendo calcaneus. *Posterolaterally* (behind the lateral malleolus). The tendons of the peroneus longus and brevis muscles. *Posteromedially* (behind the medial malleolus). The posterior tibial vessels and the tibial nerve and the long flexor tendons of the foot.

Intertarsal Joints

Subtalar Joint

Articulation. This is the posterior joint between the talus and the calcaneum. Concave inferior surface of the body and convex facet on upper surface of calcaneum.

Type. Synovial gliding joint.

Capsule. Encloses the joint.

Ligaments. **Medial and lateral talocalcaneal** ligaments strengthen the capsule. The **interosseous talocalcaneal ligament** (*Fig. 2–11*) is very strong and is attached above to the sulcus tali and below to the sulcus calcanei.

Synovial Membrane. This lines the capsule.

Movements. See calcaneocuboid joint.

Talocalcaneonavicular Joint

Articulation. The rounded head of the talus, the upper surface of the sustentaculum tali of the calcaneum, and the posterior concave surface of the navicular bone.

Type. Synovial.

Capsule. Incompletely encloses the joint.

Ligaments. The **plantar calcaneonavicular ligament** runs from the anterior border of the sustentaculum tali to the inferior surface and tuberosity of the navicular bone. The superior surface is covered with fibrocartilage and supports the head of the talus. The **bifurcated ligament** (*Fig. 2–10*) strengthens the dorsal surface of the joint (see below).

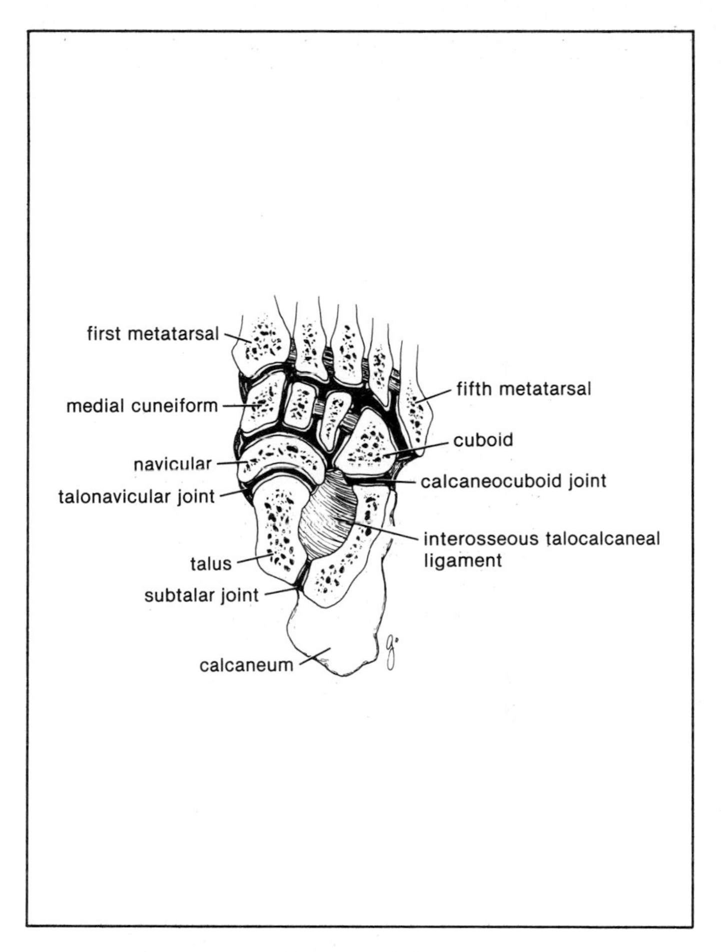

Figure 2–11. Right foot. Oblique section shows synovial cavities of the tarsal and tarsometatarsal joints, viewed from above.

Synovial Membrane. Lines the capsule.

Movements. Gliding and rotatory movements are possible. See calcaneocuboid joint.

Calcaneocuboid Joint

Articulation. The anterior end of the calcaneum and the posterior surface of the cuboid (*Fig. 2–11*).

Type. Synovial, gliding.

Capsule. Encloses the joint.

Ligaments. The **bifurcated ligament** is on the dorsal surface of the joint. The stem is attached to the dorsal surface of the calcaneum. The lateral limb is attached to the dorsal surface of the cuboid, and the medial limb to the dorsal surface of the navicular bone (*Fig. 2–10*). The **long plantar ligament** is a strong ligament on the plantar surface of the joint. It connects the calcaneum posteriorly with the cuboid, the bases of the third, fourth and fifth metatarsal bones anteriorly. The **short plantar ligament** is wide and strong and connects the plantar surface of the calcaneum to the adjoining part of the cuboid.

Synovial Membrane. This lines the capsule.

Movement. The **subtalar, talocalcaneonavicular, and calcaneocuboid joints**. The important movement of inversion and eversion take place at these joints.

Inversion. Movement of the foot so that the sole faces medially. It is performed by the tibialis anterior, extensor hallucis longus, medial tendons of extensor digitorum longus, tibialis posterior.

Eversion. The opposite movement of the foot so that the sole faces in a lateral direction. It is performed by the peroneus longus, peroneus brevis, peroneus tertius, the lateral tendons of extensor digitorum longus.

The movement of inversion is more extensive than eversion.

Cuneonavicular Joint

Articulation. Between the three cuneiform bones and the navicular bone (*Fig. 2–11*).

Type. Synovial gliding joint.

Capsule. Encloses the joint. The joint cavity is continuous with those of the intercuneiform and cuneocuboid joints, the cuneometatarsal and intermetatarsal joints, i.e., the joints between the bones of the second and third, and third and fourth, metatarsal bones.

Ligaments. **Dorsal and plantar ligaments** strengthen the capsule.

Cuboideonavicular Joint

This is a fibrous joint with the bones connected by dorsal, plantar, and interosseous ligaments. A small amount of movement is possible.

Intercuneiform and Cuneocuboid Joints

The intercuneiform and cuneocuboid joints are synovial joints of the gliding variety. The bones are connected by dorsal, plantar, and interosseous ligaments.

Tarsometatarsal and Intermetatarsal Joints

The tarsometatarsal and intermetatarsal joints are synovial joints of the gliding variety (*Fig. 2–11*). The bones are connected by dorsal, plantar, and interosseous ligaments. The tarsometatarsal joint of the big toe has a separate joint cavity.

Metatarsophalangeal and Interphalangeal Joints

The metatarsophalangeal and interphalangeal joints are similar to those of the hand (*see pp. 66 and 67*). The movements of abduction and adduction of the toes, performed by the interossei muscles, are small in amount, and take place from the midline of the second digit and not the third, as in the hand.

ARCHES OF FOOT

Bones

Medial Longitudinal Arch. Consisting of calcaneum, talus, navicular bone, three cuneiform bones, and first (medial) three metatarsal bones.

Lateral Longitudinal Arch. Consisting of calcaneum, cuboid, and fourth and fifth metatarsal bones.

Transverse Arch. Consisting of the bases of the metatarsal bones, cuboid, and the three cuneiform bones.

Mechanisms of Arch Support

Apart from the shape of the individual bones, the following muscles and ligaments support the arches.

Medial Longitudinal Arch

Muscular Support. Medial part of flexor digitorum brevis, abductor hallucis, flexor hallucis longus, medial part of flexor digitorum longus, flexor hallucis brevis, tibialis anterior, tendinous extensions of the insertion of the tibialis posterior.

Ligamentous Support. Plantar and dorsal ligaments including the important plantar calcaneonavicular ligament; the medial ligament of the ankle joint and the plantar aponeurosis.

Lateral Longitudinal Arch

Muscular Support. Abductor digiti minimi, lateral part of the flexor digitorum longus and brevis, peroneus longus and brevis.

Ligamentous Support. Long and short plantar ligaments and the plantar aponeurosis.

Transverse Arch

Muscular Support. Dorsal interossei, transverse head of adductor hallucis, peroneus longus, and peroneus brevis.

Ligamentous Support. Deep transverse ligaments, very strong plantar ligaments.

The marked wedge-shape of the cuneiform bones and the bases of the metatarsal bones plays a large part in the support of the transverse arch.

3

Skeletal Muscles

GENERAL DESCRIPTION

Skeletal muscles bring about movements by pulling on bones. The bones move toward one another by changing their position at the joints. Muscles are supported and kept in separate groups by fibrous tissue called the **deep fascia.** This should be distinguished from the superficial fascia that lies beneath the skin and is composed of loose areolar and adipose tissue. In certain regions of the body, such as the wrist and the ankle, the deep fascia is thickened to form dense fibrous bands, which pass across the tendons of muscles so as to bind them down and hold them in position. These fibrous bands, which are known as **retinacula,** prevent the tendons in front of the wrist, for example, from bowing forward when the fingers are flexed.

In the palm of the hand and the sole of the foot the deep fascia is greatly thickened to protect the underlying tendons, nerves, and blood vessels. These thickenings are known as the **palmar aponeurosis** and the **plantar aponeurosis,** respectively.

How to Learn About Skeletal Muscles

There are a large number of skeletal muscles in the body and it is important that one understands the function of each muscle and the disability the patient would suffer should the function be disturbed. In the following pages the most important muscles of the body, their origin, insertion, nerve supply, and action are described briefly in *Tables 3-1 through 3-32* (tables appear on pp. 117–164). If you have access to an articulated skeleton it is helpful to imagine that you are contracting a particular muscle and you should then move the appropriate bone or bones on the skeleton. Remember that you yourself have these muscles and skeleton and joints, so do not hesitate to reinforce your learning by performing the movements on yourself. This is particularly important in the case of the limbs. If you find it helpful, perform the muscle action in front of a mirror.

Skeletal Muscle

Skeletal muscle is so named because it is attached to the bones of the skeleton. The muscle fibers or cells are sometimes called striped or striated cells because when they are examined under a microscope they display cross striations or cross bands.

A skeletal muscle has two or more attachments. The attachment that moves the least is referred to as the **origin,** and that which moves the most, as the **insertion.** Under varying circumstances the degree of mobility of the attachments may be reversed, and therefore the terms **origin** and **insertion** are interchangeable. For example, the biceps muscle of the upper limb flexes the forearm on the arm at the elbow joint. However, if the hand grasps a heavy table, contraction of the biceps will cause the arm to be flexed on the forearm; in this situation the origin and the insertion of the biceps have been interchanged.

The fleshy part of the muscle is referred to as its **belly.** The ends of a muscle are attached to bones, cartilage, or ligaments by cords of fibrous tissue called **tendons.** Occasionally, flattened muscles are attached by a thin strong sheet of fibrous tissue called an **aponeurosis.** The individual muscle cells or fibers of a muscle are arranged either parallel or oblique to the long axis of the muscle. Because a muscle shortens by one-third or one-half of its resting length when it contracts, then it follows that muscles whose fibers run parallel to the line of pull will bring about a greater degree of movement as compared with those whose fibers are shorter and run obliquely. Examples of muscles with parallel arranged fibers are the sternocleidomastoid, the rectus abdominis, and the sartorius.

Muscles whose fibers run obliquely to the line of pull are referred to as **pennate muscles** (they resemble a feather). A **unipennate muscle** is one in which the tendon lies along one side of the muscle and the muscle fibers pass obliquely to it (e.g., extensor digitorum longus of the leg). A **bipennate muscle** is one in which the tendon lies in the center of the muscle and the muscle fibers pass to it from two sides (e.g., rectus femoris of the thigh). A **multipennate muscle** (1) may be arranged as a series of bipennate muscles lying alongside one another (e.g., the middle acromial fibers of the deltoid of the arm) or (2) may have the tendon lying within its center and the muscle fibers passing to it from all sides, converging as they go (e.g., tibialis anterior of the leg).

For a given volume of muscle substance, pennate muscles have many more fibers as compared with muscles with parallel arranged fibers, and they are therefore more powerful; therefore, range of movement has been sacrificed to strength.

Skeletal Muscle Action. Muscles produce movements by approximating the structures to which they are attached. It is important to understand that practically all movements are the result of the coordinated action of many muscles. However, to understand a muscle's action it is necessary to study it individually.

A muscle may work in the following ways: (1) as a prime mover, (2) as an antagonist, (3) as a fixator, and (4) as a synergist.

Prime Mover. A muscle is a prime mover when it is the chief muscle or member of a chief group of muscles responsible for a particular movement. For example, the extensor muscle, the quadriceps femoris, is a prime mover in the movement of extending the knee joint.

Antagonist. Any muscle that opposes the action of the prime mover is an antagonist. For example, the flexor muscle, the biceps femoris, opposes the action of the quadriceps femoris when the knee joint is extended. Before a prime mover can contract, there must be equal relaxation of the antagonist muscle; this is brought about by nervous reflex inhibition.

Fixator. This is a muscle that contracts isometrically to stabilize the origin of the prime mover so that it may act efficiently. For example, the muscles attaching the shoulder girdle to the trunk contract as fixators to allow the deltoid to act on the shoulder joint.

Synergist. There are many examples in the body where the prime mover muscle crosses a number of joints before it reaches the joint at which its main action takes place. To prevent unwanted movements in an intermediate joint, groups of muscles called synergists contract and stabilize the intermediate joints. For example, the flexor and extensor muscles of the hand contract to fix the wrist joint, and this allows the long flexor and extensor muscles of the fingers to work efficiently.

It should be understood that these are terms applied to the action of a particular muscle during a particular movement; many muscles can act as a prime mover, an antagonist, a fixator, or a synergist, depending on the movement to be accomplished.

MUSCLES OF THE HEAD AND THE NECK

The muscles of the scalp, the external ear, and the face are all derived from the second pharyngeal arch and are therefore supplied by the facial nerve (seventh cranial nerve).

Muscles of the Scalp and the External Ear

The scalp consists of five layers, the first three of which are intimately bound together and move as a whole on the skull: (1) *S*kin, (2) *C*onnective tissue of superficial fascia, (3) *A*poneurosis of occipitofrontalis muscle, (4) *L*oose connective tissue, that permits layers 1 to 3 to move on layer 5, and (5) *P*eriosteum of skull bones. Note that the first letters spell out SCALP. *(See Table 3–1, p. 117).*

Muscles of Facial Expression

The muscles of the face lie in the superficial fascia and the majority arise from the bones of the skull and are inserted into the skin. The muscles serve as sphincters and dilators to the orbit, nose, and mouth. A secondary function is to modify the expression of the face. *(See Table 3–2, p. 118)*

Muscles of Mastication

The muscles of mastication are four powerful muscles acting on the mandible *(Fig. 3–1)*. They are developed from the first pharyngeal arch and are therefore all supplied by the mandibular division of the trigeminal nerve (fifth cranial nerve). *(See Table 3–3, p. 121)*

Muscles of the Neck

The muscles of the neck are associated with fascia *(Fig. 3–1)*. The **superficial cervical fascia** is a thin layer of connective tissue uniting the dermis to the deep fascia. It contains adipose tissue and encloses the platysma muscle. Also embedded in it are the cutaneous nerves, superficial veins, and the superficial lymph nodes.

The **deep cervical fascia** supports the muscles, vessels, and viscera of the neck. In certain areas it is condensed to form well defined fibrous sheets called the investing layer, the pretracheal layer, and the prevertebral layer. It is also condensed to form the carotid sheath *(Fig. 3–1)*.

The **investing layer** encircles the neck, splitting to ensheathe the trapezius and sternocleidomastoid muscles. It is attached posteriorly to the ligamentum nuchae and above to the mandible, the zygomatic arch and the superior nuchal line of the occipital bone. Below it splits into two layers that are attached to the anterior and posterior borders of the manubrium and the clavicle. It also splits to enclose the parotid and the submandibular salivary glands.

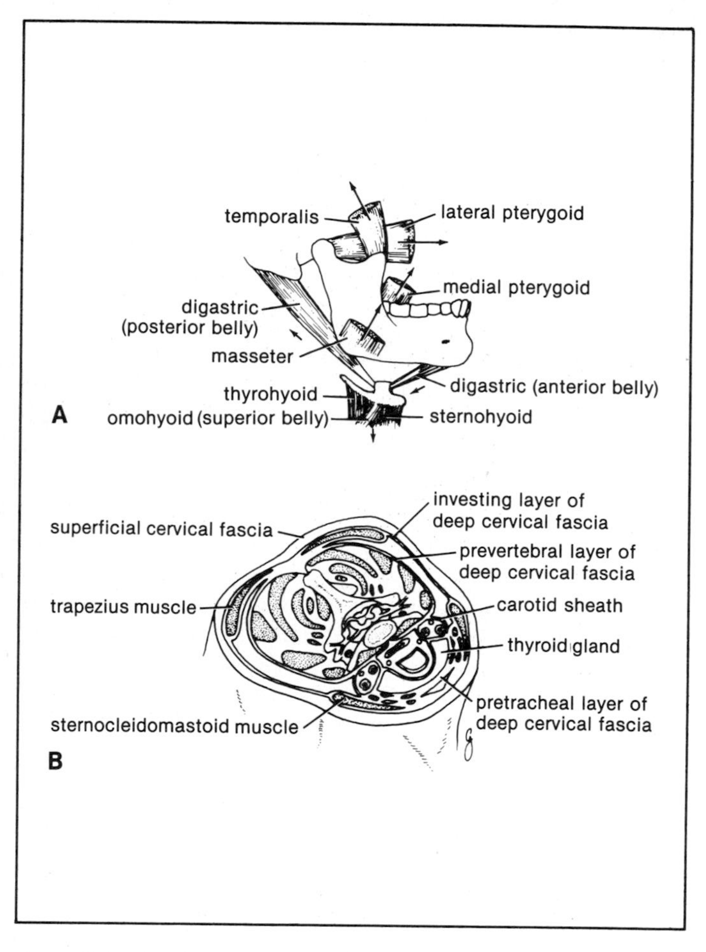

Figure 3–1. A. The attachment of the muscles of mastication to the mandible; the arrows indicate the direction of their actions. **B.** The arrangement of the layers of deep fascia in the neck.

The **pretracheal layer** is thin and is attached above to the laryngeal cartilages. Below it extends into the thorax to blend with the fibrous pericardium. Laterally it is attached to the investing layer of deep cervical fascia beneath the sternocleidomastoid muscle. It surrounds the thyroid and parathyroid glands forming a sheath for them and invests the infrahyoid muscles.

The **prevertebral layer** passes like a septum across the neck between the pharynx, the esophagus, and the vertebral column. It covers the prevertebral muscles and is attached posteriorly to the ligamentum nuchae. It forms the fascial floor of the posterior triangle and extends into the axilla to form the **axillary sheath.**

The **carotid sheath** is a condensation of deep fascia, in which are embedded the common and internal carotid arteries, the internal jugular vein, the vagus nerve, and the deep cervical group of lymph nodes. The muscles of the head and the neck are also associated with important ligaments.

- **Stylohyoid ligament** is a fibrous cord that connects the styloid process to the lesser cornu of the hyoid bone. It is the remains of the second pharyngeal arch.
- **Stylomandibular ligament** is a fibrous band that connects the styloid process to the angle of the mandible. It is a thickening of the investing layer of deep cervical fascia.
- **Sphenomandibular ligament** is a fibrous band that connects the spine of the sphenoid bone to the lingula of the mandible. It is the remains of the first pharyngeal arch.
- **Pterygomandibular ligament** is a fibrous cord that extends from the hamular process of the medial pterygoid plate to the posterior end of the mylohyoid line of the mandible. It gives attachment to the superior constrictor and buccinator muscles.

The superficial muscles of the side of the neck are shown in *Table 3–4, p. 122.*

Triangles of the Neck. The neck is commonly divided by the presence of the sternocleidomastoid muscle into **anterior and posterior triangles.** The anterior triangle is bounded by the body of the mandible above, the sternocleidomastoid muscle posteriorly, and the midline anteriorly *(Fig. 3–2).* The posterior triangle is

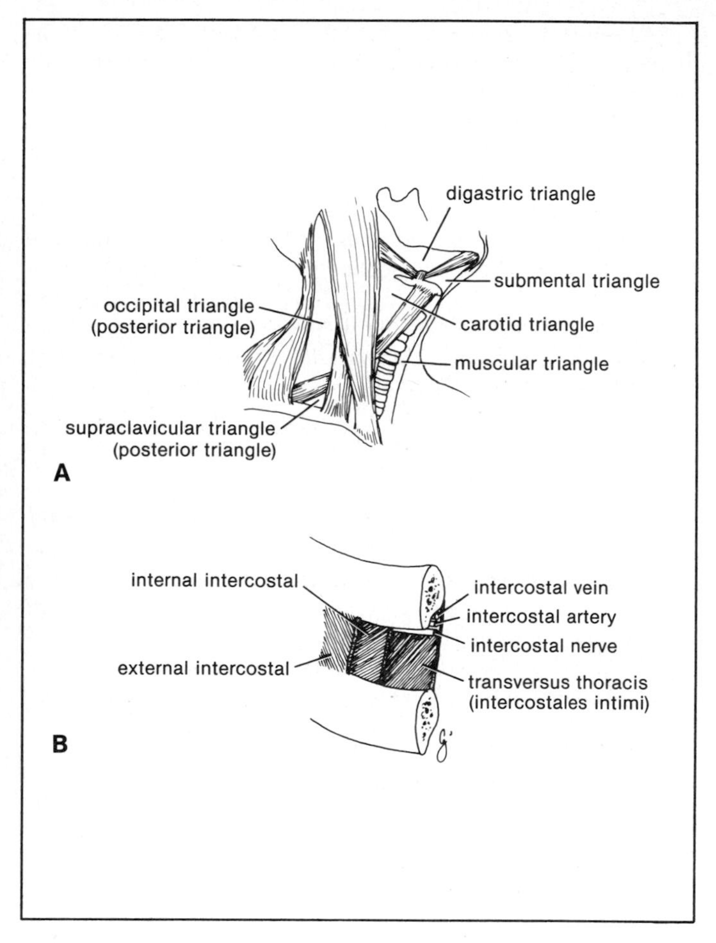

Figure 3–2. A. Muscular triangles of the neck. **B.** An intercostal space showing the position of the intercostal vessels and nerve relative to the intercostal muscles.

bounded posteriorly by the trapezius muscle, anteriorly by the sternocleidomastoid muscle, and inferiorly by the clavicle.

The anterior triangle is further subdivided into the **carotid triangle,** the **digastric triangle,** the **submental triangle,** and the **muscular triangle.** The position and boundaries of these triangles are shown in *Figure 3–2.*

The posterior triangle of the neck is subdivided by the inferior belly of the omohyoid muscle into a large **occipital triangle** above and a small **supraclavicular triangle** below *(Fig. 3–2).*

Suboccipital Triangle. The suboccipital triangle is deeply placed in the upper part of the neck just inferior to the skull. It is covered by the semispinalis capitis muscle. The three sides of the triangle are formed by the rectus capitis posterior major, and the superior and inferior oblique muscles. In the floor of the triangle are the posterior arch of the atlas, the vertebral artery, and the posterior ramus of the first cervical nerve.

The suprahyoid and infrahyoid muscles, the anterior and lateral vertebral muscles, the muscles of the back, and the suboccipital muscles are shown in *Tables 3–5 to 3–9, pp. 123–128.*

MUSCLES OF THE THORAX

Thoracic Inlet

The thorax opens into the root of the neck by a narrow aperture known as the **thoracic inlet.*** It is bounded by the superior border of the manubrium sterni, by the medial borders of the first ribs, and by the body of the first thoracic vertebra. The thoracic inlet transmits structures that pass between the thorax and the neck (esophagus, trachea, blood vessels, and so on) and for the most part lie close to the midline. On either side of these structures lies the apex of the lung with its visceral and parietal layers of pleura and protected by a deep fascial layer called the **suprapleural membrane.**

*Clinicians often loosely refer to this as the thoracic outlet because important vessels and nerves emerge from the thorax here to enter the neck and upper limbs.

Intercostal Spaces

The spaces between the ribs are called intercostal spaces. Each space contains three muscles of respiration; the external intercostal muscle, the internal intercosal muscle, and the transversus thoracis muscle. This latter muscle is related internally to the **endothoracic fascia** and parietal pleura. The intercostal nerves and blood vessels run between the intermediate and deepest layer of muscles *(Fig. 3–2)*. They are arranged in the following order from above downward: intercostal vein, intercostal artery, and intercostal nerves (i.e., VAN).

Thoracic Outlet

Inferiorly, the thorax opens into the abdomen by a wide aperture known as the **thoracic outlet.** It is bounded by the xiphisternal joint, the costal margin, and by the body of the twelfth thoracic vertebra. It is closed by the diaphragm, which is pierced by the structures that pass between the thorax and the abdomen.

Diaphragm

The diaphragm is the most important muscle of respiration *(Fig. 3–3)*. It is dome-shaped and consists of a peripheral muscular part, which arises from the margins of the thoracic outlet, and a centrally placed tendon. The origin of the diaphragm may be divided into three parts:

1. A **sternal part** consisting of small right and left portions arising from the posterior surface of the xiphoid process.
2. A **costal part** that arises from the deep surfaces of the lower six costal cartilages.
3. A **vertebral part** arising by means of vertical columns or **crura** and from the arcuate ligaments.

The **right crus** arises from the sides of the bodies of the first three lumbar vertebrae and the intervertebral discs; the **left crus** arises from the sides of the bodies of the first two lumbar vertebrae and the intervertebral disc.

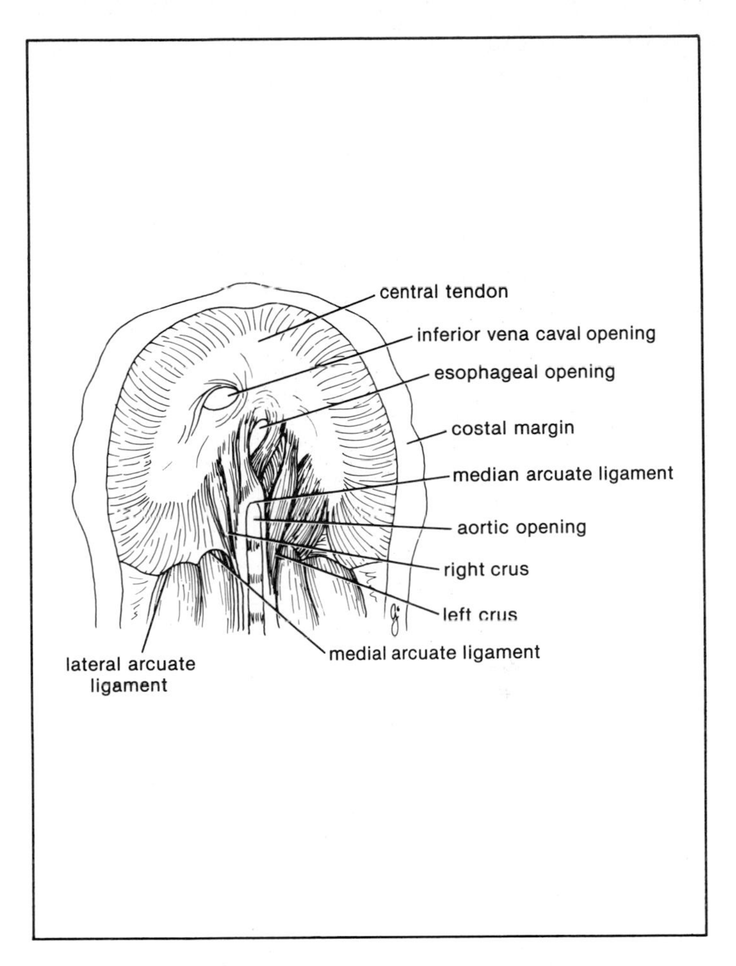

Figure 3–3. Diaphragm as seen from below.

Lateral to the crura the diaphragm arises from the **medial** and **lateral arcuate ligaments.** The diaphragm is inserted into a **central tendon.** The superior surface of the tendon is partially fused with the inferior surface of the fibrous pericardium. Some of the muscle fibers of the right crus pass up to the left and surround the esophagus in a slinglike loop. These fibers appear to act as a sphincter and assist in preventing regurgitation of the stomach contents into the thoracic part of the esophagus.

Shape of the Diaphragm. As seen in front, the diaphragm curves up into right and left domes. The right dome reaches as high as the upper border of the fifth rib, and the left dome reaches the lower border of the fifth rib. The central tendon lies at the level of the xiphisternal joint. Note that the domes support the right and left lungs whereas the central tendon supports the heart.

Nerve Supply. The motor nerve supply is from the phrenic nerve.

Openings in the Diaphragm. The diaphragm has three main openings *(Fig. 3-3).*

The **aortic opening** lies anterior to the body of the twelfth thoracic vertebra between the crura. It transmits the aorta, the thoracic duct, and the azygos vein.

The **esophageal opening** lies at the level of the tenth thoracic vertebra in a sling of muscle fibers derived from the right crus. It transmits the esophagus, the right and left vagus nerves, the esophageal branches of the left gastric vessels, and the lymphatics from the lower one-third of the esophagus.

The **caval opening** lies at the level of the eighth thoracic vertebra in the central tendon. It transmits the inferior vena cava and terminal branches of the right phrenic nerve.

In addition to these structures, the splanchnic nerves pierce the crura, the sympathetic trunk passes posterior to the medial arcuate ligament on each side, and the superior epigastric vessels pass between the sternal and costal origins of the diaphragm on each side.

The muscles of the thorax are shown in *Table 3-10, p. 129.*

MUSCLES OF THE ABDOMEN

The muscles of the anterior abdominal wall consist mainly of three broad thin sheets that are aponeurotic in front; they are the **external oblique, internal oblique,** and **transversus** from exterior to interior. On either side of the midline anteriorly there is in addition a wide vertical muscle, the **rectus abdominis.** As the aponeuroses of the three sheets pass forward they enclose the rectus abdominis to form the **rectus sheath.** In the lower part of the rectus sheath there may be present a small muscle called the **pyramidalis.**

In the lower part of the anterior abdominal wall there exists in both sexes an oblique passage called the **inguinal canal.** The posterior abdominal wall is made up of the **diaphragm,** the **quadratus lumborum,** and the **psoas muscles;** the origin of the transversus abdominis is also present.

The muscles of the anterior and lateral abdominal walls are shown in *Table 3–11, p. 131.*

Rectus Sheath

The rectus sheath *(Fig. 3–4)* is a long fibrous sheath that encloses the rectus abdominis muscle and pyramidalis muscle (if present) and contains the anterior rami of the lower six thoracic nerves and the superior and inferior epigastric vessels and lymphatics. It is formed largely by the aponeuroses of the three lateral abdominal muscles. The internal oblique aponeurosis splits at the lateral edge of the rectus abdominis to form two laminae, one passes anterior to the rectus and one passes posterior. The aponeurosis of the external oblique fuses with the anterior lamina and the transversus aponeurosis fuses with the posterior lamina. At the level of the anterior superior iliac spines, all three aponeuroses pass anterior to the rectus leaving the sheath deficient posteriorly below this level. The lower end of the posterior wall of the sheath is called the **arcuate line.** All three aponeuroses fuse with each other and with their fellows of the opposite side in the midline between the right and left recti muscles to form a fibrous band called the **linea alba.**

Note that the anterior wall of the rectus sheath is firmly attached to the tendinous intersections of the rectus abdominis

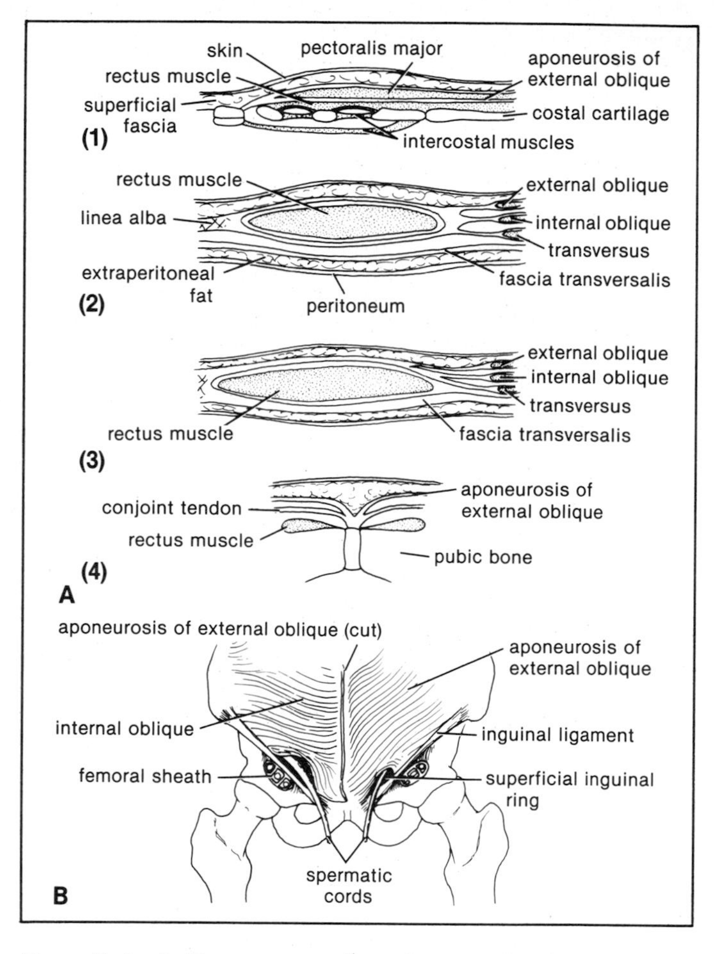

Figure 3–4. A. Transverse sections of rectus sheath seen at four levels. 1. Above costal margin. 2. Between costal margin and level of anterior superior iliac spine. 3. Below level of anterior superior iliac spine and above pubis. 4. At level of pubis. **B.** Inguinal canal and femoral sheath.

muscle whereas the posterior wall of the sheath has no attachment to the muscle. The **tendinous intersections** are usually three in number; one at the level of the xiphoid process, one at the level of the umbilicus, and one between these two.

Inguinal Ligament

The inguinal ligament *(Fig. 3–4)* connects the anterior superior iliac spine to the pubic tubercle. It is formed by the lower border of the aponeurosis of the external oblique, which is folded back upon itself. From the medial end of the ligament the **lacunar ligament** (shaped like a half moon) extends backward and upward to the pectineal line on the superior ramus of the pubis where it becomes continuous with a thickening of the periosteum called the **pectineal ligament.** The lower border of the inguinal ligament is attached to the deep fascia of the thigh, the **fascia lata.**

Fascia Transversalis

This is a thin layer of fascia that lines the transversus muscle and is continuous with a similar layer lining the diaphragm and the iliacus. The **femoral sheath** for the femoral vessels is formed from the fascia transversalis and the fascia iliaca.

Inguinal Canal

The inguinal canal *(Fig. 3–4)* is an oblique passage through the lower part of the anterior abdominal wall and is present in both sexes. It allows structures to pass to and from the testis to the abdomen in the male. In the female it permits the passage of the round ligament of the uterus from the uterus to the labium majus.

The canal is about 4 cm (1½ inches) long in the adult and extends from the deep inguinal ring, a hole in the fascia transversalis, downward and medially to the superficial inguinal ring, a hole in the aponeurosis of the external oblique muscle. It lies parallel to and immediately above the inguinal ligament. The **deep inguinal ring,** an oval opening in the fascia transversalis, lies about

1.3 cm (½ inch) above the inguinal ligament. The margins of the ring give origin to the **internal spermatic fascia.** The **superficial inguinal ring** is a triangular-shaped defect in the aponeurosis of the external oblique muscle and lies immediately above and medial to the pubic tubercle. The margins of the ring give origin to the **external spermatic fascia.**

Walls of the Inguinal Canal

Anterior Wall. External oblique aponeurosis, reinforced laterally by origin of internal oblique from inguinal ligament.

Posterior Wall. Conjoint tendon medially, fascia transversalis laterally.

Roof or Superior Wall. Arching fibers of internal oblique and transversus.

Floor or Inferior Wall. Edge of inguinal ligament.

Mechanics of the Inguinal Canal. The presence of the inguinal canal in the lower part of the anterior abdominal wall in both sexes constitutes a site of potential weakness. On coughing and straining, as in micturition, defecation, and parturition, the arching lowest fibers of the internal oblique and transversus abdominis muscles contract, flattening out the arch so that the roof of the canal is lowered toward the floor so that the canal is virtually closed.

When the great straining efforts may be necessary, as in defecation and parturition, the person naturally tends to assume the squatting position; the hip joints are flexed and the anterior surfaces of the thighs are brought up against the anterior abdominal wall. By this means the lower part of the anterior abdominal wall is protected by the thighs.

Spermatic Cord

The spermatic cord is a collection of structures that transverse the inguinal canal and pass to and from the testis. It is covered with three concentric layers of fascia derived from the layers of the anterior abdominal wall, namely the **external spermatic fascia** (from the external oblique), the **cremasteric fascia** (from the internal oblique), and the **internal spermatic fascia** (from the fascia transversalis).

The **cremaster muscle,** which is derived from the lower fibers of the internal oblique and with adjoining connective tissue forms the **cremasteric fascia,** is innervated by the genital branch of the genitofemoral nerve (L1 and L2).

The muscles of the posterior abdominal wall are shown in *Table 3–12, p. 133.*

MUSCLES OF THE PELVIS

The **piriformis muscle** lines the posterior wall of the pelvis and lies anterior to the sacrum. It leaves the pelvis through the greater sciatic foramen to enter the gluteal region and act on the femur at the hip joint. The **obturator internus** lines the lateral wall of the pelvis and lies medial to the obturator membrane. It leaves the pelvis through the lesser sciatic foramen to enter the gluteal region and act on the femur at the hip joint. The **levator ani muscles** and the **coccygeus muscles** of the two sides form with their covering fascia the **pelvic diaphragm** *(Fig. 3–5).* The pelvic diaphragm is incomplete anteriorly, to allow for the passage of the urethra and in the female, the vagina also.

The muscles of the pelvis are shown in *Table 3–13, p. 134.*

MUSCLES OF THE PERINEUM

The perineum when seen from below is diamond-shaped and is bounded anteriorly by the **symphysis pubis,** posteriorly by the **tip of the coccyx,** and laterally by the **ischial tuberosities.**

The perineum may be divided into two triangles by joining the ischial tuberosities by an imaginary line. The posterior triangle, which contains the anus, is called the **anal triangle;** the anterior triangle, which contains the urogenital orifices, is called the **urogenital triangle.**

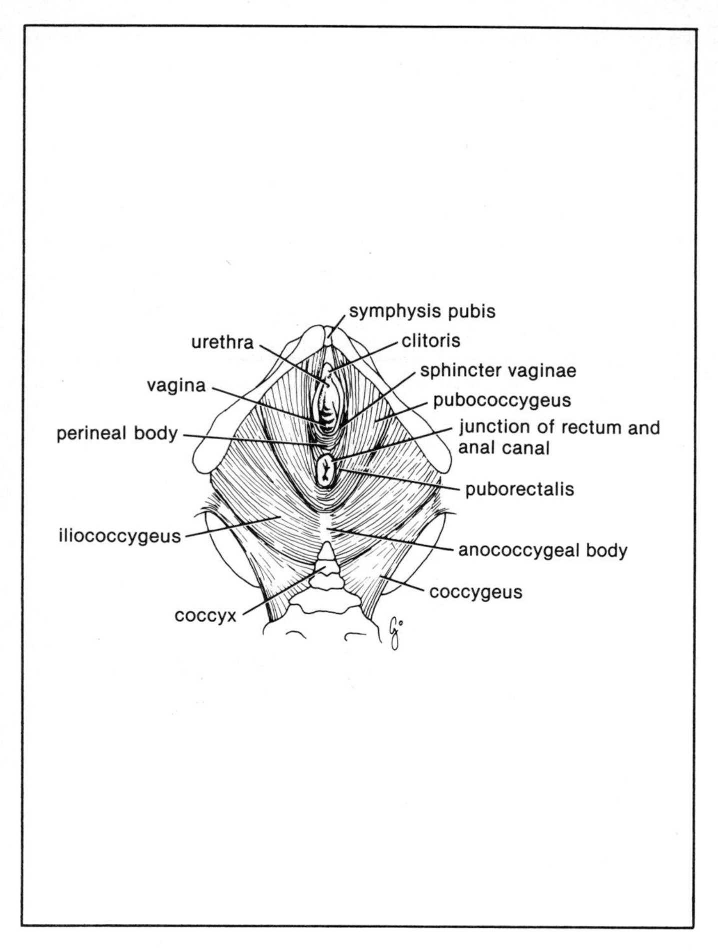

Figure 3–5. Female levator ani and coccygeus muscles.

Anococcygeal Body

This is a mass of fibrous tissue that lies between the anal canal and the coccyx.

Ischiorectal Fossa

This is a wedge-shaped space on each side of the anal canal. The base of the wedge is superficial and formed by the skin. The edge of the wedge is formed by the junction of the medial and lateral walls. The medial wall is formed by the sloping levator ani muscle and the anal canal. The lateral wall is formed by the lower part of the obturator internus muscle, covered with pelvic fascia.

The fossa is filled with fat that supports the anal canal. The pudendal nerve and internal pudendal vessels lie in a fascial canal, the **pudendal canal,** on the medial side of the ischial tuberosity.

The muscles of the anal triangle are shown in *Table 3–14, p. 135.*

Perineal Body

This is a small mass of fibrous tissue, that is attached to the center of the posterior margin of the urogenital diaphragm. It is a larger structure in the female as compared to the male and serves to support the posterior vaginal wall. In both sexes it provides a point of attachment for muscles in the perineum.

Urogenital Diaphragm

This is a musculofascial diaphragm that fills in the gap of the pubic arch. It is formed by the sphincter urethrae and the deep transverse perineal muscles, which are enclosed between a superior and an inferior layer of fascia of the urogenital diaphragm. The inferior layer of fascia is called the **perineal membrane.**

Perineal Pouches

Superficial Perineal Pouch. The **superficial perineal pouch** or space lies inferior to the urogenital diaphragm. It communicates anteriorly with the potential space that lies between the superficial fascia of the anterior abdominal wall (fascia of Scarpa) and the anterior abdominal muscles. The superficial perineal pouch is bounded inferiorly by the membranous layer of the superficial fascia (Colles' fascia). The pouch is closed off posteriorly by the attachment of this membranous layer of fascia to the posterior border of the urogenital diaphragm. The pouch contains the root of the penis (or clitoris) and its associated muscles.

Deep Perineal Pouch. The **deep perineal pouch** is the closed potential space that lies within the urogenital diaphragm. In the male the pouch contains (1) the membranous part of the urethra, (2) the sphincter urethrae, (3) the bulbourethral glands, (4) the deep transverse perineal muscles, (5) the internal pudendal vessels, and (6) the dorsal nerves of the penis.

In the female the pouch contains (1) part of the urethra, (2) part of the vagina, (3) the sphincter urethrae, (4) the deep transverse perineal muscles, (5) the internal pudendal vessels, and (6) the dorsal nerves of the clitoris.

The muscles of the urogenital triangle are shown in *Table 3–15, p. 136.*

MUSCLES OF THE UPPER LIMB

Shoulder Region

Axilla. The axilla, or armpit, is a pyramid-shaped space between the upper part of the arm and the side of the chest. The upper end, or **apex,** is directed into the root of the neck and is bounded in front by the clavicle, behind by the upper border of the scapula, and medially by the outer border of the first rib. The lower end, or **base,** is bounded in front by the anterior axillary fold (formed by the lower border of the pectoralis major muscle), behind by the posterior axillary fold (formed by the tendon of latissimus dorsi and the teres major muscle), and medially by the chest wall.

The axilla contains the principal vessels and nerves to the upper limb and many lymph nodes.

The muscles connecting the upper limb and the vertebral column, the muscles connecting the upper limb and the thoracic wall, and the scapular muscles are shown in *Tables 3–16 to 3–18, pp. 138–140.*

Rotator Cuff. The rotator cuff is the name given to the tendons of the subscapularis, supraspinatus, infraspinatus, and the teres minor muscles, which are fused to the underlying capsule of the shoulder joint. The cuff plays a very important role in stabilizing the shoulder joint.

Quadrilateral Space. The quadrilateral space is bounded above by the subscapularis in front and the teres minor behind. It is bounded below by the teres major. The space is bounded laterally by the surgical neck of the humerus and medially by the long head of the triceps.

The space is located immediately below the shoulder joint and the axillary nerve and the posterior circumflex humeral vessels, which pass through the space, may be damaged in dislocation of the shoulder joint.

Upper Arm

Fascial Compartments of the Upper Arm

The upper arm is enclosed in a sheath of deep fascia. Two fascial septa, one on the medial and one on the lateral side, extend from this sheath and are attached to the medial and lateral supracondylar ridges of the humerus, respectively. By this means the upper arm is divided into an anterior and a posterior fascial compartment, each having its muscles, nerves, and arteries.

The muscles of the upper arm are shown in *Table 3–19, p. 141.*

Cubital Fossa

The cubital fossa is a skin depression that lies in front of the elbow and is triangular in shape. It has the following boundaries: **laterally,** the brachioradialis muscle; **medially,** the pronator teres muscle. The **base** of the triangle is formed by an imaginary line drawn between the two epicondyles of the humerus. The **floor** of the fossa is formed by the supinator muscle laterally and the brachialis muscle medially. The **roof** is formed by the skin and fascia and is reinforced by the bicipital aponeurosis.

The cubital fossa contains the following structures, enumerated from the medial to the lateral side: the median nerve; the bifurcation of the brachial artery into the ulnar and radial arteries; the tendon of the biceps muscle; and the radial nerve and its deep branch.

Lying in the superficial fascia covering the cubital fossa are the important superficial veins, the cephalic and basicic veins and their tributaries.

Forearm

Fascial Compartments of the Forearm

The forearm is enclosed in a sheath of deep fascia, which is attached to the periosteum of the posterior subcutaneous border of the ulna. This fascial sheath, together with the interosseous membrane and fibrous intermuscular septa, divides up the forearm into a number of compartments, each having its own muscles, nerves, and blood supply.

Interosseous Membrane. The interosseous membrane is a thin but strong membrane uniting the radius and the ulna; it is attached to their interosseous borders. Its fibers are taut and therefore the forearm is most stable when it is in the mid-prone position, i.e., the position of function. The interosseus membrane provides attachment for neighboring muscles.

The muscles of the anterior fascial compartment of the forearm are shown in *Table 3–20, p. 143.*

Wrist

Flexor and Extensor Retinacula. The flexor and extensor retinacula are specialized bands of deep fascia seen in the region of the wrist and the hand, that hold the long flexor and extensor tendons in position. The flexor retinaculum also provides attachment for some of the short muscles of the hand. The flexor retinaculum is attached medially to the pisiform bone and the hook of the hamate, and laterally to the tubercle of the scaphoid and the trapezium. The extensor retinaculum is attached medially to the pisiform bone and the hook of the hamate, and laterally to the distal end of the radius.

Carpal Tunnel. The bones of the hand are deeply concave on their anterior surface and form a bony gutter. The gutter is converted into a tunnel by the flexor retinaculum. The median nerve lies in a **restricted** space between the flexor digitorum superficialis and the flexor carpi radialis muscles.

Hand

Fibrous Flexor Sheaths. The anterior surface of each finger, from the head of the metacarpal to the base of the distal phalanx, is provided with a strong fibrous sheath, which is attached to the sides of the phalanges. The sheath, together with the anterior surfaces of the phalanges and the interphalangeal joints, forms a blind tunnel in which the flexor tendons of the fingers lie.

Synovial Flexor Sheaths. The crowded long flexor tendons emerge from the carpal tunnel and diverge as they pass down into the hand. The tendons of the flexor digitorum superficialis and profundus invaginate a common synovial sheath from the lateral side. This common sheath extends proximally into the forearm for a short distance. Distally, the medial part of the sheath continues downward without interruption on the tendons of the little finger. The distal ends of the long flexor tendons of the index, middle, and ring fingers have **digital synovial sheaths.** The flexor pollicis longus tendon also has a synovial sheath as it passes into the thumb. The synovial flexor sheaths are essentially lubricating mechanisms, which permit the long flexor tendons to move smoothly, with the minimum of friction, beneath the flexor reticulum and the fibrous flexor sheaths.

Insertion of the Long Flexor Tendons. Each tendon of the flexor digitorum superficialis divides into two halves *(Fig. 3-6),* which pass around the profundus tendon and meet on its posterior surface, where partial decussation of the fibers takes place. The superficialis tendon, having united again, divides into two further slips that are attached to the borders of the middle phalanx. Each

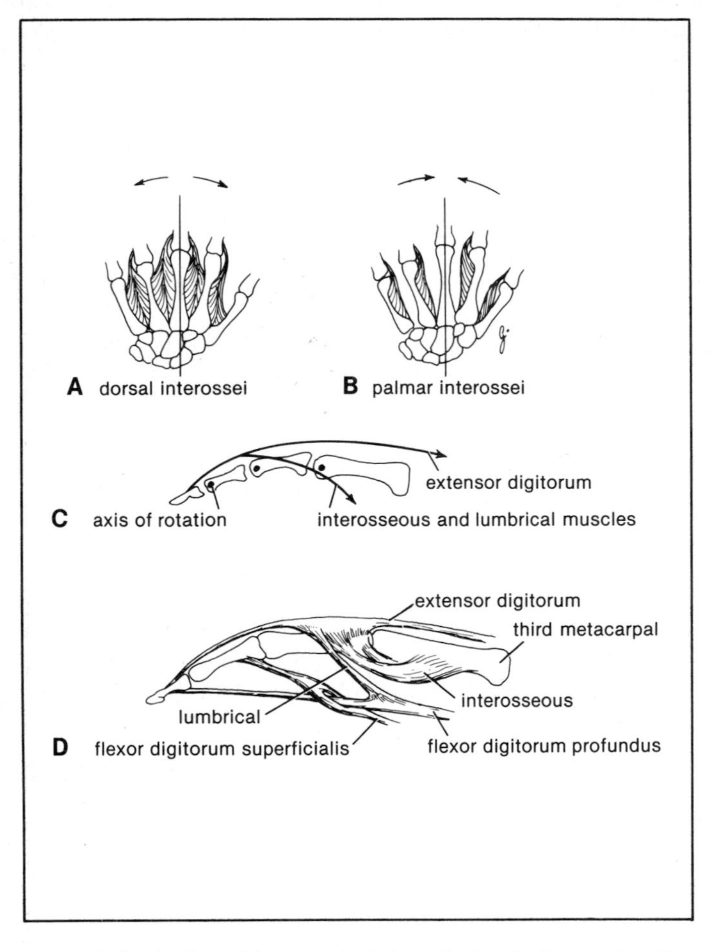

Figure 3–6. **A.** Dorsal interossei of the right hand; they abduct fingers as shown. **B.** Palmar interossei of the right hand; they adduct fingers as shown. **C.** Shows action of interosseous and lumbrical muscles in flexing metacarpophalangeal joints and extending the interphalangeal joints. **D.** Shows insertions of long flexor and extensor tendons in the fingers. Note the insertions of the lumbrical and interossei muscles.

tendon of the flexor digitorum profundus, having passed through the superficialis tendon, is inserted into the base of the distal phalanx.

The muscles of the lateral fascial compartment and the posterior fascial compartment of the forearm are shown in *Tables 3-21 and 3-22, pp. 145-147.*

Insertion of the Long Extensor Tendons. The four tendons of the extensor digitorum emerge from under the extensor retinaculum and fan out over the dorsum of the hand. The tendon to the index finger is joined on its medial side by the tendon of the extensor indicis, and the tendon to the little finger is joined on its medial side by the two tendons of the extensor digiti minimi. On the posterior surface of each finger, the extensor tendon joins the fascial expansion called the **extensor expansion.** Near the proximal interphalangeal joint, the extensor expansion splits into three parts: a central part, which is inserted into the base of the middle phalanx, and two lateral parts, which converge to be inserted into the base of the distal phalanx.

The dorsal extensor expansion receives the tendon of insertion of the corresponding interosseous muscle on each side, and further distally, receives the tendon of the lumbrical muscle on the lateral side *(Fig. 3-6).*

The small muscles of the hand are shown in *Table 3-23, p. 148.*

Fascial Spaces of the Palm. Normally, the fascial spaces of the palm are potential spaces filled with loose connective tissue. They are important clinically, because they may become infected.

Pulp Space of the Fingers. The pulp space of each finger is situated anterior to the nail and the distal phalanx; it is enclosed by deep fascia.

MUSCLES OF THE LOWER LIMB

Thigh

Deep Fascia of the Thigh (Fascia Lata). The deep fascia encloses the thigh like a trouser leg, and at its upper end is attached to the pelvis and its associated ligaments. On its lateral aspect it is thickened to form the **iliotibial tract,** which is attached above to the iliac tubercle, and below to the lateral condyle of the tibia. The iliotibial tract receives the insertion of the tensor fasciae latae and the greater part of the gluteus maximus muscle. The *saphenous opening* is a gap in the deep fascia in the front of the thigh, which transmits the great saphenous vein, some small branches of the femoral artery, and lymph vessels.

Fascial Compartments of the Thigh. Three fascial septa pass from the inner aspect of the deep fascial sheath of the thigh to the linea aspera of the femur. By this means, the thigh is divided into three compartments, each having muscles, nerves, and arteries. The compartments are anterior, medial, and posterior in position.

The muscles of the buttock or gluteal region are shown in *Table 3-24, p. 151.* The muscles of the anterior fascial compartment of the thigh are shown in *Table 3-25, p. 153.*

Femoral Triangle. The femoral triangle is a triangular area situated in the upper part of the medial aspect of the thigh. Its boundaries are as follows: base, inguinal ligament; lateral border, sartorius muscle; medial border, adductor longus muscle. The femoral triangle contains the terminal part of the femoral nerve and its branches, the femoral sheath, the femoral artery and its branches, the femoral vein and its tributaries, and the deep inguinal lymph nodes.

Femoral Sheath. The femoral sheath is a downward protrusion into the thigh of the fascia lining the abdominal walls. The sheath surrounds the femoral vessels and lymphatics for about 2.5 cm (1 inch) below the inguinal ligament. The **femoral artery,** as it enters the thigh beneath the inguinal ligament, occupies the **lateral compartment** of the sheath. The **femoral vein** occupies the **intermediate compartment** and the lymphatic vessels occupy the most **medial compartment.**

The **femoral canal** is the small medial compartment for the lymphatics. It is about 1.3 cm (½ inch) long, and its upper opening is the **femoral ring.** The femoral canal is a potentially weak area in the abdomen. A protrusion of peritoneum could be forced down the femoral canal to form a **femoral hernia.**

The muscles of the medial fascial compartment of the thigh are shown in *Table 3–26, p. 155.*

Adductor (or Subsartorial) Canal. The adductor canal is an intermuscular cleft situated on the medial aspect of the middle third of the thigh beneath the sartorius muscle. It contains the femoral artery and vein, the deep lymph vessels, and the saphenous nerve.

The muscles of the posterior fascial compartment of the thigh are shown in *Table 3–27, p. 156.*

Knee Region

Popliteal Fossa. The popliteal fossa is a diamond-shaped intermuscular space situated at the back of the knee. It contains the popliteal vessels, the small saphenous vein, the common peroneal and tibial nerves, the posterior cutaneous nerve of the thigh, connective tissue, and lymph nodes.

Boundaries. *Laterally.* The biceps femoris above and the lateral head of the gastrocnemius and plantaris below. *Medially.* The semimembranosus and semitendinosus above and the medial head of gastrocnemius below.

Leg

Deep Fascia of the Leg and the Foot. The deep fascia surrounds the leg and is continuous above with the deep fascia of the thigh. Below the tibial condyles it is attached to the anterior and medial borders of the tibia. Two intermuscular septa pass from its deep aspect to be attached to the fibula. These, together with the interosseous membrane, divide the leg into three compartments: anterior, lateral, and posterior, each having its own muscles, blood supply, and nerve supply.

Interosseous Membrane. The interosseous membrane is a strong membrane connecting the interosseous borders of the tibia and the fibula. The interosseous membrane binds the tibia and the fibula together and provides attachment for neighboring muscles.

Ankle

Retinacula of the Ankle. In the region of the ankle joint, the deep fascia is thickened to form a series of bands or retinacula, which serve to keep the long tendons in position and act as modified pulleys. The **superior extensor retinaculum** is attached to the distal ends of the anterior borders of the fibula and the tibia. The **inferior extensor retinaculum** is a Y-shaped band located in front of the ankle joint. The **flexor retinaculum** extends from the medial malleolus downward and backward to be attached to the medial surface of the calcaneum. It binds the tendons of the deep muscles to the medial side of the ankle as they pass forward from behind the medial malleolus to enter to the sole of the foot. The **superior peroneal retinaculum** extends from the lateral malleolus downward and backward to be attached to the lateral surface of the calcaneum. It binds the tendons of the peroneus longus and brevis to the lateral side of the ankle. The **inferior peroneal retinaculum** is attached to the calcaneum above and below the peroneal tendons.

The muscles of the anterior, lateral and posterior fascial compartments of the leg are shown in *Tables 3–28 to 3–30, pp. 157–159.* The muscle on the dorsum of the foot is shown in *Table 3–31, p. 161.*

MUSCLES OF THE SOLE OF THE FOOT

These are conveniently described in four layers from the inferior layer superiorly. The deep fascia is thickened to form the **plantar aponeurosis.** This is triangular in shape and attached by its apex to the medial and lateral tubercles of the calcaneum. The base of the aponeurosis divides into five slips which pass into the toes.

The muscles of the sole are shown in *Table 3–32, p. 162.*

TABLE 3–1. MUSCLES OF THE SCALP AND THE EXTERNAL EAR

Name of Muscle	Origin	Insertion	Nerve Supply	Action
Muscle of Scalp				
Occipitofrontalis				
Occipital bellies	Occipital bone	Epicranial aponeurosis	Facial nerve	Moves scalp on skull and raises eyebrows
Frontal bellies	Skin and fascia of eyebrow	Epicranial aponeurosis	Facial nerve	Moves scalp on skull and raises eyebrows
Muscles of External Ear				
Extrinsic				
Auricularis anterior	Epicranial aponeurosis	Auricle	Facial nerve	
Auricularis superior	Epicranial aponeurosis	Auricle	Facial nerve	Small amount of auricular movement in some individuals
Auricularis posterior	Epicranial aponeurosis	Auricle	Facial nerve	
Intrinsic				
Small vestigeal muscles on outer surface of auricle			Facial nerve	No action in most human ears

TABLE 3–2. MUSCLES OF FACIAL EXPRESSION

Name of Muscle	Origin	Insertion	Nerve Supply	Action
Muscles of Eyelids				
Orbicularis oculi				
Orbital part	Frontal bone Maxillary bone Medial palpebral ligament	No interruption Form concentric loops	Facial nerve	Pulls on skin of forehead, temple, and cheek like purse string "screws up the eye"
Palpebral part	Medial palpebral ligament	Lateral palpebral ligament		Closes eyelids
Lacrimal part	Lacrimal bone	Both eyelids		Compresses lacrimal sac
Corrugator supercilii	Medial part of superciliary arch of frontal bone	Skin of eyebrow	Facial nerve	Draws eyebrows medially
Muscles of Nostrils				
Compressor naris	Frontal process of maxilla	Via aponeurosis into muscle of opposite side	Facial nerve	Compresses nasal aperture
Dilator naris	Maxilla	Ala of nose	Facial nerve	Widens nasal aperture
Procerus	Nasal bone and lateral nasal cartilage	Skin between eyebrows	Facial nerve	Wrinkles skin at root of nose

Muscle	Origin	Insertion	Nerve supply	Action
Muscles of the Lips and Cheeks				
Sphincter Muscle of Lips				
Orbicularis oris	Maxilla and mandible, deep surface of skin of lips Some fibers derived from buccinator	Surround orifice of mouth	Facial nerve	Compresses lips together
Dilator Muscles of Lips				
1. Levator labii superioris alaeque nasi 2. Levator labii superioris 3. Zygomaticus minor 4. Zygomaticus major 5. Levator anguli oris 6. Risorius 7. Depressor anguli oris 8. Depressor labii inferioris 9. Mentalis	Bones and fascia around oral aperture	Substance of lips	Facial nerve	Separate lips

(continued)

TABLE 3–2. (cont.)

Name of Muscle	Origin	Insertion	Nerve Supply	Action
Muscles of the Lips and Cheeks (cont.)				
Buccinator	Alveolar margins of maxilla and mandible and pterygomandibular ligament	Fibers decussate and enter upper and lower lips	Facial nerve	Compresses cheeks and lips against teeth
Platysma	Deep fascia of upper part of chest	Lower margin of body of mandible and angle of mouth	Facial nerve	Depresses mandible and draws down lower lip and angle of mouth

TABLE 3–3. MUSCLES OF MASTICATION

Name of Muscle	Origin	Insertion	Nerve Supply	Action
Masseter	Zygomatic arch	Lateral surface ramus of mandible	Mandibular division of trigeminal nerve	Raises mandible to occlude teeth in mastication
Temporalis	Floor of temporal fossa and covering fascia	Coronoid process of mandible	Mandibular division of trigeminal nerve	Anterior and superior fibers elevate the mandible; posterior fibers retract the mandible
Lateral pterygoid	Greater wing of sphenoid and lateral pterygoid plate	Neck of mandible and articular disc of temporo-mandibular joint	Mandibular division of trigeminal nerve	Pulls neck of mandible forward
Medial pterygoid	Tuberosity of maxilla and lateral pterygoid plate	Medial surface angle of mandible	Mandibular division of trigeminal nerve	Elevates mandible

TABLE 3–4. SUPERFICIAL MUSCLES OF THE SIDE OF THE NECK

Name of Muscle	Origin	Insertion	Nerve Supply	Action
Platysma	*(See Table 3–2)*			
Trapezius	Described with Muscles of Upper Limb *(see Table 3–16)*			
Sternocleidomastoid	Manubrium sterni and medial third of clavicle	Mastoid process of temporal bone and occipital bone	Spinal part of accessory nerve and second and third cervical nerves	Muscles of the two sides acting together extend the head and flex the neck. One muscle alone rotates head to the opposite side.

TABLE 3–5. SUPRAHYOID MUSCLES

Name of Muscle	Origin	Insertion	Nerve Supply	Action
Digastric Muscle				
Posterior belly	Mastoid process of temporal bone	Intermediate tendon that is bound to hyoid bone	Facial nerve	Depresses mandible or elevates hyoid bone
Anterior belly	Lower border of mandible near midline	Intermediate tendon as above	Mylohyoid nerve Mandibular division of trigeminal nerve	
Stylohyoid	Styloid process of temporal bone	Junction of body and greater cornu of hyoid bone	Facial nerve	Elevates hyoid bone
Mylohyoid	Mylohyoid line of inner surface of body of mandible	Body of hyoid bone and raphe that extends from mandible to hyoid bone	Mandibular division of trigeminal nerve	Elevates floor of mouth and hyoid bone or depresses mandible
Geniohyoid	Inferior mental spine of back of symphysis menti of mandible	Body of hyoid bone	C1 through hypoglossal nerve	Elevates hyoid bone or depresses mandible

TABLE 3–6. INFRAHYOID MUSCLES

Name of Muscle	Origin	Insertion	Nerve Supply	Action
Sternohyoid	Manubrium sterni and medial end of clavicle	Body of hyoid bone	Ansa cervicalis (C1, C2, and C3)	Depresses the hyoid bone
Sternothyroid	Manubrium sterni	Oblique line of lamina of thyroid cartilage	Ansa cervicalis (C1, C2, and C3)	Depresses the larynx
Thyrohyoid	Oblique line on lamina of thyroid cartilage	Body of hyoid bone	C1 through hypoglossal nerve	Depresses the hyoid bone or elevates larynx
Omohyoid				
Inferior belly	Upper margin of scapula; suprascapular ligament	Intermediate tendon bound to clavicle and first rib	Ansa cervicalis (C1, C2, and C3)	Depresses hyoid bone
Superior belly	Body of hyoid bone			

TABLE 3–7. ANTERIOR AND LATERAL VERTEBRAL MUSCLES

Name of Muscle	Origin	Insertion	Nerve Supply	Action
Longus colli	Attached to anterior surface of vertebrae between atlas and third thoracic vertebra		Anterior rami of cervical nerves	Flexes cervical part of vertebral column
Longus capitis	Transverse process of lower cervical vertebrae	Occipital bone	Anterior rami of cervical nerves	Flexes head
Rectus capitis anterior	Front of lateral mass of atlas	Occipital bone	Cervical plexus	Flexes head
Rectus capitis lateralis	Transverse process of atlas	Occipital bone	Cervical plexus	Lateral flexion of head
Scalenus anterior	Transverse processes of third, fourth, fifth, and sixth cervical vertebrae	First rib	Cervical spinal nerves	Elevates first rib. Laterally flexes and rotates cervical part of vertebral column
Scalenus medius	Transverse processes of upper six cervical vertebrae	First rib	Cervical spinal nerves	Elevates first rib. Laterally flexes and rotates cervical part of vertebral column

TABLE 3–8. MUSCLES OF THE BACK[a]

Name of Muscle	Origin	Insertion	Nerve Supply	Action
Splenius capitis	Ligamentum nuchae and first four thoracic spines	Occipital bone; mastoid process of temporal bone	Posterior rami of cervical spinal nerves	Both sides together extend head. Alone pulls head to one side
Splenius cervicis	Upper thoracic spines	Upper cervical transverse processes	Posterior rami of cervical spinal nerves	Both sides together extend head. One side alone pulls head to one side
Erector Spinae (Superficial vertically running muscles)				
Iliocostalis	Sacrum, iliac crest, and lower ribs	Ribs and transverse processes of cervical vertebrae	Posterior rami of cervical, thoracic, and lumbar spinal nerves.	Extension and lateral flexion of vertebral column

Longissimus	Transverse processes of lumbar and thoracic vertebrae	Transverse processes of thoracic and cervical vertebrae, ribs and mastoid process of skull	Posterior rami of cervical, thoracic, and lumbar spinal nerves	Extension and lateral flexion of vertebral column, extension of head
Spinalis	Spines of lower thoracic and lumbar vertebrae	Spines of upper thoracic vertebrae	Posterior rami of cervical and thoracic spinal nerves	Extension of vertebral column
Transversospinales (Intermediate oblique running muscles)				
Semispinalis Multifidus Rotatores	Transverse processes	Spines	Posterior rami of spinal nerves	Rotation of vertebral column
Deepest Muscles				
Interspinales	Spine below	Spine above	Posterior rami of spinal nerves	Maintain posture
Intertransversarii	Transverse process below	Transverse process above		

[a]Detailed attachments of these muscles are not required.

TABLE 3–9. SUBOCCIPITAL MUSCLES

Name of Muscle	Origin	Insertion	Nerve Supply	Action
Rectus capitis posterior major	Spine of axis	Occipital bone	Posterior ramus of first cervical nerve	Extension of head
Rectus capitis posterior minor	Posterior arch of atlas	Occipital bone	Posterior ramus of first cervical nerve	Extension of head
Obliquus capitis inferior	Spine of axis	Transverse process of atlas	Posterior ramus of first cervical nerve	Rotates face to same side
Obliquus capitis superior	Transverse process of atlas	Occipital bone	Posterior ramus of first cervical nerve	Extends head to same side

TABLE 3–10. MUSCLES OF THE THORAX

Name of Muscle	Origin	Insertion	Nerve Supply	Action
Diaphragm	Xiphoid process; lower six costal cartilages; first three lumbar vertebrae by crura and medial and lateral arcuate ligaments	Central tendon	Phrenic nerve	Most important muscle of inspiration, increases vertical diameter of thorax by pulling central tendon downward, assists in raising lower ribs
Intercostal Muscles				
External intercostal (fibers pass downward and forward)	Inferior border of rib above	Superior border of rib below	Intercostal nerves	With first rib fixed they raise ribs during inspiration and thus increase anteroposterior and transverse diameters of thorax. With last rib fixed by addominal muscles they lower ribs during expiration

(continued)

TABLE 3–10. (cont.)

Name of Muscle	Origin	Insertion	Nerve Supply	Action
Intercostal Muscles (cont.)				
Internal intercostal (fibers pass downward and backward)	Inferior border of rib above	Superior border of rib below	Intercostal nerves	Assist external intercostal muscles
Transversus thoracis (fibers pass transversely—forms incomplete layer of muscle)	Adjacent ribs	Adjacent ribs	Intercostal nerves	Assist external and internal intercostal muscles
Levatores costarum (12 in number)	Transverse processes of seventh cervical to eleventh thoracic vertebra	Superior border of ribs	Posterior rami of thoracic spinal nerves	Elevate ribs
Serratus posterior superior	Ligamentum nuchae and upper thoracic spines	Upper ribs	Intercostal nerves	Elevates ribs
Serratus posterior inferior	Lower thoracic and upper lumbar spines	Lower ribs	Intercostal nerves	Lowers ribs

TABLE 3–11. MUSCLES OF THE ANTERIOR AND LATERAL ABDOMINAL WALLS

Name of Muscle	Origin	Insertion	Nerve Supply	Action
External oblique	Lower eight ribs	Xiphoid process, linea alba, pubic crest, pubic tubercle, iliac crest.	Lower six thoracic nerves and iliohypogastric and ilioinguinal nerves (L1)	Compresses abdominal contents; assists in flexing and rotation of trunk. Pulls down ribs in forced expiration
Internal oblique (gives rise to cremaster muscle that enters the spermatic cord)	Lumbar fascia, iliac crest, lateral two-thirds of inguinal ligament	Lower three ribs and costal cartilages, xiphoid process, linea alba, pubic symphysis. Conjoint tendon with transversus	Lower six thoracic nerves, iliohypogastric and ilioinguinal nerves (L1)	Compresses abdominal contents; assists in flexing and rotation of trunk. Pulls down ribs in forced expiration

(continued)

TABLE 3–11. (cont.)

Name of Muscle	Origin	Insertion	Nerve Supply	Action
Transversus	Lower six costal cartilages, lumbar fascia, iliac crest, lateral third of inguinal ligament	Xiphoid process, linea alba, symphysis pubis, forms conjoint tendon with internal oblique	Lower six thoracic nerves, iliohypogastric and ilioinguinal nerves (L1)	Compresses abdominal contents
Rectus abdominis	Symphysis pubis and pubic crest	Fifth, sixth and seventh costal cartilages and xiphoid process	Lower six thoracic nerves	Compresses abdominal contents and flexes vertebral column. Accessory muscle of expiration
Pyramidalis (often absent)	Anterior surface of pubis	Linea alba	Twelfth thoracic nerve	Tenses the linea alba

TABLE 3–12. MUSCLES OF THE POSTERIOR ABDOMINAL WALL

Name of Muscle	Origin	Insertion	Nerve Supply	Action
Psoas	Body of twelfth thoracic vertebra, transverse processes, bodies and intervertebral discs of the five lumbar vertebrae	With iliacus into lesser trochanter of femur	Lumbar plexus	Flexes thigh on trunk; if thigh is fixed it flexes trunk on thigh as in sitting up from lying position
Quadratus lumborum	Iliolumbar ligament, iliac crest, transverse processes of lower lumbar vertebrae	Twelfth rib	Lumbar plexus	Depresses twelfth rib during respiration; laterally flexes vertebral column same side
Iliacus	Iliac fossa	With psoas into lesser trochanter of femur	Femoral nerve	Flexes thigh on trunk; if thigh is fixed, it flexes the trunk on the thigh as in sitting up from lying position

TABLE 3–13. MUSCLES OF THE PELVIS

Name of Muscle	Origin	Insertion	Nerve Supply	Action
Piriformis	Front of sacrum	Greater trochanter of femur	Sacral plexus	Lateral rotator of femur at hip joint
Obturator internus	Obturator membrane and adjoining part of hip bone	Greater trochanter of femur	Sacral plexus	Lateral rotator of femur at hip joint
Levator ani (Levator prostatae or sphinter vaginae puborectalis pubococcygeus iliococcygeus)	Body of pubis, fascia of obturator internus, spine of ischium	Perineal body, anococcygeal body, walls of prostate, vagina, rectum and anal canal	Fourth sacral nerve, pudendal nerve	Supports pelvic viscera; sphincter to anorectal junction and vagina
Coccygeus	Spine of ischium	Lower end of sacrum; coccyx	Fourth sacral nerve	Assists levator ani to support pelvic viscera. Flexes coccyx

TABLE 3–14. MUSCLES OF THE ANAL TRIANGLE

Name of Muscle	Origin	Insertion	Nerve Supply	Action
Sphincter ani externus				
Subcutaneous part	Encircles lower end of anal canal; no bony attachments			
Superficial part	Coccyx	Perineal body	Inferior rectal nerve and perineal branch of fourth sacral nerve	Closes anal canal and anus
Deep part	Encircles upper end of anal canal; No bony attachments			

TABLE 3–15. MUSCLES OF THE UROGENITAL TRIANGLE

Name of Muscle	Origin	Insertion	Nerve Supply	Action
Male				
Superficial transverse perineal muscle	Ischial tuberosity	Perineal body	Perineal branch of pudendal nerve	Fixes perineal body
Bulbospongiosus	Perineal body and median raphe	Expansion over corpus spongiosum and cavernosum	Perineal branch of pudendal nerve	Empties urethra during micturition and ejaculation, assists in erection
Ischiocavernosus	Ischial tuberosity and ischial ramus	Expansion under side of crus	Perineal branch of pudendal nerve	Assists in erection
Deep transverse perineal muscle	Ramus of ischium	Perineal body	Perineal branch of pudendal nerve	Fixes perineal body

Sphincter urethrae	Fascia and ramus of pubis	Encircles urethra; perineal body	Perineal branch of pudendal nerve	Compresses membranous urethra
Female				
Superficial transverse perineal muscle	(See Male)			
Bulbospongiosus	Perineal body	Dorsal surface of clitoris	Perineal branch of pudendal nerve	Sphincter to vagina, assists in erection of clitoris
Ischiocavernosus	Ischial tuberosity	Sides of crus of clitoris	Perineal branch of pudendal nerve	Assists in erection of clitoris
Deep transverse perineal muscle	(See Male)			
Sphincter urethrae	(See Male)			

TABLE 3–16. MUSCLES CONNECTING THE UPPER LIMB AND THE VERTEBRAL COLUMN

Name of Muscle	Origin	Insertion	Nerve Supply	Action
Trapezius	Occipital bone, ligamentum nuchae spines of all thoracic vertebrae	Upper fibers into lateral third of clavicle; middle and lower fibers into spine of scapula	Spinal part of accessory nerve and C3 and C4	Upper fibers elevate the scapula. Middle fibers pull scapula medially. Lower fibers pull medial border of scapula downward
Latissimus dorsi	Iliac crest, lumbar fascia, spines of lower six thoracic vertebrae, lower three or four ribs and inferior angle of scapula	Floor of bicipital groove of humerus	Thoracodorsal nerve	Extends, adducts, and medially rotates the arm
Levator scapulae	Transverse process of first four cervical vertebrae	Medial border of scapula	C3 and C4 and dorsal scapular nerve	Raises medial border of scapula
Rhomboideus minor	Ligamentum nuchae and spines of seventh cervical and first thoracic vertebrae	Medial border of scapula	Dorsal scapular nerve	Raises medial border of scapula
Rhomboideus major	Second to fifth thoracic spines	Medial border of scapula	Dorsal scapular nerve	Raises medial border of scapula

TABLE 3–17. MUSCLES CONNECTING UPPER LIMB AND THORACIC WALL

Name of Muscle	Origin	Insertion	Nerve Supply	Action
Pectoralis major	Clavicle, sternum and upper six costal cartilages	Lateral lip of bicipital groove of humerus	Medial and lateral pectoral nerves from brachial plexus	Adducts arm and rotates it medially. Clavicular fibers also flex arm
Pectoralis minor	Third, fourth, and fifth ribs	Coracoid process of scapula	Medial pectoral nerve from brachial plexus	Depresses point of shoulder. If the scapula is fixed it elevates ribs of origin
Subclavius	First costal cartilage	Clavicle	Nerve to subclavius from upper trunk of brachial plexus	Depresses the clavicle and steadies this bone during movements of the shoulder girdle
Serratus anterior	Upper eight ribs	Medial border and inferior angle of scapula	Long thoracic nerve	Draws the scapula forward around the thoracic wall; rotates scapula

TABLE 3–18. SCAPULAR MUSCLES

Name of Muscle	Origin	Insertion	Nerve Supply	Action
Deltoid	Lateral third of clavicle, acromion process, spine of scapula	Middle of lateral surface of shaft of humerus	Axillary nerve	Abducts arm; anterior fibers flex arm; posterior fibers extend arm
Supraspinatus	Supraspinous fossa of scapula	Greater tuberosity of humerus	Suprascapular nerve	Abducts arm and stabilizes head of humerus in glenoid cavity of scapula
Infraspinatus	Infraspinous fossa of scapula	Greater tuberosity of humerus	Suprascapular nerve	Laterally rotates arm
Teres major	Lower third lateral border of scapula	Medial lip of bicipital groove of humerus	Lower subscapular nerve	Medially rotates and adducts arm
Teres minor	Upper two thirds lateral border of scapula	Greater tuberosity of humerus	Axillary nerve	Laterally rotates arm
Subscapularis	Subscapular fossa	Lesser tuberosity of humerus	Upper and lower subscapular nerves	Medially rotates arm

TABLE 3–19. MUSCLES OF THE UPPER ARM

Name of Muscle	Origin	Insertion	Nerve Supply	Action
Muscles of Anterior Fascial Compartment				
Biceps brachii				
Long head	Supraglenoid tubercle of scapula	Tuberosity of radius and bicipital aponeurosis into deep fascia of forearm	Musculocutaneous nerve	Supinator of forearm and flexor of elbow joint. Also weak flexor of shoulder joint
Short head	Coracoid process of scapula			
Coracobrachialis	Coracoid process of scapula	Medial aspect of shaft of humerus	Musculocutaneous nerve	Flexes arm and also weak adductor
Brachialis	Front of lower half of humerus	Coronoid process of ulna	Musculocutaneous nerve and radial nerves	Flexor of elbow joint

(continued)

TABLE 3–19. (cont.)

Name of Muscle	Origin	Insertion	Nerve Supply	Action
Muscles of Posterior Fascial Compartment				
Triceps				
Long head	Infraglenoid tubercle of scapula			
Lateral head	Upper half of posterior surface of shaft of humerus	Olecranon process of ulna	Radial nerve	Extensor of the elbow joint
Medial head	Lower half of posterior surface of shaft of humerus			

TABLE 3–20. MUSCLES OF THE ANTERIOR FASCIAL COMPARTMENT OF THE FOREARM

Name of Muscle	Origin	Insertion	Nerve Supply	Action
Pronator teres				
Humeral head	Medial epicondyle of humerus	Lateral aspect of shaft of radius	Median nerve	Pronation and flexion of forearm
Ulnar head	Coronoid process of ulna			
Flexor carpi radialis	Medial epicondyle of humerus	Bases of second and third metacarpal bones	Median nerve	Flexes and abducts hand at wrist joint
Palmaris longus (often absent)	Medial epicondyle of humerus	Flexor retinaculum and palmar aponeurosis	Median nerve	Flexes hand
Flexor carpi ulnaris				
Humeral head	Medial epicondyle of humerus	Pisiform bone, hook of the hamate, base of fifth metacarpal bone	Ulnar nerve	Flexes and adducts the hand at the wrist joint
Ulnar head	Olecranon process and posterior border of ulna			

(continued)

TABLE 3–20. (cont.)

Name of Muscle	Origin	Insertion	Nerve Supply	Action
Flexor digitorum superficialis				
Humeroulnar head	Medial epicondyle of humerus, coronoid process of ulna			Flexes middle phalanx of fingers and assists in flexing proximal phalanx and hand
Radial head	Oblique line on anterior surface of shaft of radius	Middle phalanx of middle four fingers	Median nerve	
Flexor pollicis longus	Anterior surface of shaft of radius	Distal phalanx of thumb	Anterior interosseous branch of median nerve	Flexes distal phalanx of thumb
Flexor digitorum profundus	Anterior surface of shaft of ulna; interosseous membrane	Distal phalanges of medial four fingers	Ulnar (medial half) and median (lateral half) nerves	Flexes distal phalanx of the fingers; then assists in flexion of middle and proximal phalanges and the wrist
Pronator quadratus	Anterior surface of shaft of ulna	Anterior surface of shaft of radius	Anterior interosseous branch of median nerve	Pronates forearm

TABLE 3–21. MUSCLES OF THE LATERAL FASCIAL COMPARTMENT OF THE FOREARM

Name of Muscle	Origin	Insertion	Nerve Supply	Action
Brachioradialis	Lateral supracondylar ridge of humerus	Styloid process of radius	Radial nerve	Flexes forearm at elbow joint. Rotates forearm to midprone position
Extensor carpi radialis longus	Lateral supracondylar ridge of humerus	Base of second metacarpal bone	Radial nerve	Extends and abducts hand at wrist joint

TABLE 3–22. MUSCLES OF THE POSTERIOR FASCIAL COMPARTMENT OF THE FOREARM

Name of Muscle	Origin	Insertion	Nerve Supply	Action
Extensor carpi radialis brevis	Lateral epicondyle of humerus	Base of third metacarpal bone	Deep branch of radial nerve	Extends and abducts the hand at wrist joint
Extensor digitorum	Lateral epicondyle of humerus	Middle and distal phalanges of the medial four fingers	Deep branch of radial nerve	Extends fingers and hand
Extensor digiti minimi	Lateral epicondyle of humerus	Extensor expansion of little finger	Deep branch of radial nerve	Extends metacarpal phalangeal joint of little finger
Extensor carpi ulnaris	Lateral epicondyle of humerus	Base of fifth metacarpal bone	Deep branch of radial nerve	Extends and adducts hand at the wrist joint
Anconeus	Lateral epicondyle of humerus	Olecranon process of ulna	Radial nerve	Extends elbow joint

Supinator	Lateral epicondyle of humerus, anular ligament of superior radioulnar joint and ulna	Neck and shaft of ulna	Deep branch of radial nerve	Supination of forearm
Abductor pollicis longus	Shafts of radius and ulna	Base of first metacarpal bone	Deep branch of radial nerve	Abducts and extends thumb
Extensor pollicis brevis	Shaft of radius and interosseous membrane	Base of proximal phalanx of thumb	Deep branch of radial nerve	Extends metacarpo-phalangeal joints of thumb
Extensor pollicis longus	Shaft of ulna and interosseous membrane	Base of distal phalanx of thumb	Deep branch of radial nerve	Extends distal phalanx of thumb
Extensor indicis	Shaft of ulna and interosseous membrane	Extensor expansion of index finger	Deep branch of radial nerve	Extends metacarpophalangeal joint of index finger

TABLE 3–23. SMALL MUSLCES OF THE HAND

Name of Muscle	Origin	Insertion	Nerve Supply	Action
Lumbricals	Tendons of flexor digitorum profundus	Extensor expansion of medial four fingers	First and second, i.e., lateral two, median nerve; third and fourth ulnar nerve	Flex metacarpo-phalangeal joints and extend interphalan-geal joints of fingers except thumb
Interossei				
Palmar (4)	First, second, fourth, and fifth metacarpal bones	Base of proximal phalanges of fingers; extensor expansion	Deep branch of ulnar nerve	Palmar interossei adduct fingers toward center of third finger; dorsal interossei abduct fingers from center of third finger; both palmar and dorsal flex the metacarpophalangeal joints and extend the interphalangeal joints
Palmaris brevis	Flexor retinaculum and palmar aponeurosis	Skin of palm	Superficial branch of ulnar nerve	Corrugates the skin and improves grip of palm

Short Muscles of Thumb				
Abductor pollicis brevis	Scaphoid, trapezium, flexor retinaculum	Base of proximal phalanx of thumb	Median nerve	Abduction of thumb
Flexor pollicis brevis	Flexor retinaculum	Base of proximal phalanx of thumb	Median nerve	Flexes thumb
Opponens pollicis	Flexor retinaculum	Shaft of metacarpal bone of thumb	Median nerve	Pulls thumb medially and forward across palm
Adductor pollicis	Oblique head: second and third metacarpal bones; transverse heads: third metacarpal bone	Base of proximal phalanx of thumb	Deep branch of ulnar nerve	Adduction of thumb

(continued)

TABLE 3–23. (cont.)

Name of Muscle	Origin	Insertion	Nerve Supply	Action
Short Muscles of Little Finger				
Abductor digiti minimi	Pisiform bone	Base of proximal phalanx of little finger	Deep branch of ulnar nerve	Flexes little finger
Flexor digiti minimi	Flexor retinaculum	Base of proximal phalanx of little finger	Deep branch of ulnar nerve	Flexes little finger
Opponens digiti minimi	Flexor retinaculum	Shaft of metacarpal bone of little finger	Deep branch of ulnar nerve	Flexes little finger and pulls fifth metacarpal bone forward as in cupping the hand

TABLE 3–24. MUSCLES OF THE BUTTOCK OR GLUTEAL REGION OF THE LOWER LIMB

Name of Muscle	Origin	Insertion	Nerve Supply	Action
Gluteus maximus	Outer surface of ilium, sacrum, coccyx, sacrotuberous ligament	Iliotibial tract and gluteal tuberosity of femur	Inferior gluteal nerve	Extends and laterally rotates thigh at hip joint; through iliotibial tract it extends knee joint
Gluteus medius	Outer surface of ilium	Greater trochanter of femur	Superior gluteal nerve	Abducts thigh at hip joint. Tilts pelvis when walking
Gluteus minimus	Outer surface of ilium	Greater trochanter of femur	Superior gluteal nerve	Abducts thigh at hip joint. Anterior fibers medially rotate thigh
Tensor fasciae latae	Iliac crest	Iliotibial tract	Superior gluteal nerve	Assists gluteus maximus in extending the knee joint

(continued)

TABLE 3–24. (cont.)

Name of Muscle	Origin	Insertion	Nerve Supply	Action
Piriformis	Anterior surface of sacrum	Greater trochanter of femur	First and second sacral nerves	Lateral rotator of thigh at hip joint
Obturator internus	Inner surface of obturator membrane	Greater trochanter of femur	Sacral plexus	Lateral rotator of thigh at hip joint
Gemellus superior	Spine of ischium	Greater trochanter of femur	Sacral plexus	Lateral rotator of thigh at hip joint
Gemellus inferior	Ischial tuberosity	Greater trochanter of femur	Sacral plexus	Lateral rotator of thigh at hip joint
Quadratus femoris	Ischial tuberosity	Quadrate tubercle on upper end of femur	Sacral plexus	Lateral rotator of thigh at hip joint
Obturator externus	Outer surface of obturator membrane	Greater trochanter of femur	Obturator nerve	Lateral rotator of thigh at hip joint

TABLE 3–25. MUSCLES OF THE ANTERIOR FASCIAL COMPARTMENT OF THE THIGH

Name of Muscle	Origin	Insertion	Nerve Supply	Action
Sartorius	Anterior superior iliac spine	Upper medial surface shaft of tibia	Femoral nerve	Flexes, abducts, laterally rotates thigh at hip joint; flexes and medially rotates leg at knee joint
Iliacus	Iliac fossa	With psoas into lesser trochanter of femur	Femoral nerve	Flexes thigh on trunk; if thigh is fixed, it flexes the trunk on the thigh as in sitting up from lying down
Psoas	Twelfth thoracic vertebral body; transverse processes, bodies and intervertebral discs of the five lumbar vertebrae	With iliacus into lesser trochanter of femur	Lumbar plexus	Flexes thigh on trunk; if thigh is fixed, it flexes the trunk on the thigh as in sitting up from lying down

(continued)

TABLE 3–25. (cont.)

Name of Muscle	Origin	Insertion	Nerve Supply	Action
Pectineus	Superior ramus of pubis	Upper end shaft of femur	Femoral nerve	Flexes and adducts thigh at hip joint
Quadriceps femoris				
Rectus femoris	Straight head: anterior inferior iliac spine; reflected head: ilium above acetabulum	Quadriceps tendon into patella	Femoral nerve	Extension of leg at knee joint
Vastus lateralis	Upper end and shaft of femur	Quadriceps tendon into patella	Femoral nerve	Extension of leg at knee joint
Vastus medialis	Upper end and shaft of femur	Quadriceps tendon into patella	Femoral nerve	Extension of leg at knee joint
Vastus intermedius	Shaft of femur	Quadriceps tendon into patella	Femoral nerve	Extension of leg at knee joint

TABLE 3–26. MUSCLES OF THE MEDIAL FASCIAL COMPARTMENT OF THE THIGH

Name of Muscle	Origin	Insertion	Nerve Supply	Action
Gracilis	Inferior ramus of pubis; ramus of ischium	Upper part of shaft of tibia on medial surface	Obturator nerve	Adducts thigh at hip joint; flexes leg at knee joint
Adductor longus	Body of pubis	Posterior surface of shaft of femur	Obturator nerve	Adducts thigh at hip joint and assists in lateral rotation
Adductor brevis	Inferior ramus of pubis	Posterior surface of shaft of femur	Obturator nerve	Adducts thigh at hip joint and assists in lateral rotation
Adductor magnus	Inferior ramus of pubis; ramus of ischium, ischial tuberosity	Posterior surface of shaft of femur; adductor tubercle of femur	Obturator nerve and sciatic nerve (hamstring part)	Adducts thigh at hip joint and assists in lateral rotation. Hamstring part extends thigh at hip joint

TABLE 3–27. MUSCLES OF THE POSTERIOR FASCIAL COMPARTMENT OF THE THIGH

Name of Muscle	Origin	Insertion	Nerve Supply	Action
Biceps femoris	Long head: ischial tuberosity; short head: shaft of femur	Head of fibula	Sciatic nerve (long head: tibial nerve; short head: common peroneal nerve)	Flexes and laterally rotates leg at knee joint; long head also extends thigh at hip joint
Semitendinosus	Ischial tuberosity	Upper part medial surface of shaft of tibia	Sciatic nerve (tibial portion)	Flexes and medially rotates leg at knee joint; extends thigh at hip joint
Semimembranosus	Ischial tuberosity	Medial condyle of tibia; forms oblique popliteal ligament	Sciatic nerve (tibial portion)	Flexes and medially rotates leg at knee joint; extends thigh at hip joint
Adductor magnus (hamstring portion)	Ischial tuberosity	Adductor tubercle of femur	Sciatic nerve (tibial portion)	Extends thigh at hip joint

TABLE 3–28. MUSCLES OF THE ANTERIOR FASCIAL COMPARTMENT OF THE LEG

Name of Muscle	Origin	Insertion	Nerve Supply	Action
Tibialis anterior	Shaft of tibia and interosseous membrane	Medial cuneiform and base of first metatarsal bone	Deep peroneal nerve	Extends[a] the foot at ankle joint; inverts foot at subtalar and transverse tarsal joints. Holds up medial longitudinal arch of foot
Extensor digitorum longus	Shaft of fibula and interosseous membrane	Extensor expansion of lateral four toes	Deep peroneal nerve	Extends toes; dorsiflexes foot at ankle joint
Peroneus tertius	Shaft of fibula and interosseous membrane	Base of fifth metatarsal bone	Deep peroneal nerve	Dorsiflexes foot at ankle joint; everts foot at subtalar and transverse tarsal joints
Extensor hallucis longus	Shaft of fibula and interosseous membrane	Base of distal phalanx of great toe	Deep peroneal nerve	Extends big toe; dorsiflexes foot at ankle joint; inverts foot at subtalar and transverse tarsal joints

[a]Extension, or dorsiflexion, of the ankle is the movement of the foot away from the ground.

TABLE 3–29. MUSCLES OF THE LATERAL FASCIAL COMPARTMENT OF THE LEG

Name of Muscle	Origin	Insertion	Nerve Supply	Action
Peroneus longus	Shaft of fibula	Base of first metatarsal and the medial cuneiform	Superficial peroneal nerve	Plantar flexes foot at ankle joint; everts foot at subtalar and transverse tarsal joints. Supports lateral longitudinal and transverse arches of foot
Peroneus brevis	Shaft of fibula	Base of fifth metatarsal bone	Superficial peroneal nerve	Plantar flexes foot at ankle joint; everts foot at subtalar and transverse tarsal joint. Holds up lateral longitudinal arch of foot

TABLE 3–30. MUSCLES OF THE POSTERIOR FASCIAL COMPARTMENT OF THE LEG

Name of Muscle	Origin	Insertion	Nerve Supply	Action
Superficial Group				
Gastrocnemius	Medial and lateral condyles of femur	Via Achilles tendon into calcaneum	Tibial nerve	Plantar flexes foot at ankle joint; flexes knee joint
Plantaris	Lateral supracondylar ridge of femur	Calcaneum	Tibial nerve	Plantar flexes foot at ankle joint; flexes knee joint
Soleus	Shafts of tibia and fibula	Via Achilles tendon into calcaneum	Tibial nerve	Together with gastrocnemius and plantaris is powerful plantar flexor of ankle joint. Provides main propulsive force in walking and running
Deep Group				
Popliteus	Lateral condyle of femur	Shaft of tibia	Tibial nerve	Flexes leg at knee joint. Unlocks knee joint by lateral rotation of femur on tibia and slackens ligaments of joint

(continued)

TABLE 3–30. (cont.)

Name of Muscle	Origin	Insertion	Nerve Supply	Action
Deep Group (cont.)				
Flexor digitorum longus	Shaft of tibia	Distal phalanges of lateral four toes	Tibial nerve	Flexes distal phalanges of lateral four toes; plantar flexes foot; supports medial and lateral longitudinal arches of foot
Flexor hallucis longus	Shaft of fibula	Base of distal phalanx of big toe	Tibial nerve	Flexes distal phalanx of big toe; plantar flexes foot at ankle joint; supports medial longitudinal arch of foot
Tibialis posterior	Shafts of tibia and fibula and interosseous membrane	Tuberosity of navicular bone	Tibial nerve	Plantar flexes foot at ankle joint; inverts foot at subtalar and transverse tarsal joints; supports medial longitudinal arch of foot

TABLE 3–31. MUSCLE ON THE DORSUM OF THE FOOT

Name of Muscle	Origin	Insertion	Nerve Supply	Action
Extensor digitorum brevis	Calcaneum	By four tendons into the proximal phalanx of big toe and long extensor tendons to second, third and fourth toes	Deep peroneal nerve	Extends toes

TABLE 3–32. MUSCLES OF THE SOLE

Name of Muscle	Origin	Insertion	Nerve Supply	Action
First Layer				
Abductor hallucis	Medial tubercle of calcaneum, flexor retinaculum	Medial side, base proximal phalanx big toe	Medial plantar nerve	Flexes, abducts big toe. Supports medial longitudinal arch
Flexor digitorum brevis	Medial tubercle of calcaneum	Middle phalanx of four lateral toes	Medial plantar nerve	Flexes lateral four toes. Supports medial and lateral longitudinal arches
Abductor digiti minimi	Medial and lateral tubercles of calcaneum	Lateral side base proximal phalanx fifth toe	Lateral plantar nerve	Flexes, abducts fifth toe. Supports lateral longitudinal arch
Second Layer				
Flexor accessorius (quadratus plantae)	Medial and lateral sides calcaneum	Tendon flexor digitorum longus	Lateral plantar nerve	Aids long flexor tendon to flex lateral four toes
Flexor digitorum longus	*(See Table 3–30)*	Base of distal phalanx of lateral four toes	Tibial nerve	Flexes distal phalanges of lateral four toes; plantar flexes foot; supports longitudinal arches

Lumbricals	Tendons of flexor digitorum longus	Dorsal extensor expansion of lateral four toes	First lumbrical medial plantar; remainder deep branch lateral plantar nerve	Extends toes at interphalangeal joints
Flexor hallucis longus	*(See Table 3–30)*	Base of distal phalanx of big toe	Tibial nerve	Flexes distal phalanx of big toe; plantar flexes foot; supports medial longitudinal arch
Third Layer				
Flexor hallucis brevis	Cuboid, lateral cuneiform bones; tibialis posterior insertion	Medial and lateral sides of base of proximal phalanx of big toe	Medial plantar nerve	Flexes metatarsophalangeal joint of big toe; supports medial longitudinal arch
Adductor hallucis				
Oblique head	Bases second, third, and fourth metatarsal bones	Lateral side base proximal phalanx big toe	Deep branch lateral plantar	Flexes big toe, supports transverse arch
Transverse head	Plantar ligaments			

(continued)

TABLE 3–32. (cont.)

Name of Muscle	Origin	Insertion	Nerve Supply	Action
Third Layer (cont.)				
Flexor digiti minimi brevis	Base of fifth metatarsal bone	Lateral side base of proximal phalanx big toe	Superior branch lateral plantar nerve	Flexes little toe
Fourth Layer				
Interossei				
Dorsal (4)	Adjacent sides of metatarsal bones	Bases of phalanges and dorsal expansion of corresponding toes	Lateral plantar nerve	Abduct toes from second toe; flex metatarsophalangeal joints; extend interphalangeal joints
Plantar (3)	Third, fourth, and fifth metatarsal bones	Bases of phalanges and dorsal expansion of corresponding toes	Lateral plantar nerve	Adduct toes top second toe; flex metatarso-phalangeal joints; extend interphalangeal joints
Peroneus longus	*(See Table 3–29)*			
Tibialis posterior	*(See Table 3–30)*			

4

Vascular System

HEART

The heart is a hollow muscular organ that is somewhat pyramidal in shape and lies within the pericardium in the mediastinum. It is connected at its base to the great blood vessels but otherwise lies free within the pericardium.

The heart has three surfaces: sternocostal (anterior), diaphragmatic (inferior), and a base (posterior). It also has an apex that is directed downward, and forward, and to the left.

Surface Markings of the Heart

Superior. Line from the second left costal cartilage to the third right costal cartilage.

Inferior. Line from the sixth right costal cartilage to the apex.

Right Border. Line drawn from a point on the third right costal cartilage 1.3 cm (½ inch) from the edge of the sternum downward to a point on the sixth right costal cartilage 1.3 cm (½ inch) from the edge of the sternum.

Left Border. Line drawn from second left costal cartilage 1.3 cm (½ inch) from edge of the sternum to the apex of the heart.

The **apex** of the heart lies at the level of the fifth left intercostal space, 9 cm (3½ inches) from the midline (*Fig. 4–1*).

Heart Valves

Surface Markings of the Heart Valves. The surface markings of the heart valves are of academic value only; the clinician is interested in listening to the valves in action.

The **tricuspid valve** lies behind the right half of the sternum opposite the fourth intercostal space.

The **mitral valve** lies behind the left half of the sternum opposite the fourth costal cartilage.

The **pulmonary valve** lies behind the medial end of the third left costal cartilage and the adjoining part of the sternum.

The **aortic valve** lies behind the left half of the sternum opposite the third intercostal space.

Auscultation of the Heart Valves. The **tricuspid valve** is best heard over the right half of the lower end of the body of the sternum.

The **mitral valve** is best heard over the apex of the heart.

The **pulmonary valve** is best heard over the medial end of the second left intercostal space.

The **aortic valve** is best heard over the medial end of the second right intercostal space.

Structure of the Heart

The heart is divided longitudinally by a vertical septum and transversely by a vertical septum producing four chambers, the right and left atria and the right and left ventricles. The right atrium lies anterior to the left atrium and the right ventricle lies anterior to the left ventricle (*Figs. 4–2 and 4–3*).

The walls of the heart are composed of cardiac muscle, the **myocardium,** covered externally with serous pericardium, called

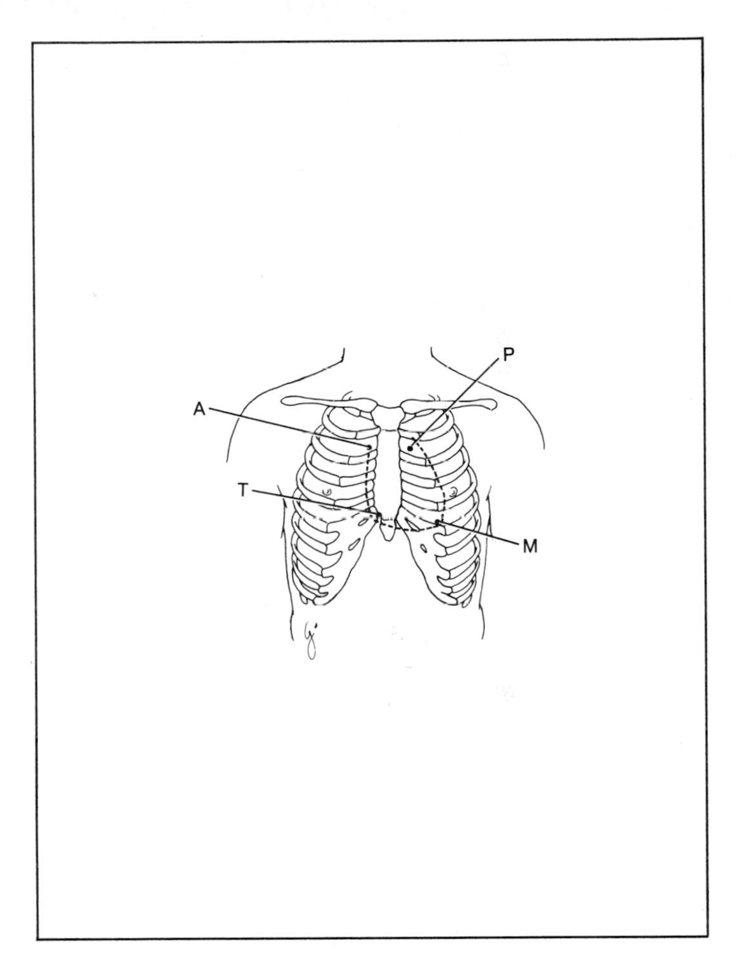

Figure 4–1. *A,* aortic valve sound; *P,* pulmonary valve sound; *M,* mitral valve sound; and *T,* tricuspid valve sound. Black dots indicate positions where heart valves may be heard with least interference.

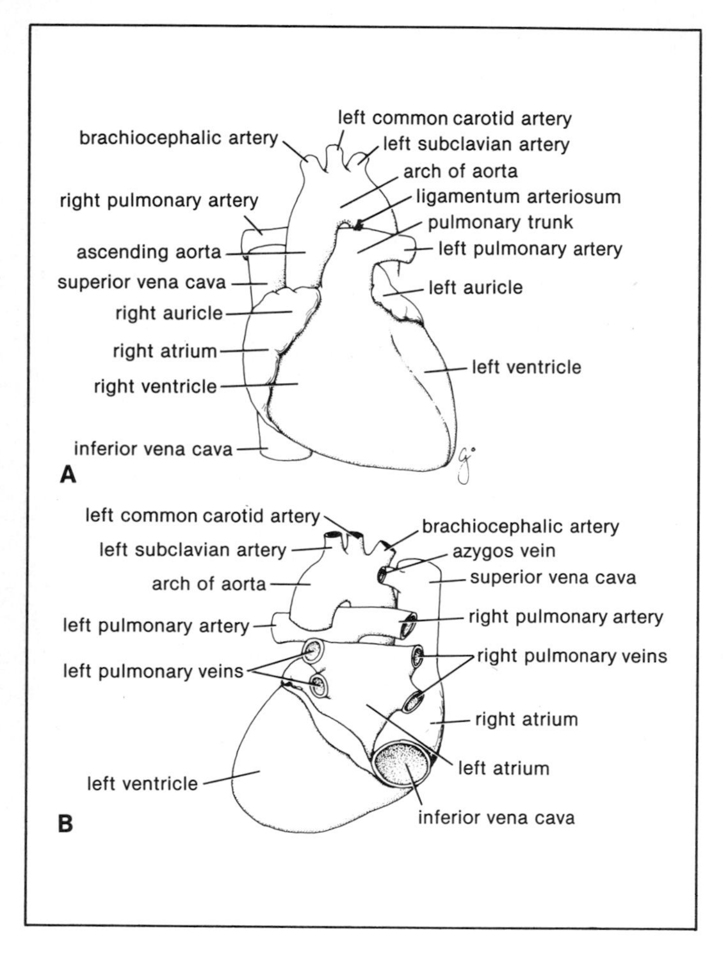

Figure 4–2. **A.** Anterior surface of heart. **B.** Posterior surface or base of heart.

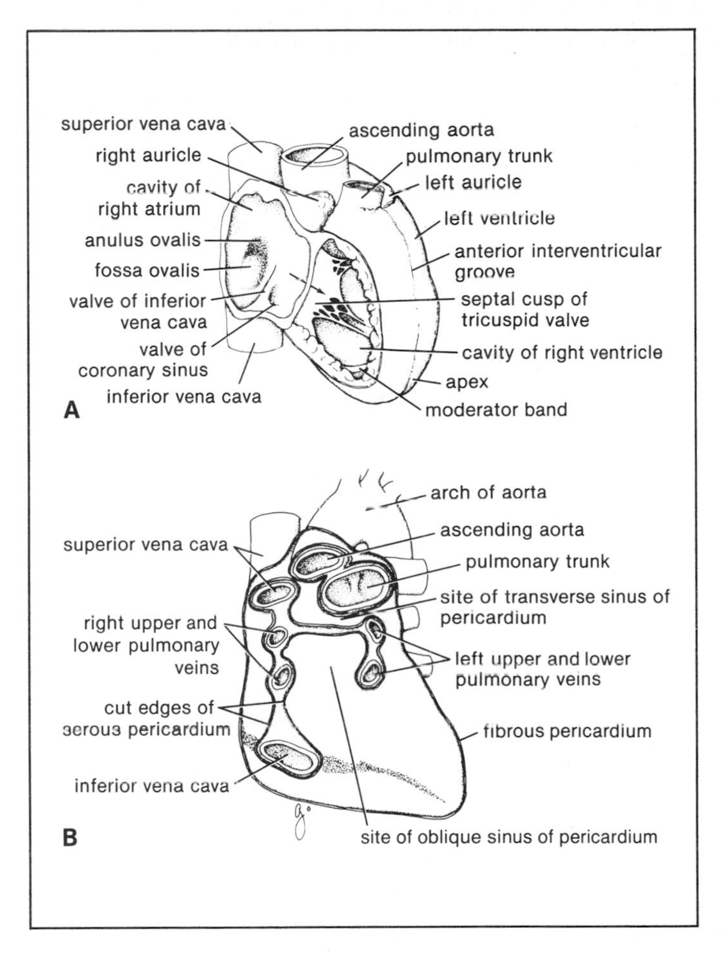

Figure 4–3. A. Interior of right atrium and right ventricle; the anterior walls of these heart chambers have been removed. **B.** Interior of the pericardial cavity, anterior view; the heart has been removed.

the **epicardium,** and lined internally with a layer of endothelium, the **endocardium.**

The **skeleton of the heart** consists of fibrous rings that surround the atrioventricular, pulmonary, and aortic orifices and are continuous with the membranous upper part of the ventricular septum.

Chambers of the Heart

Right Atrium. The right atrium consists of a main cavity and an auricle (*Fig. 4–3*). Externally at the juction of the two parts is a vertical groove, the **sulcus terminalis,** which on the inside forms a ridge, the *crista terminalis.* The main part of the atrium that lies posterior to the ridge is smooth-walled whereas the interior of the auricle is trabeculated by bundles of muscle fibers, the **musculi pectinati.**

Openings. The **superior vena cava** opens into the upper part of the right atrium; it is devoid of any valve. The **inferior vena cava** (larger than that of the superior vena cava) opens into the lower part of the right atrium; it is guarded by a rudimentary valve.

The **coronary sinus** opens into the right atrium between the inferior vena cava and the atrioventricular orifice; it is guarded by a rudimentary valve.

The **right atrioventricular orifice** lies anterior to the inferior vena caval opening and is guarded by the tricuspid valve.

There are also many small orifices of small veins that drain the wall of the heart and open directly into the right atrium.

Fetal Remnants. In addition to the rudimentary valve of the inferior vena cava, there are the **fossa ovalis** and **anulus ovalis.** These latter structures lie on the atrial septum that separates the right atrium from the left atrium. The fossa ovalis is a shallow depression that is the site of the foramen ovale in the fetus. The anulus ovalis forms the upper margin of the fossa.

Right Ventricle. The right ventricle (*Fig. 4–3*) communicates with the right atrium through the atrioventricular orifice, and with the pulmonary trunk through the pulmonary orifice. As the cavity approaches the pulmonary orifice it becomes funnel-shaped and is here referred to as the **infundibulum.**

The walls of the right ventricle are much thicker than those of the right atrium and show a number of internal projecting ridges formed of muscle bundles. The projecting ridges are known as **trabeculae carneae** and are of three types. The first type are the **papillary muscles,** which project inward being attached by their bases to the ventricular wall and their apices are connected by fibrous chords (the **chordae tendineae**) to the cusps of the tricuspid valve. The second type are attached at their ends to the ventricular wall, being free in the middle. Of these, one, the **moderator band,** crosses the ventricular cavity from the septal to the anterior wall. It conveys the right branch of the atrioventricular bundle, part of the conducting system of the heart. The third type are simply prominent ridges.

The **tricuspid valve** guards the atrioventricular orifice. It consists of three cusps formed by a fold of endocardium with some fibrous tissue enclosed. The bases of the cusps are attached to the fibrous ring of the skeleton of the heart, whereas to their free edges and ventricular surfaces are attached the chordae tendineae. The chordae tendineae connect the cusps to the papillary muscles. The cusps are anterior, septal, and inferior. The anterior cusp lies anteriorly, the septal cusp lies against the ventricular septum, and the inferior cusp lies inferiorly.

The **pulmonary valve** guards the pulmonary orifice. It consists of three semilunar cusps formed by folds of endocardium with some fibrous tissue enclosed. The curved lower margin of each cusp is attached to the arterial wall. The open mouths of the cusps are directed upward into the pulmonary trunk. At the root of the pulmonary trunk are three dilations called the **sinuses,** and one is situated external to each cusp (see aortic valve).

The three semilunar cusps are arranged with one posterior and two anterior.

Left Atrium. The left atrium (*Fig. 4–2*) consists of a main cavity and an auricle. The interior of the auricle possesses muscular ridges as on the right side.

Openings. The four **pulmonary veins,** two from each lung, open through the posterior wall and are devoid of valves. The left **atrioventricular** orifice is guarded by the mitral valve.

Left Ventricle. The left ventricle (*Figs. 4–2 and 4–3*) communicates with the left atrium through the atrioventricular orifice, and with the aorta through the aortic orifice. The walls of the left ventricle are three times thicker than those of the right ventricle. There are well-developed trabeculae carneae, two papillary muscles, but there is no moderator band. The part of the ventricle below the aortic orifice is called the **aortic vestibule.**

The **mitral valve** guards the atrioventricular orifice. It consists of two cusps, one anterior and one posterior, which have a similar structure to those of the tricuspid valve. The attachment of the chordae tendineae to the cusps and the papillary muscles is similar to the tricuspid valve.

The **aortic valve** guards the aortic orifice and is precisely similar in structure to the pulmonary valve. One cusp is situated on the anterior wall and two are located on the posterior wall. Behind each cusp the aortic wall bulges to form an **aortic sinus.** The anterior aortic sinus gives origin to the right coronary artery, and the left posterior sinus gives origin to the left coronary artery.

Conducting System of the Heart

The **sinoatrial node** is situated at the upper part of the sulcus terminalis just to the right of the opening of the superior vena cava into the right atrium.

The **atrioventricular node** is situated in the lower part of the atrial septum just above the attachment of the septal cusp of the tricuspid valve.

The **atrioventricular bundle** is continuous above with the atrioventricular node and below with the fibers of the Purkinje plexus. It descends behind the septal cusp of the tricuspid valve on the membranous part of the ventricular septum. On reaching the muscular part of the septum it divides into two branches, one for each ventricle. The atrioventricular bundle is the only muscular connection between the myocardium of the atria and the myocardium of the ventricles.

Blood Supply of the Heart

Arterial Supply. The right and left coronary arteries arise from the ascending aorta immediately above the aortic valve (*Fig. 4–4*).

The **right coronary artery** arises from the anterior aortic sinus and runs forward between the pulmonary trunk and the right auricle (*Fig. 4–4*). It descends in the right atrioventricular groove, giving branches to the right atrium and right ventricle. At the inferior border of the heart it continues posteriorly along the atrioventricular groove to anastomose with the left coronary artery in the posterior interventricular groove. It gives off a **marginal branch,** which supplies the right ventricle, and a **posterior interventricular branch,** which supplies both ventricles. The posterior interventricular branch anastomoses with the anterior interventricular branch of the left coronary artery.

The **left coronary artery** arises from the left posterior aortic sinus and passes forward between the pulmonary trunk and the left auricle (*Fig. 4–4*). It then divides into an anterior interventricular branch and a circumflex branch. The **anterior interventricular branch** runs downward to the apex of the heart in the anterior interventricular groove, supplying both ventricles. It then passes around the apex of the heart to anastomose with the posterior interventricular branch of the right coronary artery. The **circumflex branch** winds around the left margin of the heart in the atrioventricular groove and ends by anastomosing with the right coronary artery.

Anastomoses between the terminal branches of the right and left coronary arteries are not large enough to provide an adequate supply of blood to the cardiac muscle should one of the larger

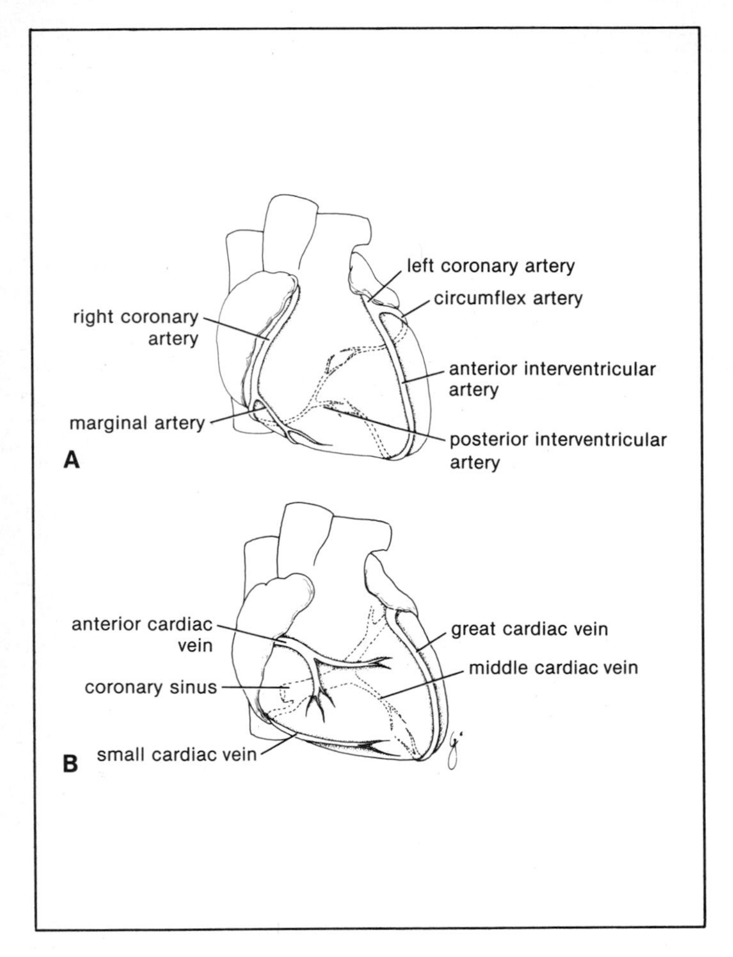

Figure 4–4. **A.** Coronary arteries and their main branches. **B.** Coronary veins and their main tributaries.

branches become blocked by disease. A sudden block of one of the larger branches of either coronary artery will inevitably lead to necrosis of the cardiac muscle (**myocardial infarction**) in that vascular area and often death to the patient.

Venous Drainage. Most of the venous blood from the heart wall drains into the right atrium through the coronary sinus (*Fig. 4–4*). The remainder drains directly into the right atrium through the **anterior cardiac vein** and small veins, the **venae cordis minimae.**

The **coronary sinus** lies in the posterior part of the atrioventricular groove and is a continuation of the great cardiac vein (*Fig. 4–4*).

The **great cardiac vein** ascends from the apex of the heart in the anterior interventricular groove. It then enters the atrioventricular groove curving to the left side and back of the heart to empty into the coronary sinus.

The **middle cardiac vein** runs from the apex of the heart in the posterior interventricular groove and empties into the coronary sinus.

The **small cardiac vein** accompanies the marginal artery along the inferior border of the heart and empties into the coronary sinus.

The **anterior cardiac vein** drains the anterior surface of the right atrium and the right ventricle and empties directly into the right atrium.

Blood Circulation Through the Heart

Blood is continuously returning to the right atrium through the superior and inferior venae cavae and the coronary sinus. From here the blood passes through the right atrioventricular orifice into the right ventricle. It is then conveyed to the lungs by the pulmonary trunk. After passing through the lungs the blood is returned to the left atrium by the right and left pulmonary veins. The blood then passes into the left ventricle through the left atrioventricular orifice. It is conveyed to the remainder of the body through the aorta.

PERICARDIUM

The pericardium is a sac that encloses the heart and the roots of the great vessels. It lies in the middle mediastinum.

The **fibrous pericardium** is the fibrous part of the pericardium. It is attached anteriorly to the sternum by the **sternopericardial ligaments,** above to the walls of the great blood vessels, namely the aorta, the pulmonary trunk, the superior vena cava, and the pulmonary veins. Below it is firmly attached to the central tendon of the diaphragm. The fibrous pericardium limits unnecessary movement of the heart.

The **serous pericardium** lines the fibrous pericardium and coats the heart. It is divided into parietal and visceral layers. The parietal layer lines the fibrous pericardium and is reflected around the roots of the great vessels to become continuous with the visceral layer that closely covers the heart **(epicardium)**. The slitlike space between the parietal and visceral layers is called the **pericardial cavity.** The cavity contains a small amount of tissue fluid, which acts as a lubricant to facilitate movements of the heart.

The **transverse sinus** (*Fig. 4–3*) is a passage on the posterior surface of the heart that lies between the reflection of serous pericardium around the aorta and pulmonary trunk and the reflection around the great veins.

The **oblique sinus** (*Fig. 4–3*) is a recess formed by the reflection of the serous pericardium around the venae cavae and the four pulmonary veins.

Nerve Supply. The fibrous pericardium and the parietal layer of serous pericardium are supplied by the phrenic nerves.

ARTERIES OF THE BODY

Pulmonary Trunk

The pulmonary trunk conveys deoxygenated blood from the right ventricle to the lungs (*Fig. 4–2*). It leaves the upper part of the right ventricle and runs upward, backward, and to the left. It is about 5 cm (2 inches) long and terminates in the concavity of the aortic arch by dividing into right and left pulmonary arteries.

The pulmonary trunk is contained within the fibrous pericardium and together with the ascending aorta, it is enclosed in a sheath of serous pericardium.

The **ligamentum arteriosum** is a fibrous band that connects the bifurcation of the pulmonary trunk to the lower surface of the aortic arch (*Fig. 4–2*). It is the remains of the **ductus arteriosus,** which in the fetus conducts blood from the pulmonary trunk to the aorta, thus bypassing the lungs. Following birth, the ductus closes.

Branches. The **right pulmonary artery** runs to the right behind the ascending aorta and superior vena cava to enter the root of the right lung where it divides into three primary branches, one for each lobe.

The **left pulmonary artery** runs to the left in front of the descending aorta to enter the root of the left lung, where it divides into two primary branches, one for each lobe.

Aorta

The aorta is the main arterial trunk that delivers oxygenated blood from the left ventricle of the heart to the tissues of the body (*Fig. 4–2 and 4–5*). It is divided for purposes of description into the following parts: ascending aorta, arch of aorta, descending thoracic aorta, and abdominal aorta.

Ascending Aorta. The ascending aorta begins at the base of the left ventricle and runs upward and forward to come to lie behind the right half of the sternum at the level of the sternal angle, where it becomes continuous with the arch of the aorta. It is enclosed with the pulmonary trunk in a sheath of serous pericardium. At its root it possesses three bulges, the **sinuses of the aorta,** one behind each aortic cusp.

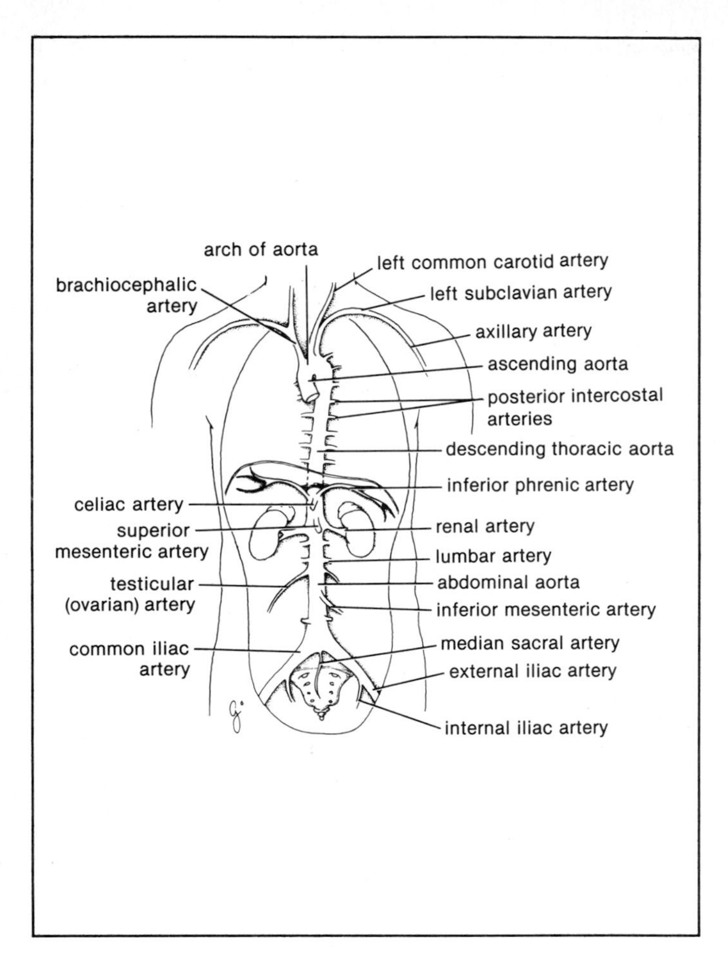

Figure 4–5. The aorta and its major branches.

Branches. The **right coronary artery** arises from the anterior aortic sinus, and the **left coronary artery** arises from the left posterior aortic sinus (*see p. 173*).

Arch of Aorta. The arch of the aorta is a continuation of the ascending aorta (*Fig. 4–5*). It lies behind the manubrium sterni and arches upward, backward, and to the left in front of the trachea. (Its main direction is backward.) It then passes downward to the left of the trachea, and at the level of the sternal angle becomes continuous with the descending thoracic aorta.

Inferiorly, the aortic arch is related to the bifurcation of the pulmonary trunk, the root of the left lung, the ligamentum arteriosum, and the left recurrent laryngeal nerve.

Branches. The **brachiocephalic artery** arises from the convex surface of the aortic arch and passes upward and to the right of the trachea. It terminates behind the right sternoclavicular joint by dividing into the right subclavian and right common carotid arteries.

The **left common carotid artery** arises from the convex surface of the aortic arch on the left side of the brachiocephalic artery. It runs upward and to the left of the trachea and enters the neck behind the left sternoclavicular joint.

The **left subclavian artery** arises from the aortic arch behind the left common carotid artery. It runs upward along the left side of the trachea and the esophagus to enter the root of the neck. It lies in contact with the apex of the left lung over which it arches.

Descending Thoracic Aorta. The descending thoracic aorta begins as a continuation of the arch of the aorta on the left side of the lower border of the body of the fourth thoracic vertebra (i.e., opposite the sternal angle) (*Fig. 4–5*). It runs downward in the posterior mediastinum inclining forward and medially to reach the anterior surface of the vertebral column. At the level of the twelfth thoracic vertebra it passes behind the diaphragm (through the aortic opening) in the midline and becomes continuous with the abdominal aorta.

Branches. **Posterior intercostal arteries*** are given off to the lower nine intercostal spaces on each side. The arteries of the right side cross the vertebral column and pass behind the esophagus. On both sides the arteries run behind the sympathetic trunks. On entering the intercostal space each artery runs forward in the costal groove between the vein above and the corresponding nerve below. The arteries anastomose with the anterior intercostal arteries from the internal thoracic and musculophrenic arteries. The lowest two intercostal arteries pass forward into the anterior abdominal wall. Each posterior intercostal artery gives off a posterior branch that supplies the muscles and skin of the back and supplies the spinal cord and the meninges.

Subcostal arteries are given off on each side and run along the lower border of the twelfth rib to enter the abdominal wall.

Pericardial, esophageal, and **bronchial arteries** are small branches that are distributed to these organs.

ARTERIES OF THE HEAD AND THE NECK

Common Carotid Artery

The right common carotid artery arises from the brachiocephalic artery behind the right sternoclavicular joint (*Fig. 4–6*). The left artery arises from the arch of the aorta in the superior mediastinum. The common carotid artery runs upward and backward through the neck, under cover of the anterior border of the sternocleidomastoid muscle. At the upper border of the thyroid cartilage it divides into the external and internal carotid arteries.

The common carotid artery at its point of division or the beginning of the internal carotid artery, shows a localized dilatation, called the **carotid sinus** (*Fig. 4–6*). Here the tunica media is thin and the tunica adventitia is thick and contains nerve endings from the glossopharyngeal nerve. The carotid sinus is a presso-

*The first and second posterior intercostal arteries arise from the costocervical branch of the subclavian.

receptor and assists in the regulation of the blood pressure in the cerebral arteries.

The **carotid body** is a small structure that lies posterior to the common carotid artery at its point of bifurcation. It is innervated by the glossopharyngeal nerve and is a chemoreceptor, being sensitive to excess carbon dioxide and reduced oxygen tensions in the blood circulating through the organ. When the chemoreceptors are stimulated thc heart and respiratory rates are increased.

The common carotid artery ascends the neck in the carotid fascial sheath and is closely related to the internal jugular vein laterally; the vagus nerve lies between these two structures.

Branches. External and internal carotid arteries.

External Carotid Artery

The external carotid artery is one of the terminal branches of the common carotid artery (*Fig. 4–6*). It begins at the level of the upper border of the thyroid cartilage undercover of the anterior border of the sternocleidomastoid muscle. It terminates in the substance of the parotid gland behind the neck of the mandible by dividing into the superficial temporal and maxillary arteries.

Close to its origin the artery emerges from under cover of the sternocleidomastoid muscle and it lies within the carotid triangle where its pulsations can be felt. At first it lies medial to the internal carotid artery, but as it ascends it passes backward and lies lateral to it.

Branches

1. Superior thyroid artery
2. Ascending pharyngeal artery
3. Lingual artery
4. Facial artery
5. Occipital artery
6. Posterior auricular artery
7. Superficial temporal artery
8. Maxillary artery

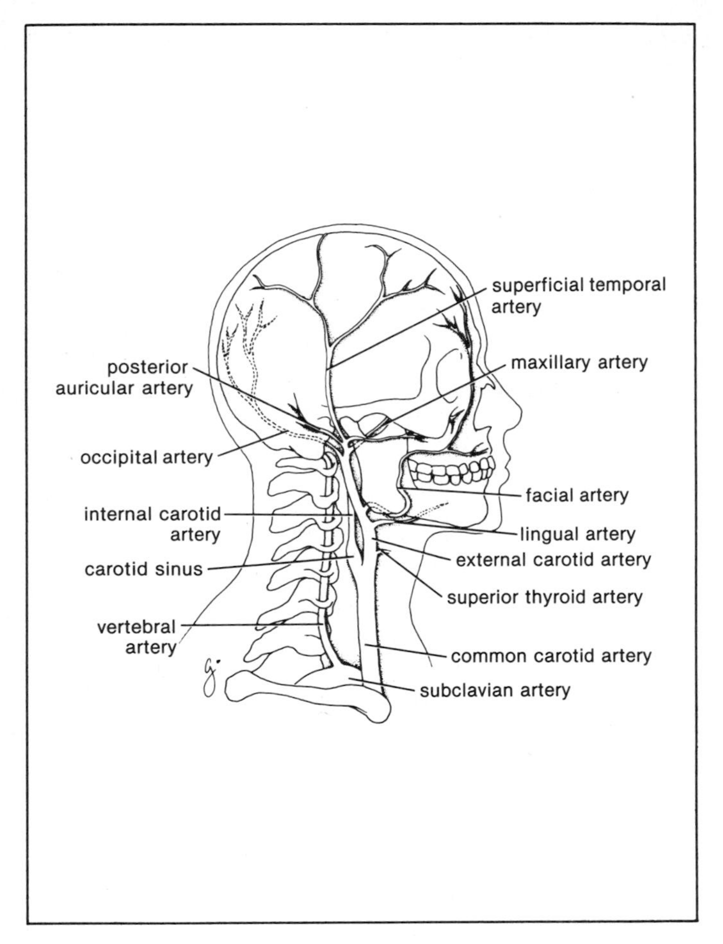

Figure 4–6. The major arteries of the head and the neck.

The **superior thyroid artery** arises from the front of the external carotid artery near its origin (*Fig. 4–6*). It curves downward and forward to the upper pole of the thyroid gland. It is accompanied by the external laryngeal nerve. It gives off a branch to the sternocleidomastoid muscle and the superior laryngeal artery.

The **ascending pharyngeal artery** is the smallest branch and arises close to the origin of the external carotid artery. It ascends along the wall of the pharynx and supplies this structure.

The **lingual artery** arises from the anterior surface of the external carotid artery, opposite the tip of the greater cornu of the hyoid bone (*Fig. 4–6*). It loops upward and then passes forward to supply the tongue. The loop of the artery is crossed superficially by the hypoglossal nerve.

The **facial artery** arises from the anterior surface of the external carotid artery, just above the tip of the greater cornu of the hyoid bone (*Fig. 4–6*). It loops upward on the lateral surface of the superior constrictor muscle close to the tonsil. It grooves the deep surface of the submandibular salivary gland then passes downward between the gland and the body of the mandible. It loops around the lower border of the mandible and ascends over the face close to the anterior border of the masseter muscle. Its pulsations can be felt against the mandible. The facial artery has a tortuous course as it ascends across the buccinator muscle lateral to the mouth. It terminates at the medial angle of the orbit.

The facial artery gives off branches to the tonsil, the submandibular salivary gland, and numerous branches to the muscles and the skin of the face.

The **occipital artery** arises from the posterior surface of the external carotid artery (*Fig. 4–6*). It follows the lower border of the posterior belly of the digastric muscle to reach the back of the scalp. The hypoglossal nerve loops around it.

The **posterior auricular artery** arises from the posterior surface of the external carotid artery and follows the upper border of the posterior belly of the digastric muscle to reach the groove between the auricle and the back of the scalp (*Fig. 4–6*).

The **superficial temporal artery,** the smaller terminal branch of the external carotid artery, ascends over the root of the zygomatic arch where it may be palpated (*Fig. 4–6*). It lies in front of the auricle in company with the auriculotemporal nerve. It divides into anterior and posterior branches that supply the scalp.

The **maxillary artery,** the larger terminal branch of the external carotid artery, arises behind the neck of the mandible in the parotid salivary gland (*Fig. 4–6*). It runs forward medial to the neck of the mandible to reach the lateral pterygoid muscle. The artery then inclines upward and forward lying either superficially or deep to the lower head of the lateral pterygoid muscle. The artery leaves the infratemporal fossa by entering the pterygopalatine fossa. Here it splits up into branches, which accompany the branches of the maxillary nerve.

The maxillary artery supplies the upper and lower jaws, the muscles of mastication, the nose, the palate, and the meninges.

Important Branch. The **middle meningeal artery** arises from the maxillary artery medial to the neck of the mandible. It ascends between the roots of the auriculotemporal nerve and enters the skull through the foramen spinosum. The artery runs laterally and forward on the greater wing of the sphenoid and divides into anterior and posterior branches that supply the dura mater and the bones of the skull.

The anterior branch is particularly important because it lies close to the motor area of the cerebral cortex. This branch is accompanied by its vein grooves or tunnels through the upper part of the greater wing of the sphenoid bone and the anterior inferior angle of the parietal bone. Here the skull is particularly thin in most individuals so that the artery or vein may be damaged at this site following a blow to the head. The artery and the vein lie between the meningeal layer of dura and the periosteal layer of dura so that the resulting hemorrhage would be extradural. The pressure of the extravasated blood on the cerebral cortex would result in paralysis of muscles on the opposite side of the body.

Internal Carotid Artery

The internal carotid artery begins at the bifurcation of the common carotid artery at the level of the upper border of the thyroid cartilage (*Fig. 4–6*). It ascends the neck in the carotid sheath accompanied by the internal jugular vein and the vagus nerve. At first lying superficially in the carotid triangle, it then passes deep to the parotid gland, the styloid process, and the muscles attached to it.

The internal carotid artery enters the cranial cavity by passing forward through the carotid canal in the petrous part of the temporal bone. It passes upward and then runs forward in the cavernous sinus. At the anterior end of the sinus the artery passes upward through the roof medial to the anterior clinoid process. The artery then inclines backward lateral to the optic chiasma to terminate by dividing into the anterior and middle cerebral arteries.

Branches. The cervical portion of the internal carotid artery has no branches, but the remainder of the artery gives off numerous small branches and four important branches.

1. The **ophthalmic artery** arises as the internal carotid artery emerges from the cavernous sinus. It enters the orbit through the optic canal, below and lateral to the optic nerve. Its branches supply the contents of the orbit and the skin above the orbital margin. The **central artery of the retina** is an important branch that enters the substance of the optic nerve and runs forward to enter the eyeball. This is an end artery and is the only blood supply to the retina.
2. The **posterior communicating artery** runs backward to join the posterior cerebral artery, thus contributing to the arterial circle (circulus arteriosus) at the base of the brain.
3. The **anterior cerebral artery** is a terminal branch of the internal carotid artery and arises at the medial end of the lateral cerebral sulcus. It passes forward in the longitudinal fissure of the cerebrum and winds round the genu of the corpus callosum to supply the medial and superolateral surfaces of the cerebral hemisphere. It is joined to the artery of the opposite side by the **anterior communicating artery.**
4. The **middle cerebral artery** is the largest branch of the internal carotid artery and runs laterally in the lateral cerebral

sulcus. It supplies the entire lateral surface of the cerebral hemisphere, except for the narrow strip supplied by the anterior cerebral artery, the occipital pole, and the inferolateral surface of the hemisphere, which are supplied by the posterior cerebral artery. This artery thus supplies all the "motor area" of the cerebral cortex except the "leg area." It also gives off important central branches that supply masses of gray matter and the internal capsule.

Circulus Arteriosus

The circulus arteriosus lies in the subarachnoid space in the interpeduncular fossa at the base of the brain. The circle is formed by the anastomosis between the branches of the two internal carotid arteries and the two vertebral arteries. The anterior communicating, the anterior cerebral, the internal carotid, the posterior communicating, the posterior cerebral, and the basilar (formed by the junction of the two vertebral arteries) all contribute to the circle. Cortical and central branches arise from the arterial circle and supply the brain.

Subclavian Arteries

The **right subclavian artery** arises from the brachiocephalic artery, behind the right sternoclavicular joint (*Fig. 4–6*). It arches upward and laterally over the pleura and comes to lie on the first rib between the scalenus anterior and medius. At the outer border of the first rib it becomes continuous with the axillary artery.

The **left subclavian artery** arises from the arch of the aorta behind the origin of the left common carotid artery (*Fig. 4–5*). It ascends to the root of the neck and then arches laterally in a manner similar to that of the right subclavian artery.

The scalenus anterior muscle passes anterior to the artery on each side and divides it into three parts. First part, from the origin of the artery to the medial border of the scalenus anterior. Second part is the portion that lies posterior to the scalenus anterior. Third part, from the lateral border of scalenus anterior to the lateral border of the first rib.

First Part of the Subclavian Artery

Important Relations. Anteriorly. From medial to lateral are the common carotid artery, the ansa subclavia, the vagus nerve, the internal jugular and vertebral veins, and on the left side the phrenic nerve. *Posteriorly.* The suprapleural membrane, the dome of the cervical pleura, the apex of the lung, the ansa subclavia, and, on the right side, the right recurrent laryngeal nerve.

Branches: Vertebral Artery. The vertebral artery arises from the upper margin of the subclavian artery and ascends in the neck in front of the transverse process of the seventh cervical vertebra (*Fig. 4–6*). It then ascends through the foramina in the transverse processes of the upper six cervical vertebrae. It passes medially above the posterior arch of the atlas, pierces the dura mater, and enters the vertebral canal. The artery then ascends into the skull through the foramen magnum and passes upward, forward, and medially on the medulla oblongata. On reaching the anterior aspect of the medulla it joins the vessel of the opposite side at the lower border of the pons to form the basilar artery.

The **basilar artery** ascends in a groove on the anterior surface of the pons, giving off branches to the pons, cerebellum, and internal ear and finally divides into the two posterior cerebral arteries. The **posterior cerebral artery** on each side curves laterally and backward around the midbrain. Cortical branches supply the inferolateral surfaces of the temporal lobe and the visual cortex on the lateral and medial surfaces of the occipital lobe.

BRANCHES OF THE VERTEBRAL ARTERY IN THE NECK. Spinal and muscular branches. The spinal branches enter the vertebral canal through the intervertebral foramina.

BRANCHES OF THE VERTEBRAL ARTERY IN THE SKULL. These include (1) meningeal arteries, (2) anterior and posterior spinal arteries, (3) posterior inferior cerebellar artery, and (4) medullary arteries.

Thyrocervical Trunk. The thyrocervical trunk is a wide, short trunk that arises from the front of the first part of the subclavian artery, near the medial border of scalenus anterior.

BRANCHES. The **inferior thyroid artery** ascends along the medial border of the scalenus anterior muscle to the level of the cricoid cartilage. It then turns medially and downward behind the carotid sheath and its contents, namely, the internal jugular vein, the vagus nerve, and the common carotid artery, and passes in front of the vertebral artery and the sympathetic trunk. It then reaches the posterior border of the thyroid gland where it is closely related to the recurrent laryngeal nerve. The inferior thyroid artery supplies not only the thyroid gland but the inferior parathyroid glands, the larynx, the pharynx, the trachea, and the esophagus.

The **superficial cervical artery** passes laterally over the scaleni muscles and the phrenic nerve. In the posterior triangle it crosses the trunks of the brachial plexus and disappears under the trapezius muscle.

The **suprascapular artery** runs laterally and downward over the scalenus anterior and the phrenic nerve. It crosses in front of the brachial plexus and subclavian artery in the posterior triangle of the neck to reach the upper border of the scapula. It then follows the suprascapular nerve into the supraspinous fossa and takes part in the anastomosis around the scapula.

Internal Thoracic Artery. The internal thoracic artery arises from the lower border of the first part of the subclavian artery. It enters the thorax behind the first costal cartilage and in front of the pleura. It is crossed from the lateral to the medial side by the phrenic nerve. It descends vertically on the pleura behind the costal cartilages, a finger-breadth lateral to the sternum. The artery ends in the sixth intercostal space by dividing into the superior epigastric and musculophrenic arteries.

BRANCHES

1. **Anterior intercostal arteries** for the upper six intercostal spaces

2. **Perforating arteries** pierce the intercostal muscles and supply the skin and the mammary gland
3. **Pericardiophrenic artery** supplies the pericardium and the diaphragm
4. **Mediastinal arteries** to the mediastinum, including the thymus gland
5. **Superior epigastric artery** enters the rectus sheath and supplies the upper part of the rectus muscle
6. **Musculophrenic artery** follows the costal margin on the upper surface of the diaphragm; it supplies the diaphragm and the lower intercostal spaces.

Second Part of the Subclavian Artery

Important Relations. Anteriorly. The scalenus anterior muscle. *Posteriorly.* The suprapleural membrane, the dome of the cervical pleura, and the apex of the lung.

Branch. The **costocervical trunk** arises from the back of the second part of the subclavian artery. It runs backward over the dome of the pleura to the neck of the first rib. Here, it divides into the **superior intercostal artery,** which gives rise to the posterior intercostal arteries for the first and second intercostal spaces, and the **deep cervical artery,** which supplies the muscles of the back of the neck.

Third Part of the Subclavian Artery. The third part of the subclavian artery, with the nerves of the brachial plexus, is surrounded by a sheath of fascia, the **axillary sheath.** This sheath is derived from the prevertebral layer of the deep cervical fascia.

Important Relations. Anteriorly. The skin, superficial fascia, platysma, supraclavicular nerves, deep cervical fascia, external jugular vein, suprascapular artery, and the clavicle. At first the artery is covered by the sternocleidomastoid muscle. In the intermediate part of its course, it is comparatively superficial, and its pulsations can be easily felt. The last part of the artery passes behind the clavicle. *Posteriorly.* The lower trunk of the brachial plexus and the scalenus medius muscle.

Branches. The third part of the subclavian artery usually has no branches. Occasionally the superficial cervical or suprascapular artery, or both, arise from this part of the subclavian artery.

ARTERIES OF THE UPPER LIMB

Axillary Artery

The axillary artery begins at the lateral border of the first rib as a continuation of the subclavian artery, and at the lower border of the teres major muscle it becomes the brachial artery (*Fig. 4–7*). Throughout its course, the artery is closely related to the cords of the brachial plexus and their branches and is enclosed with them in a connective tissue sheath, called the **axillary sheath.**

The pectoralis minor muscle crosses in front of the artery dividing it for purposes of description into three parts. First part, from the lateral border of the first rib to the upper border of the pectoralis minor. Second part is the portion that lies posterior to the pectoralis minor. Third part, from the lower border of pectoralis minor to the lower border of the teres major.

Important Relations of the First Part of the Artery. *Anteriorly.* Skin, fasciae and pectoralis major. *Posteriorly.* Long thoracic nerve. *Laterally.* Three cords of brachial plexus. *Medially.* Axillary vein.

Important Relations of the Second Part of the Artery. *Anteriorly.* Skin, fasciae, pectoralis major and minor. *Posteriorly.* Posterior cord of brachial plexus. *Laterally.* Lateral cord of brachial plexus. *Medially.* Medial cord of brachial plexus and axillary vein.

Important Relations of the Third Part of the Artery. *Anteriorly.* Skin, fasciae, pectoralis major. The artery is crossed by the

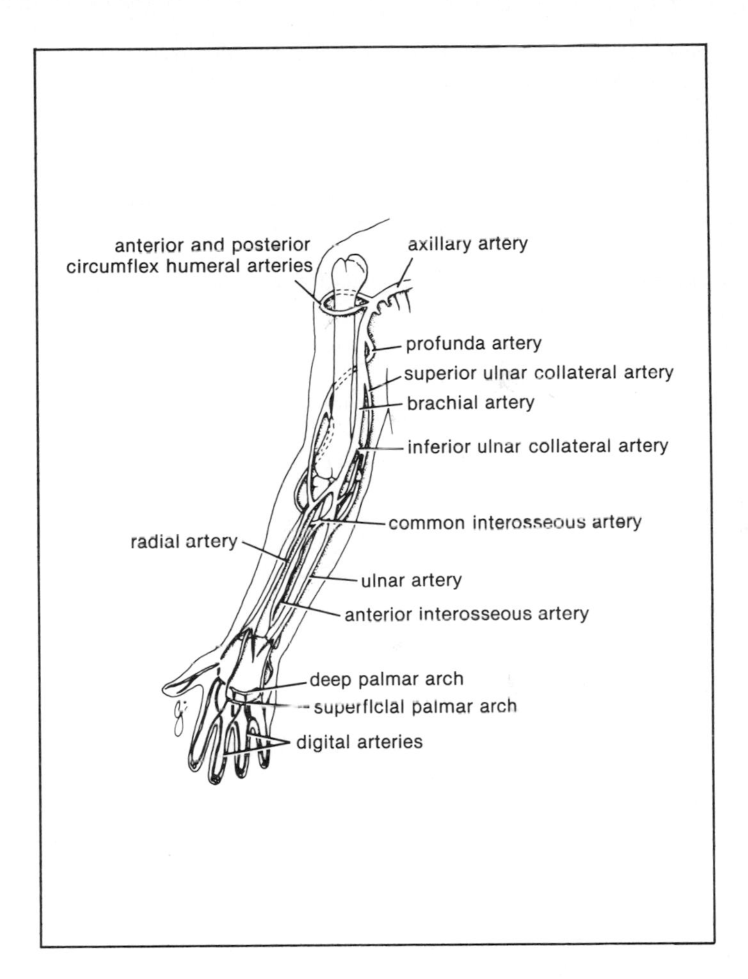

Figure 4–7. The major arteries of the upper limb.

medial root of the median nerve. *Posteriorly.* Axillary and radial nerves. *Laterally.* Lateral root of median nerve and the musculocutaneous nerve. *Medially.* Ulnar nerve, medial cutaneous nerve of arm and axillary vein.

Branches. The first part of the artery gives off one branch, the second part two branches, and the third part three branches.

Branch of the First Part. The **highest thoracic artery** is small and runs to the chest wall along the upper border of pectoralis minor.

Branches of the Second Part. The **thoracoacromial artery** immediately divides into four terminal branches. The **lateral thoracic artery** runs to the chest wall along the lower border of the pectoralis minor. In the female it supplies the mammary gland.

Branches of the Third Part. The **subscapular artery** runs along the lower border of the subscapularis muscle to the inferior angle of the scapula. The **anterior** and **posterior circumflex humeral** arteries wind around the front and back of the surgical neck of the humerus, respectively (*Fig. 4–7*).

Arterial Anastomosis Around the Shoulder Joint

To compensate for temporary occlusion of the axillary artery during movements of the shoulder joint, and to ensure adequate blood flow to the upper limb, the following arteries anastomose with one another. The suprascapular and superficial cervical arteries (branches of the thyrocervical trunk from the first part of the subclavian artery) with the subscapular and anterior and posterior circumflex humeral arteries (branches of the third part of the axillary artery).

Brachial Artery

The brachial artery begins at the lower border of the teres major as a direct continuation of the axillary artery (*Fig. 4–7*). It ends at the level of the neck of the radius by dividing into the radial and ulnar arteries.

Relations. *Anteriorly.* The artery is superficial and is overlapped from the lateral side by the coracobrachialis and the biceps. The median nerve crosses its middle part and the bicipital aponeurosis its lower part. *Posteriorly.* The triceps, the insertion of coracobrachialis and the brachialis. *Medially.* The ulnar nerve and the basilic vein in the upper part of the arm and the median nerve in the lower part of the arm. *Laterally.* The median nerve, the coracobrachialis, and the biceps in the upper part of the arm and the tendon of the biceps in the lower part of its course.

Branches

1. **Muscular branches.**
2. **Nutrient artery** to the humerus.
3. **Profunda artery** is a large branch that follows the radial nerve into the posterior compartment of the arm (*Fig. 4–7*).
4. **Superior ulnar collateral artery** follows the ulnar nerve.
5. **Inferior ulnar collateral artery** takes part in the anastomosis around the elbow joint.

Radial Artery

The radial artery, one of the terminal branches of the brachial artery, begins in the cubital fossa at the level of the neck of the radius (*Fig. 4–7*). It descends through the forearm lying superficially throughout most of its course. In the region of the wrist it winds backward, round the lateral side of the carpus to the proximal end of the space between the first and second metacarpal bones. Here, it passes anteriorly into the palm between the two heads of the first dorsal interosseous muscle and joins the deep branch of the ulnar artery forming the **deep palmar arch.**

In the lower part of the forearm the radial artery lies on the anterior surface of the radius and is covered only by skin and fascia. Here, the artery has on its lateral side the tendon of brachioradialis and on its medial side the tendon of the flexor carpi radialis. It is at this site just above the wrist that the radial pulse can be palpated in the living subject.

In the middle third of its course the radial nerve lies on its lateral side.

Branches

1. **Muscular branches.**
2. **Recurrent branch** takes part in the arterial anastomosis around the elbow joint.
3. **Superficial palmar branch** arises just above the wrist. It enters the palm and frequently joins the ulnar artery to form the **superficial palmar arch** (*Fig. 4–7*).
4. **First dorsal metacarpal artery** arises just before the radial artery passes between the two heads of the first dorsal interosseous muscle. It supplies the adjacent sides of the thumb and index finger.
5. **Arteria princeps pollicis** arises from the radial artery in the palm. It divides into two branches which supply the sides of the thumb.
6. **Arteria radialis indicis** arises from the radial artery in the palm. It supplies the lateral side of the index finger.

Deep Palmar Arch. This arch is deeply placed in the palm and extends from the proximal end of the first interosseous space to the base of the fifth metacarpal bone (*Fig. 4–7*). It is formed by the radial artery and terminates by anastomosing with the deep branch of the ulnar artery.

Branches. Palmar, metacarpal, perforating, and recurrent.

Ulnar Artery

The ulnar artery is the larger of the two terminal branches of the brachial artery (*Fig. 4–7*). It begins in the cubital fossa at the level of the neck of the radius. It descends to the wrist and enters the palm in front of the flexor retinaculum in company with the ulnar nerve. It ends by forming the superficial palmar arch, often uniting with the superficial palmar branch of the radial artery.

In the upper part of its course it passes deep to the flexor muscles that arise from the common flexor origin on the medial epicondyle of the humerus. In the lower part of its course it becomes superficial and lies between the tendons of flexor carpi ulnaris and the tendons of flexor digitorum superficialis. As the artery passes in front of the flexor retinaculum it is covered only by skin and fasciae and lies just lateral to the pisiform bone. It is here that the pulsations can be palpated in the living subject.

Branches

1. **Muscular branches.**
2. **Recurrent branches** take part in the arterial anastomosis around the wrist joint.
3. **Common interosseous artery** arises from the upper part of the ulnar artery and quickly divides into **anterior and posterior interosseous arteries.** These latter arteries descend on the anterior and posterior surfaces of the interosseous membrane, respectively.
4. **Deep palmar branch** arises in front of the flexor reticulum and joins the radial artery to complete the deep palmar arch.

Superficial Palmar Arch. This arterial arch lies immediately beneath the palmar aponeurosis and on the long flexor tendons (*Fig. 4–7*). It is a direct continuation of the ulnar artery. The arch is often completed on the lateral side by the superficial palmar branch of the radial artery. The curve of the arch lies across the palm, level with the distal border of the extended thumb. Four digital arteries arise from the arch and supply the medial side of the little finger and the adjacent sides of the little, ring, middle, and index fingers, respectively.

ARTERIES OF THE ABDOMEN AND THE PELVIS

Abdominal Aorta

The aorta enters the abdomen through the aortic opening of the diaphragm in front of the twelfth thoracic vertebra (*Fig. 4–5*). It descends on the anterior surfaces of the bodies of the lumbar vertebrae and in front of the fourth lumbar vertebra it divides into the two common iliac arteries. On the right side of the aorta lie the inferior vena cava, the cisterna chyli, and the beginning of the azygos vein. On its left side lies the left sympathetic trunk. On the anterior surface it is related to the stomach, the celiac plexus, the pancreas, the splenic vein, the left renal vein, the third part of duodenum, the coils of small intestine, and the peritoneum.

Branches

1. Three anterior visceral branches
   a. Celiac artery
   b. Superior mesenteric artery
   c. Inferior mesenteric artery
2. Three lateral visceral branches
   a. Suprarenal artery
   b. Renal artery
   c. Testicular or ovarian artery
3. Five lateral abdominal wall branches
   a. Inferior phrenic artery
   b. Four lumbar arteries
4. Three terminal arteries
   a. Two common iliac arteries
   b. Median sacral artery

Celiac Artery

The celiac artery or trunk is very short and arises from the commencement of the abdominal aorta (*Fig. 4–5*). It is the artery that supplies the foregut. It is surrounded by the celiac plexus and lies behind the lesser sac of peritoneum. It has three terminal branches: the left gastric, the splenic, and the hepatic arteries.

Left Gastric Artery. This small artery runs to the cardiac end of the stomach, gives off a few esophageal branches, then turns to the right along the lesser curvature of the stomach. It anastomoses with the right gastric artery.

Splenic Artery. This large artery runs to the left in a wavy course along the upper border of the pancreas and behind the stomach. On reaching the left kidney the artery enters the lienorenal ligament and runs to the hilum of the spleen.

Branches

1. **Pancreatic branches.**
2. **Left gastroepiploic artery** arises near the hilum of the spleen and reaches the greater curvature of the stomach in the gastrosplenic omentum. It passes to the right along the greater curvature of the stomach between the layers of the greater omentum. It anastomoses with the right gastroepiploic artery.
3. **Short gastric arteries,** five or six in number, arise from the end of the splenic artery and reach the fundus of the stomach in the gastrosplenic omentum. They anastomose with the left gastric artery and the left gastroepiploic artery.

Hepatic Artery. * This medium-sized artery runs forward and to the right and then ascends between the layers of the lesser omentum. It lies in front of the opening into the lesser sac and is placed to the left of the common bile duct and in front of the portal

*For purposes of description the hepatic artery is sometimes divided into the **common hepatic artery,** which extends from its origin to the gastroduodenal branch, and the **hepatic artery proper,** which is the remainder of the artery.

vein. At the porta hepatis it divides into right and left branches to supply the corresponding lobes of the liver.

Branches

1. **Right gastric artery** arises from the hepatic artery at the upper border of the pylorus and runs to the left in the lesser omentum along the lesser curvature of the stomach. It anastomoses with the left gastric artery.
2. **Gastroduodenal artery** is a large branch that descends behind the first part of the duodenum. It divides into the **right gastroepiploic artery** that runs along the greater curvature of the stomach between the layers of the greater omentum, and the **superior pancreaticoduodenal artery** that descends between the second part of the duodenum and the head of the pancreas.
3. **Right and left hepatic arteries.** The right hepatic artery usually gives off the **cystic artery,** which runs to the neck of the gallbladder.

Superior Mesenteric Artery

The superior mesenteric artery arises from the front of the abdominal aorta just below the celiac artery (*Fig. 4–5*). It runs downward and to the right behind the neck of the pancreas and in front of the third part of the duodenum. It continues downward to the right between the layers of the mesentery of the small intestine and ends by anastomosing with the ileal branch of its own ileocolic branch.

The superior mesenteric artery is the artery of the midgut and supplies the intestine from halfway along the duodenum to just proximal to the left colic flexure.

Branches

1. **Inferior pancreaticoduodenal artery** passes to the right as a single or double branch along the upper border of the third part of the duodenum and the head of the pancreas.

2. **Middle colic artery** runs forward in the transverse mesocolon to supply the transverse colon.
3. **Right colic artery** is often a branch of the ileocolic artery. It passes to the right to supply the ascending colon.
4. **Ileocolic artery** passes downward and to the right. It gives rise to a **superior branch** that anastomoses with the right colic artery and an **inferior branch** that anastomoses with the end of the superior mesenteric artery. The inferior branch gives rise to the **anterior and posterior cecal arteries;** the **appendicular artery** is a branch of the posterior cecal artery.
5. **Jejunal and ileal branches.** These are 12 to 15 in number and arise from the left side of the superior mesenteric artery. Each artery divides into two, which unite with adjacent branches to form a series of arcades. Branches from the arcades divide and unite to form a second, third, and fourth series of arcades. Fewer arcades supply the jejunum as compared with the ileum. From the terminal arcades small straight vessels supply the intestine.

Inferior Mesenteric Artery

The inferior mesenteric artery arises from the abdominal aorta about 3.8 cm (1½ inches) above its bifurcation (*Fig. 4-5*). The artery runs downward and to the left and crosses the left common iliac artery. Here, it changes name and becomes the superior rectal artery.

The inferior mesenteric artery is the artery of the hindgut and supplies the large intestine from the distal one-third of the transverse colon to halfway down the anal canal.

Branches

1. **Superior left colic artery** (left colic artery) runs upward and to the left and supplies the distal third of the transverse colon, the left colic (splenic) flexure, and the upper part of the descending colon.

2. **Inferior left colic arteries** (sigmoid arteries). These are two or three in number and supply the descending and sigmoid (pelvic) colon.
3. **Superior rectal artery** is a continuation of the inferior mesenteric artery as it crosses the left common iliac artery. It descends into the pelvis behind the rectum. The artery supplies the rectum and the upper half of the anal canal and anastomoses with the middle rectal and inferior rectal arteries.

Marginal Artery

The anastomosis of the colic arteries around the concave margin of the large intestine form a single arterial trunk called the marginal artery. This begins at the ileocolic junction, where it anastomoses with the ileal branches of the superior mesenteric artery, and it ends where it anastomoses less freely with the superior rectal artery.

Middle Suprarenal Arteries

The middle suprarenal artery arises on each side of the aorta and runs horizontally laterally to the suprarenal gland. It anastomoses with the superior suprarenal artery of the phrenic and the inferior suprarenal artery from the renal.

Renal Arteries

The renal artery arises on each side of the aorta just below the origin of the superior mesenteric artery (*Fig. 4–5*). The right artery is longer and passes behind the inferior vena cava. The left artery is higher. The renal artery gives off the inferior suprarenal artery.

Testicular Arteries

The testicular arteries arise from the front of the aorta just below the origin of the renal arteries (*Fig. 4–5*). Each artery is long and slender and passes obliquely downward and laterally behind the peritoneum. It crosses the ureter and the external iliac artery to reach the deep inguinal ring where it joins the spermatic cord. Having passed through the inguinal canal it enters the scrotum and supplies the testis and the epididymis.

In the female the **ovarian artery** has a similar abdominal course. Having crossed the external iliac artery at the pelvic inlet, it enters the suspensory ligament of the ovary. It then passes into the broad ligament and enters the ovary by way of the mesovarium.

Inferior Phrenic Arteries

The inferior phrenic arteries arise from the aorta just beneath the diaphragm. They run upward and laterally supplying the diaphragm and each gives off a superior suprarenal artery.

Lumbar Arteries

The four pairs of lumbar arteries arise from the back of the aorta and pass round the bodies of the upper four lumbar vertebrae. They end between the muscle layers of the abdominal wall and supply the muscles and the skin of the back. The first lumbar artery gives off branches to the lower part of the spinal cord.

Median Sacral Artery

The median sacral artery is a small branch that arises at the bifurcation of the aorta. It descends into the pelvis in the midline and runs over the anterior surface of the sacrum and the coccyx.

Common Iliac Arteries

The right and left common iliac arteries are the terminal branches of the abdominal aorta (*Fig. 4–5*). They arise at the level of the fourth lumbar vertebra and run downward and laterally to end opposite the sacroiliac joint by dividing into external and internal iliac arteries. At the bifurcation, the common iliac artery on each side is crossed anteriorly by the ureter.

Branches of Common Iliac Arteries

External Iliac Artery. The external iliac artery runs along the medial border of the psoas muscle, following the pelvic brim, and gives off the inferior epigastric and deep circumflex iliac branches. The artery enters the thigh by passing under the inguinal ligament, to become the femoral artery. The **inferior epigastric artery** arises just above the inguinal ligament. It passes upward and medially along the medial margin of the deep inguinal ring and enters the rectus sheath behind the rectus abdominis muscle. The **deep circumflex iliac artery** arises close to the inferior epigastric artery. It ascends laterally to the anterior superior iliac spine and the iliac crest supplying the muscles of the abdominal wall.

Internal Iliac Artery. The internal iliac artery passes down into the pelvis to the upper margin of the greater sciatic foramen where it divides into anterior and posterior divisions. The branches of these divisions supply the pelvic viscera, the perineum, the buttock, and the sacral canal.

Branches of Anterior Division of Internal Iliac Artery

Superior Vesical Artery. This is the proximal patent part of the umbilical artery and supplies the upper portion of the bladder. It gives off the **artery to the vas deferens.**

Inferior Vesical Artery. This supplies the base of the bladder and the prostate and seminal vesicles in the male.

Middle Rectal Artery. Commonly this artery arises with the inferior vesical artery. It supplies the muscle of the lower rectum and anastomoses with the superior rectal and inferior rectal arteries.

Uterine Artery. This runs medially on the floor of the pelvis and crosses the ureter superiorly. It passes above the lateral fornix of the vagina to reach the uterus. Here, it ascends between the layers of the broad ligament along the lateral margin of the uterus. It ends by following the uterine tube laterally where it anastomoses with the ovarian artery. The uterine artery gives off a vaginal branch.

Vaginal Artery. This artery usually takes the place of the inferior vesical artery present in the male. It supplies the vagina and the base of the bladder.

Obturator Artery. This artery runs forward along the lateral wall of the pelvis in company with the obturator nerve. It leaves the pelvis through the obturator canal and divides into two branches that pass around the margin of the outer surface of the obturator membrane. It gives off muscular branches and an articular branch to the hip joint; the latter sends an important twig to the head of the femur.

In about 25 percent of people the obturator artery is replaced by the pubic branch of the inferior epigastric artery. This **abnormal obturator artery** may descend into the pelvis on the medial side of the femoral ring lying on the lacunar ligament. Here, it would lie close to the neck of a femoral hernial sac.

Internal Pudendal Artery. This artery leaves the pelvis through the greater sciatic foramen and enters the gluteal region below the piriformis muscle. It enters the perineum by passing through the lesser sciatic foramen. The artery then passes forward in the pudendal canal with the pudendal nerve and by means of its branches supplies the musculature of the anal canal and the skin and muscles of the perineum.

BRANCHES

1. The **inferior rectal artery** leaves the pudendal canal, crosses the ischiorectal fossa and supplies the lower half of the anal canal.

2. Branches to the scrotum and penis in the male and to the labia and clitoris in the female. These vessels include the artery of the bulb of the penis (artery of bulb of vestibule), the deep artery of penis (clitoris), and the dorsal artery of penis (clitoris).

Inferior Gluteal Artery. This artery leaves the pelvis through the lower part of the greater sciatic foramen below the piriformis. It divides into numerous branches that supply the gluteal region.

Branches of the Posterior Division of the Internal Iliac Artery

Superior Gluteal Artery. This artery leaves the pelvis through the upper part of the greater sciatic foramen above the piriformis. It divides into numerous branches that supply the gluteal region.

Iliolumbar Artery. This artery ascends across the pelvic inlet posterior to the external iliac vessels, psoas and iliacus muscles.

Lateral Sacral Arteries. These arteries descend in front of the sacral plexus giving off branches to neighboring structures.

ARTERIES OF THE LOWER LIMB

The chief artery of the lower limb is the femoral artery (*Fig. 4–8*). This artery is supplemented in the gluteal region by the superior and inferior gluteal arteries and in the adductor region by the obturator artery.

Femoral Artery

The femoral artery is a continuation of the external iliac artery. It begins behind the inguinal ligament where it lies midway between the anterior superior iliac spine and the symphysis pubis. It is here that its pulsations may be palpated in the living subject.

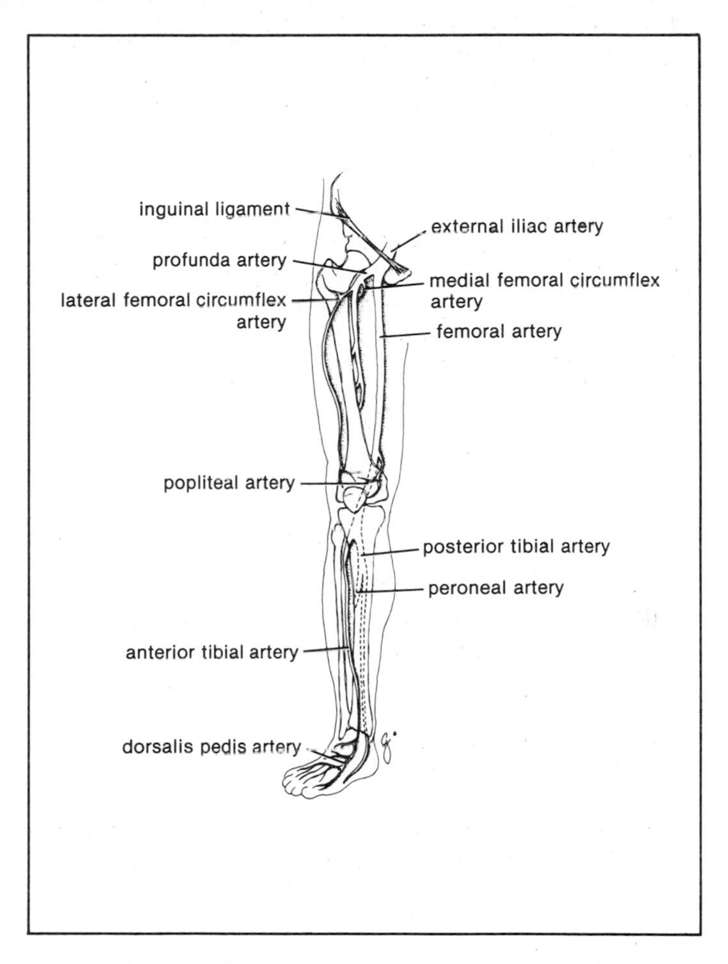

Figure 4–8. The major arteries of the lower limb.

The artery descends almost vertically toward the adductor tubercle of the femur and ends at the opening of the adductor magnus muscle by entering the popliteal space as the popliteal artery.

Relations. *Anteriorly.* In the upper part of its course in the femoral triangle, it is superficial and is covered by skin and fascia. In the lower part of its course, it lies in the adductor canal (subsartorial canal) and passes behind the sartorius muscle. *Posteriorly.* The psoas muscle, which separates the artery from the hip joint. *Medially.* The artery is related to the femoral vein and the femoral canal in the upper part of its course. *Laterally.* The femoral nerve and its branches.

The femoral artery is accompanied by the femoral vein, which lies on the medial side at the inguinal ligament and posterior to it at the apex of the femoral triangle. At the opening in the adductor magnus the vein lies on the lateral side of the artery.

Branches

1. **Superficial circumflex iliac artery.** This arises just below the inguinal ligament and runs laterally to the region of the anterior superior iliac spine.
2. **Superficial epigastric artery.** This arises just below the inguinal ligament and ascends on to the abdominal wall as high as the umbilicus.
3. **Superficial external pudendal artery.**
4. **Deep external pudendal artery.** This artery and the preceding artery arise just below the inguinal ligament and run medially to supply the skin of the scrotum (or labium majus).
5. **Profunda femoris artery.** This is a large and important branch that arises from the lateral and posterior surface of the femoral artery about 4 cm (1½ inches) below the inguinal ligament (*Fig. 4–8*). It supplies structures in the anterior, medial, and posterior fascial compartments of the thigh by means of the following branches. At its origin it gives off the **medial and lateral femoral circumflex arteries,**

and during its course it gives off **three perforating arteries.** The artery ends by becoming the **fourth perforating artery.**

6. **Descending genicular artery.** This is a small branch that arises from the femoral artery in the subsartorial canal.

Popliteal Artery

The popliteal artery is a continuation of the femoral artery and extends from the opening in the adductor magnus to the lower border of the popliteous muscle, where it divides into anterior and posterior tibial arteries (*Fig. 4–8*). It is deeply placed in the popliteal fossa.

Relations. *Anteriorly.* The popliteal surface of the femur, the knee joint, and the popliteous muscle. *Posteriorly.* The popliteal vein, the tibial nerve, fascia, and skin.

Branches

1. Muscular branches
2. Articular branches to the knee joint
3. Terminal branches: anterior and posterior tibial arteries.

Arterial Anastomosis Around the Knee Joint

To compensate for the narrowing of the popliteal artery, which occurs during flexion of the knee, a profuse arterial anastomosis exists. The vessels involved are the descending genicular of the femoral, the lateral femoral circumflex of the profunda femoris, the articular branches of the popliteal, and branches from the anterior and posterior tibial arteries.

Anterior Tibial Artery

The anterior tibial artery arises at the bifurcation of the popliteal artery at the level of the lower border of the popliteus muscle (*Fig. 4–8*). It passes forward between the tibia and the fibula through the upper part of the interosseous membrane to enter the anterior compartment of the leg. It descends with the deep peroneal nerve to the front of the ankle joint where it becomes the dorsalis pedis artery.

In the upper part of its course it lies deep beneath the muscles of the anterior compartment. In the lower part of the course it lies superficial in front of the lower end of the tibia. At the ankle it has the tendon of the extensor hallucis longus on the medial side and the tendons of extensor digitorum longus on its lateral side; it is here that pulsations can easily be felt in the living subject.

Branches

1. Muscular branches
2. Anastomotic branches, which anastomose with branches of other arteries around the knee and ankle joints.

Dorsalis Pedis Artery

The dorsalis pedis artery begins in front of the ankle joint as a continuation of the anterior tibial artery (*Fig. 4–8*). It ends by passing downward into the sole through the proximal part of the space between the first and second metatarsal bones. Having passed between the two heads of the first dorsal interosseous muscle, it joins the lateral plantar artery and completes the plantar arch.

The artery is superficial in position. On its lateral side lie the tendons of the extensor digitorum longus and on the medial side the tendon of extensor hallucis longus. Its pulsations can be easily felt in the living subject.

Branches

1. **Lateral tarsal artery.** Supplies the dorsum of the foot.
2. **Arcuate artery.** Runs laterally across the bases of the metatarsal bones and gives off metatarsal branches to the toes.

3. **First dorsal metatarsal artery.** Supplies both sides of the big toe.

Posterior Tibial Artery

The posterior tibial artery arises at the bifurcation of the popliteal artery at the level of the lower border of the popliteus muscle (*Fig. 4–8*). It descends in the posterior compartment of the leg accompanied by the tibial nerve. The artery terminates behind the medial malleolus by dividing into medial and lateral plantar arteries.

In the upper two-thirds of its course it lies deep to the gastrocnemius and soleus muscles. In the lower third of the leg it becomes superficial and lies on the posterior surface of the tibia and is here covered only by fascia and skin. The pulsations of the artery may be palpated at a spot midway between the medial malleolus and the heel in the living subject.

Branches

1. **Peroneal artery.** This is a large artery that arises close to the origin of the posterior tibial artery (*Fig. 4–8*). It descends behind the fibula in close association with the flexor hallucis longus muscle. The peroneal artery has **muscular branches,** a **nutrient artery** to the fibula, and **anastomotic** branches around the ankle joint. A **perforating branch** pierces the lower part of the interosseous membrane to reach the front of the ankle.
2. **Muscular branches.**
3. **Nutrient artery** to the tibia.
4. **Anastomotic branches** around the ankle joint.
5. **Medial and lateral plantar arteries.**

Medial Plantar Artery. The medial plantar artery is the smaller of the terminal branches of the posterior tibial artery. It arises behind the medial malleolus and runs forward along the medial border of the foot with the medial plantar nerve. It lies deep to the abductor hallucis muscle and ends by supplying the big toe. It gives off many muscular and cutaneous branches.

Lateral Plantar Artery. The lateral plantar artery is the larger of the terminal branches of the posterior tibial artery. It arises behind the medial malleolus and runs forward deep to the abductor hallucis and the flexor digitorum brevis with the lateral plantar nerve. At the base of the fifth metatarsal bone, the artery curves medially to form the plantar arch. It anastomoses with the dorsalis pedis artery. The lateral plantar artery gives off many muscular and cutaneous branches.

Plantar Arch. The plantar arch extends from the base of the fifth metatarsal bone to the proximal part of the space between the first and second metatarsal bones. It is formed by the terminal part of the lateral plantar artery and is completed medially by joining the dorsalis pedis artery. The plantar arch gives off numerous branches that include perforating and metatarsal arteries. The metatarsal arteries give rise to digital arteries that supply the lateral four toes.

VEINS OF THE BODY

Veins of the Head and the Neck

The veins of the head and the neck may be divided for purposes of description into (1) the veins of the brain, the venous sinuses, the diploic and emissary veins and (2) the veins of the scalp, the face, and the neck (*Fig. 4–9*).

Veins of the Brain. These veins have no valves and no muscle tissue in their walls. They are very thin. They are made up of the cerebral veins, the cerebellar veins, and the veins of the brainstem.

Cerebral Veins. These veins may be divided into external veins that lie on the surface of the cerebral hemispheres with the ce-

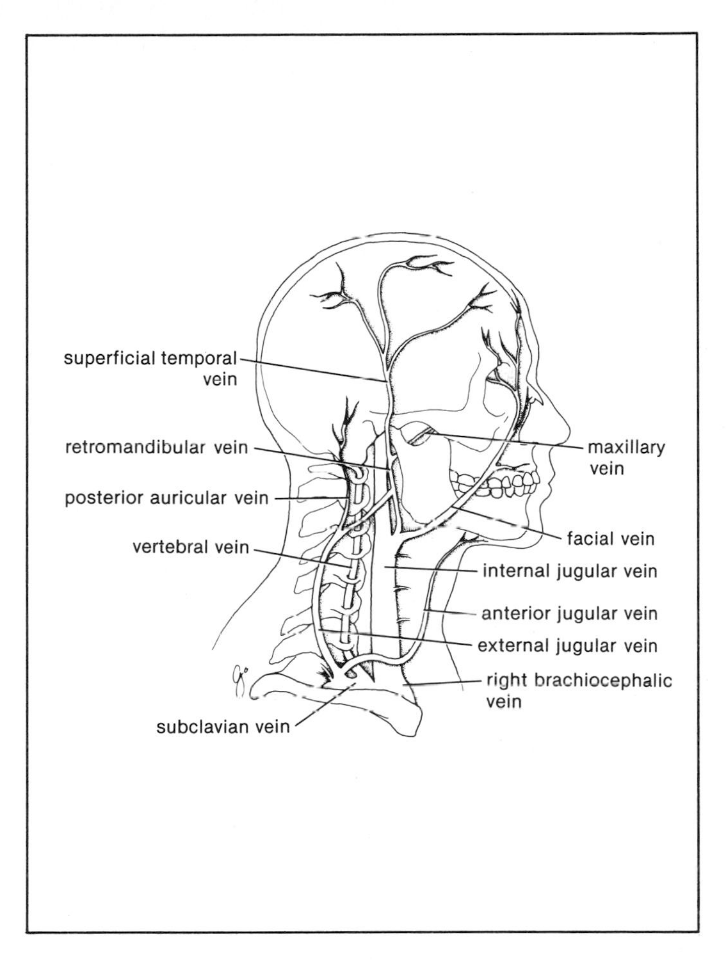

Figure 4–9. The major veins of the head and the neck.

rebral arteries in the subarachnoid space, and the internal veins that drain the interior of the hemispheres.

The **internal cerebral veins** are formed on each side by the union of the **thalamostriate vein** and the **choroid vein,** which drains the choroid plexus of the lateral ventricle. The internal cerebral veins run backward in the roof of the third ventricle and below the splenium of the corpus callosum unite to form the **great cerebral vein.** The great cerebral vein drains into the straight sinus.

Cerebellar Veins. The superior and inferior cerebellar veins lie on the surface of each cerebellar hemisphere. The former drains into the straight sinus and the latter into the transverse and sigmoid sinuses.

Veins of the Brainstem. These form a superficial plexus that drains into adjacent venous sinuses.

Venous Sinuses. The venous sinuses of the cranial cavity are placed between the layers of the dura mater. Their walls are thick and composed of fibrous tissue; there is no muscular tissue. The sinuses possess no valves. They receive tributaries from the brain, from the diploë of the skull, from the orbit, and from the internal ear.

Superior Sagittal Sinus. This sinus lies in the upper fixed border of the falx cerebri. It begins in front at the foramen cecum, where it occasionally receives a vein from the nasal cavity. It runs backward and at the internal occipital protuberance it usually becomes continuous with the right transverse sinus. On each side the superior sagittal sinus communicates through small openings with cavities called the **venous lacunae.** Numerous arachnoid villi and granulations project into the lacunae.

The superior sagittal sinus receives the superior cerebral veins. At the internal occipital protuberance it is dilated to form the **confluence of the sinuses.** It is here that the sinus becomes continuous with the right transverse sinus and is connected to the left transverse sinus and the occipital sinus.

Inferior Sagittal Sinus. This sinus lies in the lower free margin of the falx cerebri. It runs backward to join the greater cerebral vein to form the straight sinus. It receives cerebral veins from the medial surface of the cerebral hemisphere.

Straight Sinus. This sinus lies at the junction of the falx cerebri with the tentorium cerebelli. It is formed by the union of the inferior sagittal sinus with the great cerebral vein. It drains usually into the left transverse sinus. It receives some of the superior cerebellar veins.

Transverse Sinuses. The right sinus begins at the internal occipital protuberance as a continuation of the superior sagittal sinus; the left sinus is usually a continuation of the straight sinus. Each sinus runs forward and laterally along the attached margin of the tentorium cerebelli. They end on each side by becoming the sigmoid sinus. The transverse sinuses receive the superior petrosal sinuses, the inferior cerebral, and the inferior cerebellar veins.

Sigmoid Sinuses. These sinuses are a direct continuation of the transverse sinuses. Each sinus curves downward behind the mastoid antrum and leaves the skull through the posterior part of the jugular foramen, where it becomes the internal jugular vein.

Occipital Sinus. This sinus occupies the attached margin of the falx cerebelli. It begins near the foramen magnum where it communicates with the vertebral veins and drains into the confluence of sinuses.

Cavernous Sinuses. Each sinus is situated on the side of the body of the sphenoid bone and extends from the superior orbital fissure in front to the apex of the petrous part of the temporal bone behind. The sinus receive the inferior ophthalmic vein (through this vein it communicates with the facial vein), the sphenoparietal sinus, the central vein of the retina, and the cerebral veins. The cavernous sinus drains into the transverse sinus through the superior petrosal sinus.

Intercavernous sinuses connect the two cavernous sinuses through the sella turcica.

Traveling through the cavernous sinus on each side are the internal carotid artery and the sixth cranial nerve. In the lateral wall lie the third and fourth cranial nerves, and the ophthalmic and maxillary divisions of the fifth cranial nerve.

Superior and Inferior Petrosal Sinuses. The superior petrosal sinus runs along the superior border of the petrous part of the temporal bone. It drains the cavernous sinus into the transverse sinus. The inferior petrosal sinus runs along the inferior border of the petrous part of the temporal bone and drains the cavernous sinus into the internal jugular vein.

Diploic Veins. Diploic veins occupy channels within the bones of the vault of the skull. They communicate with the venous sinuses, the meningeal veins, and the veins of the pericranium.

Emissary Veins. The emissary veins are valveless veins that pass through foramina in the skull and connect the venous sinuses with the veins outside the skull. They are a route through which infection can travel from outside the skull to the venous sinuses within.

Veins of the Scalp, the Face, and the Neck

Facial Vein. The facial vein is formed at the medial angle of the eye by the union of the **supraorbital** and **supratrochlear veins** (*Fig. 4–9*). It is connected through the superior ophthalmic vein with the cavernous sinus. The facial vein descends behind the facial artery to the lower margin of the body of the mandible. It crosses superficially to the submandibular salivary gland, and is joined by the anterior division of the retromandibular vein, and drains into the internal jugular vein.

The facial vein receives tributaries that correspond to the

branches of the facial artery. It is joined to the pterygoid venous plexus by the **deep facial vein.**

Superficial Temporal Vein. This vein is formed as a network on the side of scalp (*Fig. 4–9*). It passes downward over the zygomatic arch in the company of the superficial temporal artery and the auriculotemporal nerve. It enters the substance of the parotid gland and joins the maxillary vein to form the retromandibular vein.

Maxillary Vein. This short vein is formed in the infratemporal fossa by the union of the veins that form the pterygoid venous plexus (*Fig. 4–9*). The **pterygoid venous plexus** is made up of veins that correspond to branches of the maxillary artery and are closely associated with the lateral pterygoid muscle. The maxillary vein joins the superficial temporal vein at the neck of the mandible to form the retromandibular vein.

Retromandibular Vein. This vein is formed by the union of the superficial temporal and the maxillary veins (*Fig. 4–9).* It descends in the parotid gland superficial to the external carotid artery. It divides into an anterior branch that joins the facial vein and a posterior branch that joins the **posterior auricular vein** to form the external jugular vein.

External Jugular Vein. This vein is formed just behind the angle of the mandible by the union of the posterior auricular vein with the posterior division of the retromandibular vein (*Fig. 4–9*). It descends obliquely across the sternocleidomastoid muscle, beneath the platysma muscle, and drains into the subclavian vein behind the middle of the clavicle. The external jugular vein receives the following tributaries: the **posterior external jugular vein** from the back of the scalp, the **superficial cervical vein,** the **suprascapular vein,** and the **anterior jugular vein.**

Anterior Jugular Vein. This vein starts in the submental region and descends in the neck close to the midline (*Fig. 4–9*). Just above the sternum the veins of the two sides are united by a transverse trunk, called the **jugular arch.** The anterior jugular vein passes laterally under the sternocleidomastoid muscle to join the external jugular vein.

Internal Jugular Vein. This is a large vein that drains blood from the brain, from the face, and from the neck (*Fig. 4–9*). It begins at the jugular foramen in the skull as a continuation of the sigmoid sinus. It descends through the neck in the carotid sheath lateral to the vagus nerve, the internal and common carotid arteries. It ends by joining the subclavian vein to form the brachiocephalic vein behind the medial end of the clavicle. It is closely related to the **deep cervical lymph nodes** throughout its course.

TRIBUTARIES

1. Inferior petrosal sinus
2. Facial vein
3. Pharyngeal veins
4. Lingual veins
5. Superior thyroid vein
6. Middle thyroid vein.

Vertebral Vein. The vertebral vein is formed from a plexus of veins beneath the skull, which descend through the foramina in the transverse processes of the cervical vertebrae around the vertebral artery (*Fig. 4–9*). On reaching the sixth or seventh cervical transverse process the vertebral vein emerges and descends to open into the brachiocephalic vein.

Subclavian Vein. This vein is a continuation of the axillary vein at the outer border of the first rib (*Fig. 4–9*). It joins the internal jugular vein behind the medial end of the clavicle to form the brachiocephalic vein. It is separated from the subclavian artery by the scalenus anterior muscle. It receives the external jugular vein and it often receives on the left side the **thoracic duct** and on the right side the **right lymphatic duct.**

Veins of the Upper Limb

The veins of the upper limb may be divided into superficial and deep groups (*Fig. 4–10*). The superficial veins are of great clinical importance and lie in the superficial fascia. The deep veins accompany the main arteries.

Superficial Veins

Dorsal Venous Arch or Network. This venous network of superficial veins lies on the dorsum of the hand. It receives most of the blood from the hand and the fingers. It is drained on the lateral side by the cephalic vein and on the medial side by the basilic vein.

Cephalic Vein. This vein arises from the lateral side of the dorsal venous network (*Fig. 4–10*). It ascends round the lateral border of the forearm, just lateral to the styloid process of the radius, to reach the anterior aspect of the forearm. It then ascends into the arm and runs along the lateral border of the biceps to reach the interval between the deltoid and pectoralis major muscles. The cephalic ends by piercing the deep fascia to enter the axillary vein.

Basilic Vein. This vein arises from the medial side of the dorsal venous network and ascends on the posterior surface of the forearm (*Fig. 4–10*). Just below the elbow it inclines forward to reach the anterior aspect of the forearm. The vein then runs upward medial to the biceps and pierces the deep fascia at about the middle of the arm. The **median cubital vein** links the cephalic and basilic veins in the cubital fossa (*Fig. 4–10*). The basilic vein joins the venae comitantes of the brachial artery at the lower border of the teres major muscle to form the axillary vein.

Median Vein of the Forearm. This is a small vein that arises in the palm and ascends on the front of the forearm. It drains into the basilic vein, or the median cubital vein, or divides into two

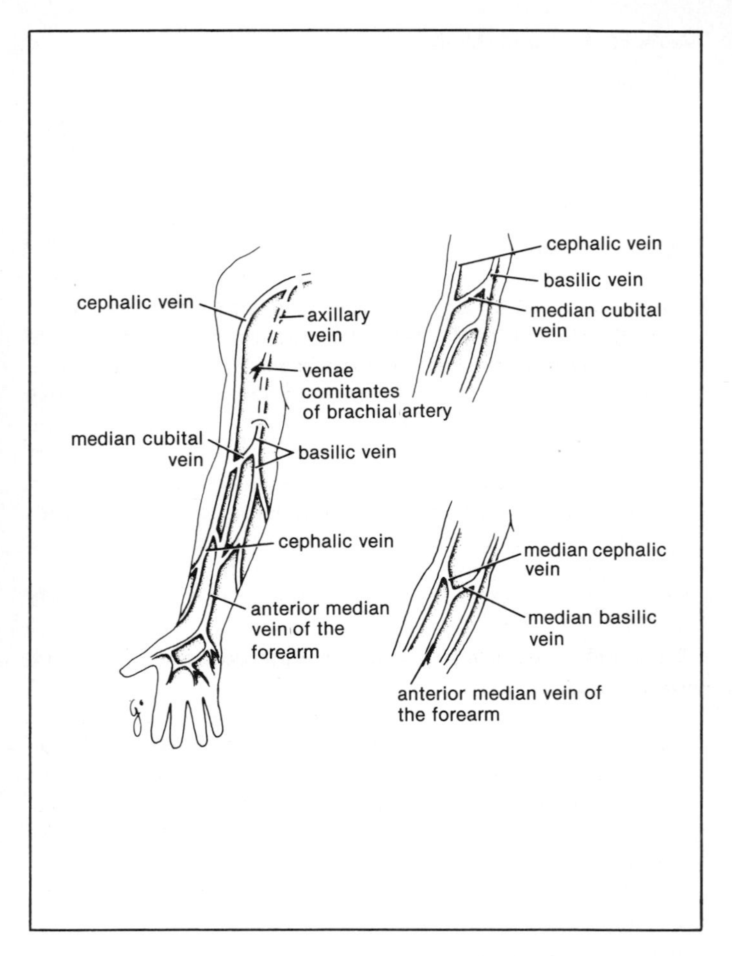

Figure 4–10. Superficial veins of the upper limb. On right note common variations seen in the region of the elbow.

branches one of which joins the basilic (**median basilic vein**) and the other joins the cephalic (**median cephalic vein**).

The superficial veins of the upper limbs are used for taking blood samples or for blood transfusion. They are also used for the introduction of cardiac catheters.

Deep Veins

Venae Comitantes. The deep veins accompany the respective arteries as venae comitantes. The two venae comitantes of the brachial artery join the basilic vein at the lower border of the teres major to form the axillary vein.

Axillary Vein. This vein is formed at the lower border of the teres major muscle by the union of the venae comitantes of the brachial artery with the basilic vein (*Fig. 4–10*). It becomes the subclavian vein at the outer border of the first rib. It receives tributaries that correspond to the branches of the axillary artery.

Veins of the Thorax

Brachiocephalic Veins. The **right brachiocephalic vein** is formed at the root of the neck by the union of the right subclavian and the right internal jugular veins. The **left brachiocephalic vein** has a similar origin on the left side of the root of the neck. It passes obliquely downward and to the right behind the manubrium sterni and in front of the large branches of the aortic arch. It joins the right brachiocephalic vein to form the superior vena cava (*Fig. 4–11*).

The tributaries of each vein include the **vertebral vein,** the **internal thoracic vein,** the **inferior thyroid vein,** and the **first posterior intercostal vein.**

Superior Vena Cava. The superior vena cava collects all the venous blood from the head and the neck and both upper limbs and is formed by the union of the two brachiocephalic veins. It descends vertically to end in the right atrium of the heart. The azy-

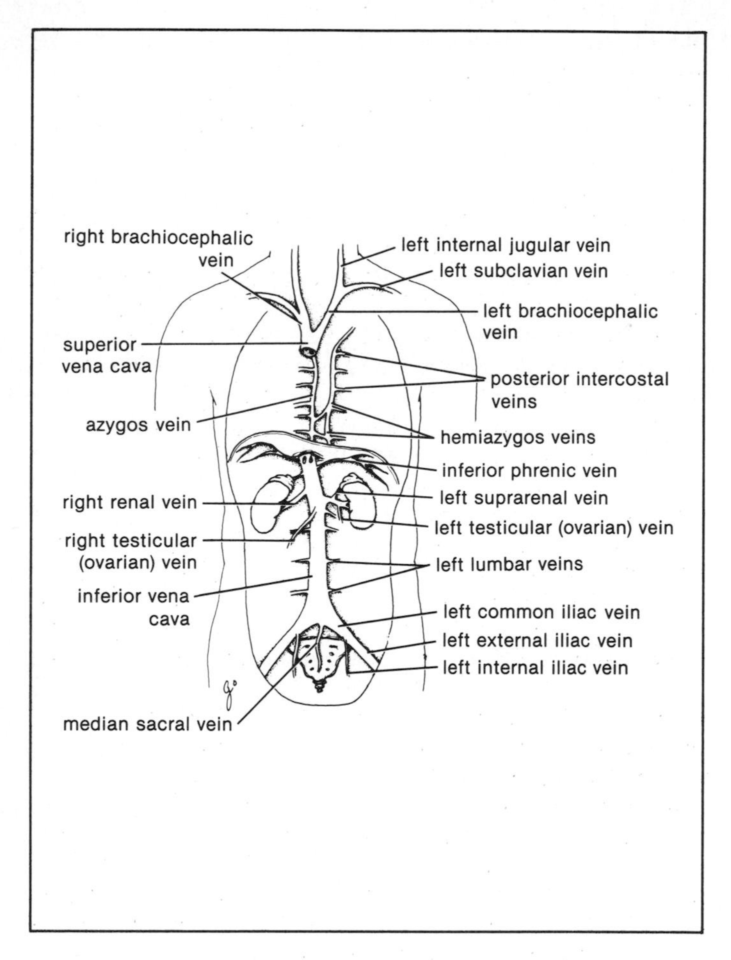

Figure 4–11. Major veins draining into the superior vena cava and the inferior vena cava.

gos vein joins the posterior aspect of the superior vena cava just before it enters the pericardium (*Fig. 4–10*).

Azygos Vein. This vein has a variable origin. It is often formed by the union of the **right ascending lumbar vein** and the **right subcostal vein.** It ascends through the aortic opening in the diaphragm on the right side of the aorta to the level of the fifth thoracic vertebra. Here it arches forward above the root of the right lung to empty into the posterior surface of the superior vena cava (*Fig. 4–11*). The azygos vein has numerous tributaries that include the eight lower right intercostal veins, the right superior intercostal vein, the superior and inferior hemiazygos veins, and numerous mediastinal veins.

Inferior Hemiazygos Vein. It is often formed by the union of the left ascending lumbar vein and the left subcostal vein. It ascends through the left crus of the diaphragm and at about the level of the eighth thoracic vertebra turns to the right and joins the azygos vein (*Fig. 4–11*). It receives some lower left intercostal veins and mediastinal veins.

Superior Hemiazygos Vein. This vein is formed by the union of the fourth to the eighth intercostal veins. It joins the azygos vein at the level of the seventh thoracic vertebra (*Fig. 4–11*).

Inferior Vena Cava. The inferior vena cava is formed in the abdomen (*see p. 222*) and perforates the tendinous part of the diaphragm. It passes through the pericardium and opens into the lower and back part of the right atrium of the heart. The valve of the inferior vena cava is important in the fetus but rudimentary in the adult.

Pulmonary Veins. There are four pulmonary veins (two on each side). They carry oxygenated blood from the lungs and empty into the left atrium. There are no valves.

Cardiac Veins. These are described on *p. 175*.

Veins of the Abdomen and the Pelvis

External Iliac Vein. This vein begins on each side as a continuation of the femoral vein behind the inguinal ligament. It ascends along the brim of the pelvis to the sacroiliac joint where it unites with the internal iliac vein to form the common iliac vein (*Fig. 4–11*). It receives the **inferior epigastric vein,** the **deep circumflex iliac vein,** and the **pubic vein;** the latter connects the external iliac with the obturator vein.

Internal Iliac Vein. This vein is formed on each side in the pelvis by the union of all the veins that correspond to the branches of the internal iliac artery.

Common Iliac Veins. These veins are formed by the union of the external and internal iliac veins in front of the sacroiliac joint (*Fig. 4–11*). They receive the **iliolumbar veins** and sometimes the **lateral sacral veins.** The **median sacral vein** sometimes opens into the left common iliac vein.

Inferior Vena Cava. The inferior vena cava is formed by the union of the common iliac veins behind the right common iliac artery at the level of the fifth lumbar vertebra (*Fig. 4–11*). It ascends on the right side of the aorta, pierces the central tendon of the diaphragm at the level of the eighth thoracic vertebra, and drains into the right atrium of the heart.

Relations. As the inferior vena cava passes up the posterior abdominal wall it has the following important relations. *Anteriorly.* Coils of small intestine, third part of the duodenum, head of pancreas, first part of duodenum, entrance into the lesser sac (which separates the inferior vena cava from the portal vein, common bile duct, and hepatic artery), and liver. *Laterally.* The right sympathetic trunk and the right ureter. *Medially.* Abdominal aorta.

Tributaries. The inferior vena cava has the following tributaries.

1. Two anterior visceral tributaries—the hepatic veins
2. Three lateral visceral tributaries
   a. Right suprarenal (the left vein drains into the left renal)
   b. Renal veins
   c. Right testicular or ovarian vein (the left vein drains into the left renal vein)
3. Five lateral abdominal wall tributaries
   a. Inferior phrenic vein
   b. Four lumbar veins
4. Three veins of origin
   a. Two common iliac veins
   b. Median sacral vein (sometimes enters left common iliac vein)

Portal Vein

The portal vein is about 5 cm (2 inches) long and is formed behind the neck of the pancreas by the union of the superior mesenteric and the splenic veins (*Fig. 4–12*). It runs upward and to the right, posterior to the first part of the duodenum, and enters the lesser omentum. Here, the portal vein lies posterior to the common bile duct and the hepatic artery. It then passes up in front of the opening into the lesser sac to the porta hepatis, where it divides into right and left terminal branches (*Fig. 4–12*).

The portal venous system begins as a capillary network in the organs it drains and ends in capillary-like vessels, termed sinusoids in the liver. The portal vein drains blood from the alimentary canal from the lower end of the esophagus to half way down the anal canal; from the pancreas, the gallbladder, and the bile ducts; and from the spleen.

The tributaries of the portal vein are: the splenic vein, the superior mesenteric vein, the left gastric vein, the right gastric vein, and the cystic veins.

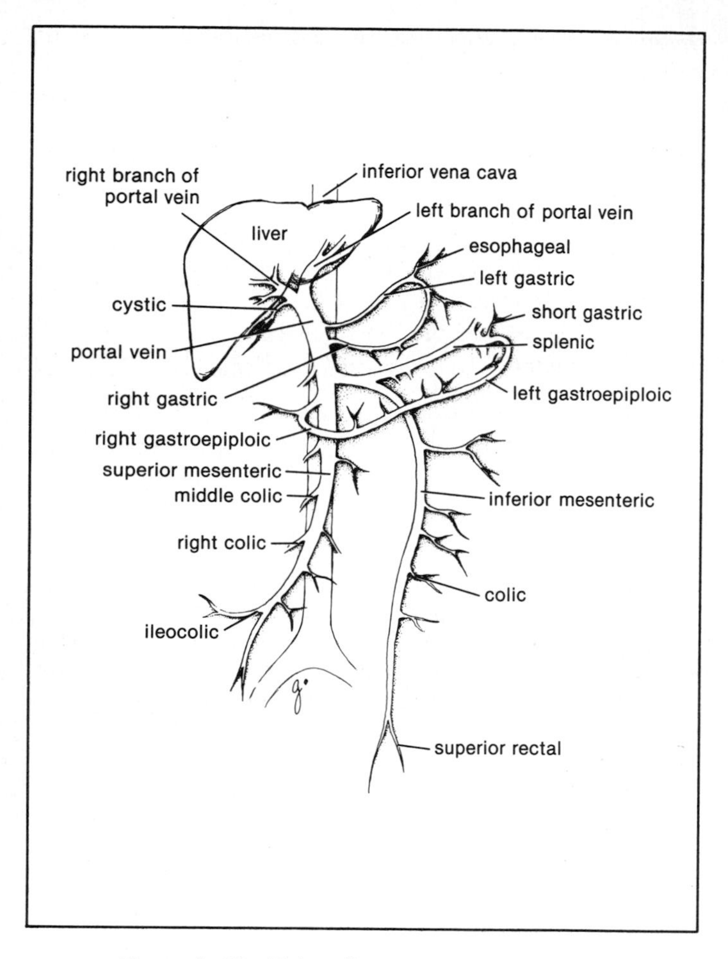

Figure 4–12. Major tributaries of the portal vein.

Splenic Vein. This vein arises in the hilum of the spleen and passes to the right in the lienorenal ligament. It crosses in front of the left kidney and behind the body of the pancreas. It lies below the splenic artery. It ends behind the neck of the pancreas by uniting with the superior mesenteric vein to form the portal vein (*Fig. 4–12*). It receives the short gastric, left gastroepiploic, inferior mesenteric, and pancreatic veins.

Inferior Mesenteric Vein. This vein drains blood from the upper half of the anal canal, the rectum, the pelvic (sigmoid) colon, the descending colon, the left colic (splenic) flexure, and the distal third of the transverse colon. It ascends on the posterior abdominal wall on the left side of its artery. It passes close to the duodenojejunal flexure and joins the splenic vein behind the body of the pancreas (*Fig. 4–12*). It receives the superior rectal veins, the sigmoid (lower left colic) veins, and the left colic (superior left colic) vein.

Superior Mesenteric Vein. The superior mesenteric vein drains the jejunum, the ileum, the cecum and appendix, and the ascending and proximal two-thirds of the transverse colon. It ascends in the root of the mesentery of the small intestine on the right of the superior mesenteric artery. It then passes in front of the third part of the duodenum and joins the splenic vein behind the neck of the pancreas to form the portal vein (*Fig. 4–12*). It receives jejunal, ileal, ileocolic, right colic, middle colic, inferior pancreaticoduodenal, and right gastroepiploic veins.

Left Gastric Vein. This vein drains the left portion of the lesser curvature of the stomach and the distal part of the esophagus. It opens into the portal vein (*Fig. 4–12*).

Right Gastric Vein. This vein drains the right portion of the lesser curvature of the stomach and opens into the portal vein (*Fig. 4–12*).

Cystic Veins. These veins drain the gallbladder either directly into the liver or join the portal vein.

Portal Systemic Anastomoses. These important communications are as follows:

1. At the lower third of the esophagus, the esophageal branches of the left gastric vein (portal tributary) anastomose with the esophageal veins draining the middle third of the esophagus into the azygos veins (systemic tributary).
2. Halfway down the anal canal, the superior rectal veins (portal tributary) draining the upper half of the anal canal anastomose with the middle and inferior rectal veins (systemic tributaries).
3. The **paraumbilical veins** connect the left branch of the portal vein with the superficial veins of the anterior abdominal wall (systemic tributaries). The paraumbilical veins travel in the falciform ligament and accompany the ligamentum teres.
4. The veins of the ascending colon, descending colon, duodenum, pancreas, and liver (portal tributaries) anastomose with the renal, lumbar, and phrenic veins (systemic tributaries).

Veins of the Lower Limb

The veins of the lower limb may be divided into superficial and deep groups (*Fig. 4–13*). The superficial veins are of great clinical importance and lie in the superficial fascia. The deep veins accompany the main arteries.

Superficial Veins

Dorsal Venous Arch. This venous arch lies on the dorsum of the foot (*Fig. 4–13*). The greater part of the blood from the whole foot drains into the arch via digital veins, and communicating veins that pass through the interosseous spaces. The dorsal venous arch is drained on the medial side by the great saphenous vein and on the lateral side by the small saphenous vein.

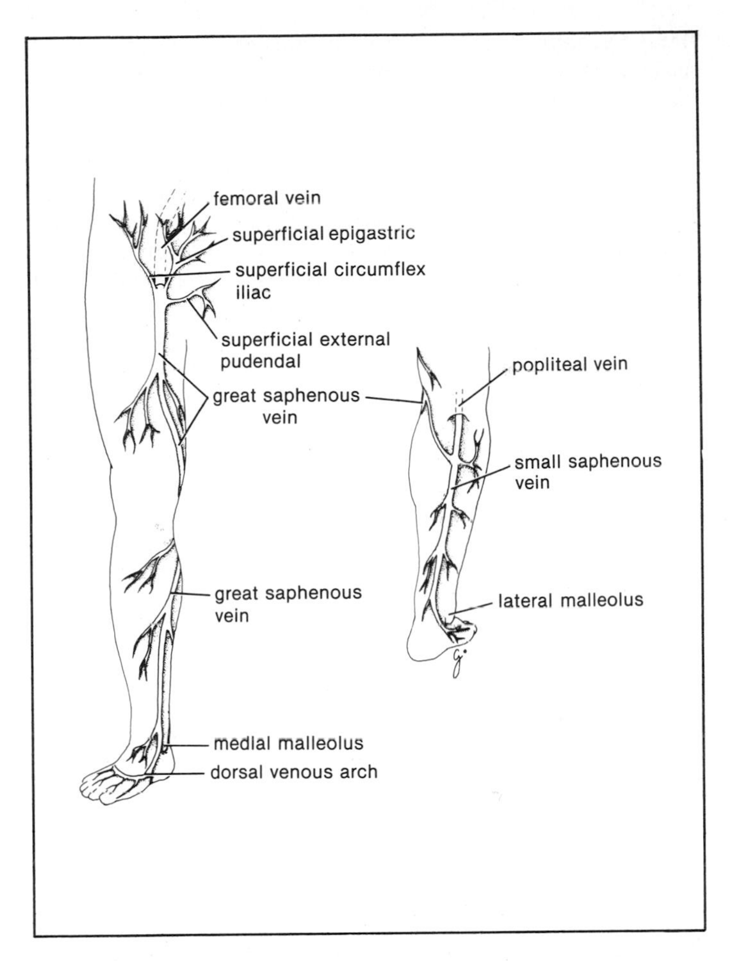

Figure 4–13. Superficial veins of the right lower limb.

Great Saphenous Vein. The great saphenous vein arises from the medial end of the dorsal venous arch of the foot and ascends directly in front of the medial malleolus (*Fig. 4–13*). It is accompanied by the saphenous nerve and lies in the superficial fascia. It passes up the medial side of the leg, behind the knee and curves forward around the medial side of the thigh. It then runs through the saphenous opening in the deep fascia and joins the femoral vein about 4 cm (1½ inches) below and lateral to the pubic tubercle. The great saphenous vein possesses numerous valves and is connected to the small saphenous vein by branches that pass behind the knee. A number of **perforating veins** connect the great saphenous vein with the deep veins along the medial side of the calf.

The great saphenous vein receives near its termination the following small named tributaries: (1) the **superficial circumflex iliac vein,** (2) the **superficial epigastric vein,** and (3) the **superficial external pudendal vein.**

Small Saphenous Vein. The small saphenous vein arises from the lateral end of the dorsal venous arch of the foot (*Fig. 4–13*). It ascends behind the lateral malleolus in company with the sural nerve. It then passes up the back of the leg and pierces the deep fascia. It enters the popliteal fossa by passing between the two heads of the gastrocnemius muscle and drains into the popliteal vein. The small saphenous vein communicates with the deep veins and the great saphenous vein.

The superficial veins of the lower limbs are commonly used for blood transfusion; they are common sites for varicosities.

Deep Veins

Venae Comitantes. The deep veins accompany the respective arteries as venae comitantes. The venae comitantes of the anterior and posterior tibial arteries unite in the popliteal fossa to form the popliteal vein.

Popliteal Vein. The popliteal vein, formed by the union of the venae comitantes of the anterior and posterior tibial arteries, ascends through the popliteal space and at the opening in the adductor magnus becomes the femoral vein. It receives numerous tributaries including the small saphenous vein.

Femoral Vein. The femoral vein is a continuation of the popliteal vein at the opening in the adductor magnus. It ascends through the thigh and comes to lie on the medial side of the femoral artery in the intermediate compartment of the femoral sheath. It ends by becoming continuous with the external iliac vein behind the inguinal ligament. The femoral vein receives the great saphenous vein and veins that correspond to branches of the femoral artery.

5

Lymphatic System

GENERAL DESCRIPTION

The lymphatic system consists of lymphatic vessels and lymphatic tissues. The lymphatic vessels assist the capillaries and the venules of the cardiovascular system in the removal of tissue fluid from the tissues and return it to the blood. Lymphatic vessels are found in most tissues and organs in the body but are absent from the central nervous system, the eyeball, the internal ear, the epidermis, the cartilage, and the bone.

The lymphatic tissue is a type of connective tissue that contains large numbers of lymphocytes. It is organized into the following principal structures (1) the lymph nodes, (2) the thymus, and (3) the spleen. The lymphatic tissue is essential for the immunological defenses of the body against bacteria and viruses.

The lymphatic capillaries begin as blind-ended tubes. They differ from blood capillaries in that they can absorb proteins and large particulate matter from the tissue spaces, whereas the fluid absorbed by the blood capillaries is an aqueous solution of inorganic salts and sugar. **Lymph** is the name given to tissue fluid once it has entered a lymphatic vessel.

Lymph from the peripheral capillary plexuses passes into larger collecting vessels. At strategic points along the course of these vessels are small, ovoid masses of lymphatic tissue called **lymph nodes**. The direction of the flow of lymph is determined by the valves of the lymphatic vessels. Lymphatic vessels tend to run alongside blood vessels. In the limbs, the superficial lymphatic vessels of the skin and subcutaneous tissue tend to follow the superficial veins; the deeper lymphatic vessels follow the deep arteries and veins.

The lymph from the greater part of the body reaches the blood via the thoracic duct. The lymph from the right side of the head and neck, the right upper limb, and the right side of the thorax, however, reaches the blood via the right lymphatic duct (*Fig. 5–1*).

Major Lymphatic Ducts

Thoracic Duct. The thoracic duct begins below in the abdomen at the level of the second lumbar vertebra as a dilated sac, the **cisterna chyli** (*Fig. 5–1*). It ascends through the aortic opening in the diaphragm, on the right side of the descending aorta. It gradually crosses the median plane behind the esophagus and reaches the left border of the esophagus at the level of the fourth thoracic vertebra (sternal angle). It then ascends along the left border of the esophagus to the root of the neck. Here it turns laterally to the left behind the carotid sheath and then turns down in front of the phrenic nerve and crosses the subclavian artery to enter the beginning of the left brachiocephalic vein. It often enters the vein as several branches. At the termination the thoracic duct receives the left jugular, subclavian, and mediastinal lymph trunks, although these trunks may drain independently into neighboring large veins in this region.

The thoracic duct thus conveys to the blood all the lymph from the lower limbs, pelvic cavity, abdominal cavity, left side of the thorax, and left side of the head, neck, and left upper limb (*Fig. 5–1*).

Right Lymphatic Duct. The right jugular, subclavian, and bronchomediastinal trunks, which drain the right side of the head

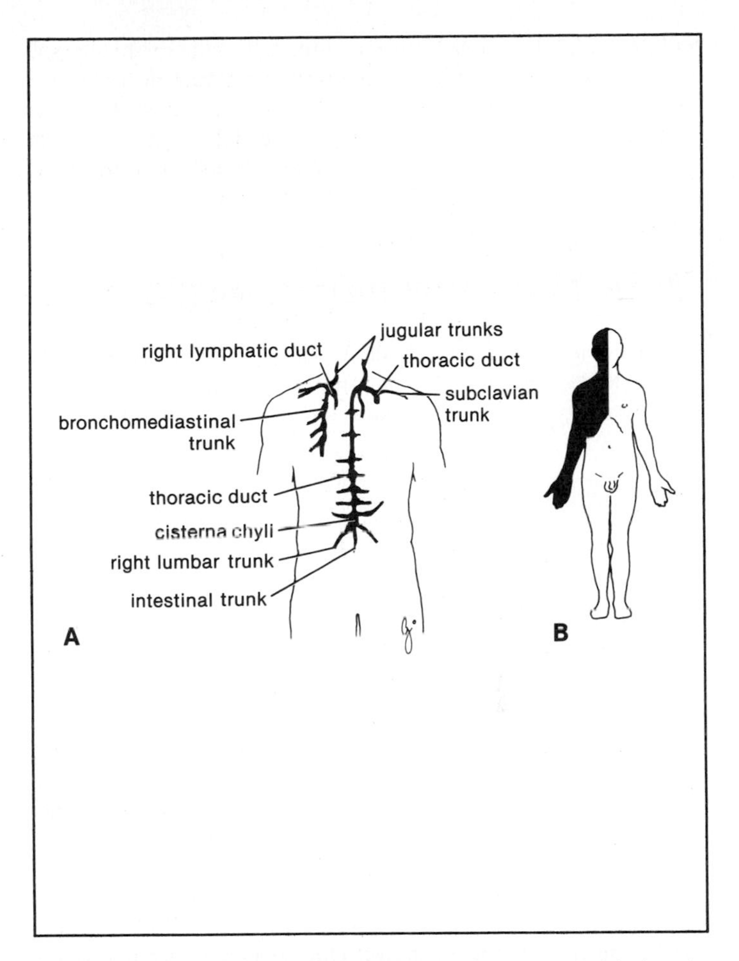

Figure 5–1. **A.** Thoracic duct and right lymphatic duct and their main tributaries. **B.** The regions of the body draining lymph into the thoracic duct (*clear*) and right lymphatic duct (*black*).

and neck, the right upper limb, and the right side of the thorax, respectively, may join to form the right lymphatic duct (*Fig. 5–1*). This common duct if present is short, about 1.3 cm (½ inch) long, and opens into the beginning of the right brachiocephalic vein. Alternatively, the trunks open independently into the great veins at the root of the neck.

LYMPHATIC DRAINAGE OF THE HEAD AND THE NECK

The lymph nodes of the head and the neck (*Fig. 5–2*) are arranged as a regional collar that extends from below the chin to the occipital region and as a deep vertical terminal group that is embedded in the carotid sheath.

Regional Nodes

The regional nodes are arranged as follows:

1. The **occipital nodes** are situated over the occipital bone at the apex of the posterior triangle. They receive lymph from the back of the scalp.
2. The **retroauricular (mastoid) nodes** lie behind the ear over the mastoid process. They receive lymph from the scalp above the ear, the auricle, and the posterior wall of the external auditory meatus.
3. The **parotid nodes** are situated on or within the parotid salivary gland. They receive lymph from the scalp above the parotid gland, the eyelids, the parotid gland, the auricle, and the anterior wall of the external auditory meatus.
4. The **buccal (facial) nodes**. One or two nodes lie on the buccinator muscle. They drain lymph that ultimately passes into the submandibular nodes.
5. The **submandibular nodes** lie superficial to the submandibular salivary gland. They receive lymph from the front of the scalp, the nose, the cheek, the upper and lower lip

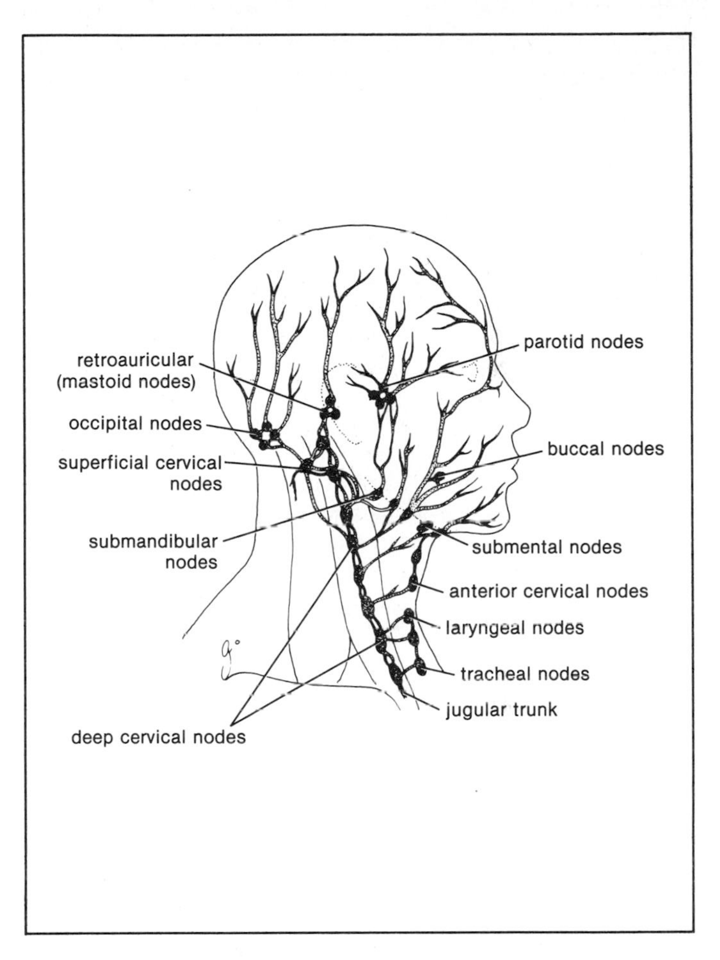

Figure 5–2. Lymphatic drainage of the head and the neck.

(except the central part), the frontal, maxillary, and ethmoid sinuses, the upper and lower teeth (except the lower incisors), the anterior two-thirds of the tongue (except the tip), the floor of the mouth, and vestibule and the gums.

6. The **submental nodes** lie in the submental triangle just below the chin. They receive lymph from the tip of the tongue, the floor of the anterior part of the mouth, the incisor teeth, the center part of the lower lip, and the skin over the chin.
7. The **anterior cervical lymph nodes** lie along the course of the anterior jugular veins. They receive lymph from the skin and superficial tissues of the front of the neck.
8. The **superficial cervical lymph nodes** lie along the course of the external jugular vein. They receive lymph from the skin over the angle of the jaw, the skin over the lower part of the parotid gland, and the lobe of the ear.
9. The **retropharyngeal lymph nodes** lie behind the pharynx and in front of the vertebral column. They receive lymph from the nasal pharynx, the auditory tube, and the vertebral column.
10. The **laryngeal lymph nodes** lie in front of the larynx. They receive lymph from the larynx.
11. The **tracheal (paratracheal) lymph nodes** lie alongside the trachea. They receive lymph from neighboring structures, including the thyroid gland.

Deep Cervical Lymph Nodes

The deep cervical lymph nodes form a vertical chain along the course of the internal jugular vein within the carotid sheath (*Fig. 5–2*). They receive lymph from all the above 11 groups of regional nodes. The **jugulodigastric node**, which is located below and behind the angle of the mandible, is mainly concerned with the drainage of the tonsil and the tongue. The **jugulo-omohyoid node** is mainly associated with drainage of the tongue.

The efferent lymph vessels join to form the **jugular trunk**, which drains into the thoracic duct or the right lymphatic duct (*Fig. 5-2*).

LYMPHATIC DRAINAGE OF THE UPPER LIMB AND THE BREAST

The lymph vessels of the upper limb are arranged as superficial and deep sets. The superficial vessels ascend the limb in the superficial fascia and accompany the superficial veins. The deep lymph vessels lie deep to the deep fascia and follow the deep arteries and veins. All the lymph vessels of the upper limb ultimately drain into lymph nodes that are situated in the axilla.

Axillary Lymph Nodes

The axillary lymph nodes drain lymph vessels from the entire upper limb. In addition they drain vessels from the lateral part of the breast, and the superficial lymph vessels from the thoracoabdominal walls above the level of the umbilicus.

The lymph nodes are 20 to 30 in number and are located as follows:

1. **Anterior (pectoral) nodes** lie along the lower border of pectoralis minor and behind pectoralis major. They receive lymph from the lateral part of the breast and the superficial vessels from the thoracoabdominal wall above the level of the umbilicus.
2. **Posterior (subscapular) nodes** lie in front of the subscapularis muscle. They receive superficial lymph vessels from the back, down as far as the level of the iliac crests.
3. **Lateral nodes** lie along the medial side of the axillary vein. They receive most of the lymph vessels of the upper limb (except the superficial vessels draining the lateral side (see infraclavicular nodes).
4. **Central nodes** lie in the center of the axilla embedded in fat. They receive lymph from the above three groups.

5. **Infraclavicular (deltopectoral) nodes** lie in the interval between the deltoid and pectoralis major muscles along the course of the cephalic vein. They receive lymph from the superficial vessels from the lateral side of the hand, the forearm and the arm; the lymph vessels accompany the cephalic vein.
6. **Apical group** lies at the apex of the axilla. They receive lymph from all the other axillary nodes. The apical nodes drain into the subclavian trunk, which on the left side, drains into the thoracic duct but on the right side drains into the right lymphatic trunk.

The **supratrochlear (cubital) lymph node** lies in the superficial fascia in the cubital fossa close to the trochlea. It receives lymph from the third, fourth, and fifth fingers, the medial part of the hand, and the medial side of the forearm. The efferent lymph vessels ascend to the lateral axillary lymph nodes.

Lymphatic Drainage of the Breast

The lateral quadrants of the breast drain into the anterior axillary or pectoral nodes. The medial quadrants drain into lymph nodes lying along the internal thoracic artery within the thorax. Some superficial lymph vessels communicate with those of the opposite breast and with those of the anterior abdominal wall. A few lymph vessels drain posteriorly into the posterior intercostal nodes.

The lymphatic drainage of the breast is of great clinical importance because of the high incidence of cancer of the breast and the subsequent spread to the lymph nodes that drain the breast.

LYMPHATIC DRAINAGE OF THE THORAX

The thoracic wall and the thoracic viscera are drained into the following groups of lymph nodes:

1. **Axillary nodes**. The superficial lymph vessels from the skin and the subcutaneous tissues of the anterior and posterior

walls of the chest are drained into the axillary nodes. The anterior chest wall drains into the anterior axillary nodes and the posterior chest wall drains into the posterior nodes.

2. **Internal thoracic (parasternal) nodes** are five in number and lie alongside the internal thoracic artery. They receive lymph from the medial quadrants of the breast, the deep structures of the anterior thoracic and abdominal walls down as far as the umbilicus. They also receive lymph from the upper surface of the liver. The efferent vessels drain into the bronchomediastinal trunk, which drains on the right side into the right lymphatic duct and on the left into the thoracic duct.
3. **Intercostal nodes** lie close to the heads of the ribs. They receive lymph from the intercostal spaces and a few from the breast. The efferent lymph vessels drain into the thoracic and right lymphatic ducts.
4. **Diaphragmatic nodes** lie on the upper surface of the diaphragm. They drain lymph from the diaphragm and the upper surface of the liver. The efferent vessels drain into the internal thoracic and posterior mediastinal nodes.
5. **Brachiocephalic nodes** lie with the brachiocephalic veins in the superior mediastinum. They drain lymph from the thyroid and the pericardium and their efferents drain into the bronchiomediastinal trunks.
6. **Posterior mediastinal nodes** lie close to the descending thoracic aorta. They drain lymph from the esophagus, the pericardium, and the diaphragmatic nodes. Their efferent vessels drain into the thoracic duct.
7. **Tracheobronchial nodes** lie alongside (1) the trachea, (2) the main bronchi at the hilus of lungs, and (3) the bronchi within the lungs. They drain the lymph from the lungs, the trachea, and the heart. Their efferent vessels drain into the bronchiomediastinal trunk.

Lymphatic Drainage of the Lungs

The lymph vessels begin in **superficial and deep plexuses**. The superficial plexus lies beneath the visceral pleura and then drains into the bronchopulmonary nodes at the hilus of the lung. The deep plexus accompanies the bronchi and drains via pulmonary nodes into bronchopulmonary nodes at the hilus. There is no free communication between the plexuses.

Lymphatic Drainage of the Esophagus

The cervical part drains into the deep cervical nodes, the middle part drains into the posterior mediastinal nodes, and the lower abdominal part drains into the left gastric and then the celiac nodes.

LYMPHATIC DRAINAGE OF THE ABDOMEN AND THE PELVIS

The lymph from most of the abdominal wall and from all the viscera (except a small part of the liver) drains into the thoracic duct. The lymph from the gastrointestinal tract, including the liver, the gallbladder, the pancreas, and the spleen first drains into the preaortic lymph nodes. The lymph from the remaining organs and the abdominal and pelvic walls first drains into the **paraaortic (lateral aortic or lumbar) nodes**. The afferent lymph vessels to these nodes tend to accompany arteries and there are a number of outlying small groups of nodes that the lymph passes through; these nodes take the names of the arteries along which they lie.

Preaortic lymph nodes lie along the anterior surface of the abdominal part of the aorta. Their efferent vessels form the **intestinal trunk** that drains into the cisterna chyli. These nodes may be divided into the **celiac**, **superior mesenteric**, and **inferior mesenteric groups** that lie close to the origins of these arteries.

Paraaortic (lateral aortic or lumbar) nodes are right and left groups that lie alongside the abdominal part of the aorta. Their efferent vessels form the **right and left lumbar trunks** that drain into the cisterna chyli. Lymph from the pelvis passes first through a number of peripheral nodes that are associated with arteries.

These nodes include **internal iliac, external iliac**, and **common iliac nodes.**

Lymphatic Drainage of the Abdominal Part of the Esophagus, the Stomach, and the First Half of the Duodenum

Abdominal Part of the Esophagus. Drains into left gastric nodes.

Stomach. The left half of the lesser curvature drains into left gastric nodes. The right half of the lesser curvature drains into the right gastric nodes. The fundus and the left half of the greater curvature drains into the left gastroepiploic nodes and the pancreaticosplenic nodes. The right half of the greater curvature drains into the right gastroepiploic nodes and the gastroduodenal nodes.

First Half of the Duodenum. The upper half of the duodenum drains into the pyloric nodes (superior pancreaticoduodenal nodes) and the gastroduodenal nodes. All these nodes drain into the celiac nodes.

Lymphatic Drainage of the Lower Half of the Duodenum, the Jejunum, the Ileum, the Cecum, the Appendix, the Ascending Colon, and the Proximal Two-thirds of the Transverse Colon

The lymph passes through lymph nodes that lie along the terminal branches of the superior mesenteric artery. All these nodes drain finally into the superior mesenteric nodes.

Lymphatic Drainage of the Distal Third of the Transverse Colon, the Descending Colon, the Sigmoid Colon, the Rectum, and the Upper Half of the Anal Canal

The lymph passes through lymph nodes that lie along the terminal branches of the inferior mesenteric artery. All these nodes drain finally into the inferior mesenteric nodes.

Lymphatic Drainage of the Liver

The lymph passes to the **hepatic nodes** in the porta hepatis and then to the celiac nodes. The bare areas of the liver drain through the diaphragmatic nodes to the posterior mediastinal nodes.

Lymphatic Drainage of the Pancreas

The lymph drains to nodes that lie along the arterial supply, namely the pancreaticoduodenal nodes, the splenic nodes, and finally the celiac nodes.

Lymphatic Drainage of the Spleen

The lymph drains into the pancreaticosplenic nodes.

Lymphatic Drainage of the Suprarenal Gland and the Kidney

The lymph drains into the lateral aortic nodes.

Lymphatic Drainage of the Bladder

The lymph drains into the internal and external iliac nodes.

Lymphatic Drainage of the Testis and the Ovary

The lymph ascends on the posterior abdominal wall in company with the gonadal blood vessels to drain into the lateral aortic nodes at the level of the first lumbar vertebra.

Lymphatic Drainage of the Prostate

The lymph drains into the internal iliac nodes.

Lymphatic Drainage of the Penis and the Scrotum

The lymph from the superficial tissues drains into the superficial inguinal nodes. The lymph from the glans penis passes to the deep inguinal and external iliac nodes. The lymph from the erectile tissue passes to the internal iliac nodes.

Lymphatic Drainage of the Uterus

The lymph from the body and cervix pass to the internal and external iliac nodes. Lymph vessels from the fundus accompany the ovarian vessels to the lateral aortic lymph nodes at the level of the first lumbar vertebra. A few lymph vessels pass with the round ligament of the uterus to the superficial inguinal nodes.

Lymphatic Drainage of the Vagina

Lymph from the upper part of the vagina drains into the internal and external iliac nodes. The lymph from the orifice and from the vulva drains into the superficial inguinal nodes.

Lymph Drainage of the Lower Half of the Anal Canal

Lymph descends to the anus and then drains into the superficial inguinal nodes.

LYMPHATIC DRAINAGE OF THE LOWER LIMB

The lymph vessels of the lower limb are arranged as superficial and deep sets. The superficial vessels ascend the limb with the superficial veins. The deep lymph vessels lie deep to the deep fascia and follow the deep arteries and veins. All the lymph vessels from the lower limb ultimately drain into the deep inguinal group of nodes that are situated in the groin.

Superficial Inguinal Lymph Nodes

These nodes lie in the superficial fascia just below the inguinal ligament. They may be divided into a horizontal and a vertical group. The **horizontal group** receives lymph from the superficial lymph vessels of the anterior abdominal wall below the umbilicus, from the perineum, the external genitalia of both sexes (but not the testes), and the lower half of the anal canal. They also receive lymph from the superficial lymph vessels of the buttocks.

The **vertical group** lies along the terminal part of the great saphenous vein and receives the majority of the superficial lymph vessels of the lower limb except from the back and lateral side of the calf and lateral side of the foot.

The superficial inguinal nodes all drain into the deep inguinal nodes.

Deep Inguinal Lymph Nodes

There are usually three nodes situated along the medial side of the femoral vein and in the femoral canal. They receive all the lymph from the superficial inguinal nodes and from all the deep structures of the lower limb. The efferent lymph vessels ascend through the femoral canal into the abdominal cavity and drain into the external iliac nodes.

Popliteal Lymph Nodes

These nodes lie in the popliteal fossa. They receive superficial lymph vessels that accompany the small saphenous vein from the lateral side of the foot and the back and lateral side of the calf. They also receive lymph from the deep structures of the leg below the knee. The efferent vessels drain upward to the deep inguinal nodes.

THYMUS

The thymus is a flat bilobed structure lying in the superior mediastinum and the anterior mediastinum of the thorax. In the newborn infant it reaches its largest size relative to the size of the body. It continues to grow until puberty but thereafter undergoes involution. Lymphatic vessels do not drain into it but large numbers leave it. It is one of the most important organs concerned with the defense against infection.

Blood Supply. Inferior thyroid and internal thoracic arteries.

SPLEEN

The spleen is the largest single mass of lymphatic tissue in the body. It lies in the abdomen just beneath the left half of the diaphragm. Unlike lymph nodes, it does not lie along the course of lymphatic vessels but along the course of the systemic circulation, having a large splenic artery and a splenic vein.

The spleen is reddish in color, ovoid in shape, with a notched anterior border. It is surrounded by peritoneum that passes from the hilus to the stomach as the **gastrosplenic omentum** (ligament) and to the left kidney as the **lienorenal ligament**. The gastrosplenic omentum contains the short gastric and left gastroepiploic vessels and the lienorenal ligament contains the splenic vessels and the tail of the pancreas.

The spleen is related **anteriorly** to the stomach, the tail of the pancreas, and the left colic flexure; the left kidney lies along its medial border. **Posteriorly** lie the diaphragm, the left lung, and the ninth, tenth and eleventh ribs.

Arterial Supply. Splenic artery branch of the celiac artery. The splenic vein joins the superior mesenteric vein to form the portal vein.

6

Central Nervous System

The nervous system is divided into two main parts, the **central nervous system,** which consists of the brain and the spinal cord, and the **peripheral nervous system,** which consists of the cranial and spinal nerves and their associated ganglia. The brain is that part of the nervous system that lies inside the cranial cavity. It is continuous with the spinal cord through the foramen magnum.

The central nervous system is composed of large numbers of nerve cells and their processes, which are supported by specialized tissue called **neuroglia. Neuron** is the name given to the nerve cell and all its processes. The long processes of a nerve cell are called **axons** or **nerve fibers.**

The interior of the central nervous system is organized into gray and white matter. **Gray matter** consists of nerve cells and the proximal portions of their processes embedded in neuroglia. **White matter** consists of nerve fibers embedded in neuroglia.

SPINAL CORD

The spinal cord is cylindrical in shape and begins superiorly at the foramen magnum where it is continuous with the medulla oblongata of the brain. It terminates inferiorly in the adult at the level of the lower border of the first lumbar vertebra. The spinal cord thus occupies the upper two-thirds of the vertebral canal and is surrounded by three meninges, the **dura mater,** the **arachnoid mater,** and the **pia mater** (*Fig. 6–1*). Further protection is provided by the **cerebrospinal fluid,** which surrounds the spinal cord in the **subarachnoid space.**

Enlargements of the Spinal Cord

In the cervical region, where the spinal cord gives origin to the brachial plexus, and in the lower thoracic and lumbar regions, where it gives origin to the lumbosacral plexus, there are fusiform enlargements, called the **cervical and lumbar enlargements.** Inferiorly the lumbar enlargement tapers off into the **conus medullaris** from the apex of which a prolongation of the pia mater called the **filum terminale** descends to be attached to the back of the coccyx.

Fissures of the Spinal Cord

The spinal cord possesses, in the midline anteriorly, a deep longitudinal fissure, the **anterior median fissure,** and, on the posterior surface, a shallow furrow, the **posterior median sulcus.**

Roots of the Spinal Nerves

Along the entire length of the spinal cord are attached 31 pairs of spinal nerves by the **anterior or motor roots** and the **posterior or sensory roots** (*Fig. 6–1*). Each root is attached to the cord by a series of **rootlets,** which extend the whole length of the corresponding segment of the cord. Each posterior nerve root possesses a **posterior root ganglion.**

In the upper cervical region the spinal nerve roots are short and run almost horizontally, but the roots of the lumbar and sacral nerves below the level of the termination of the cord (lower border of the first lumbar vertebra in the adult) form a vertical leash of nerves around the filum terminale, called the **cauda equina.**

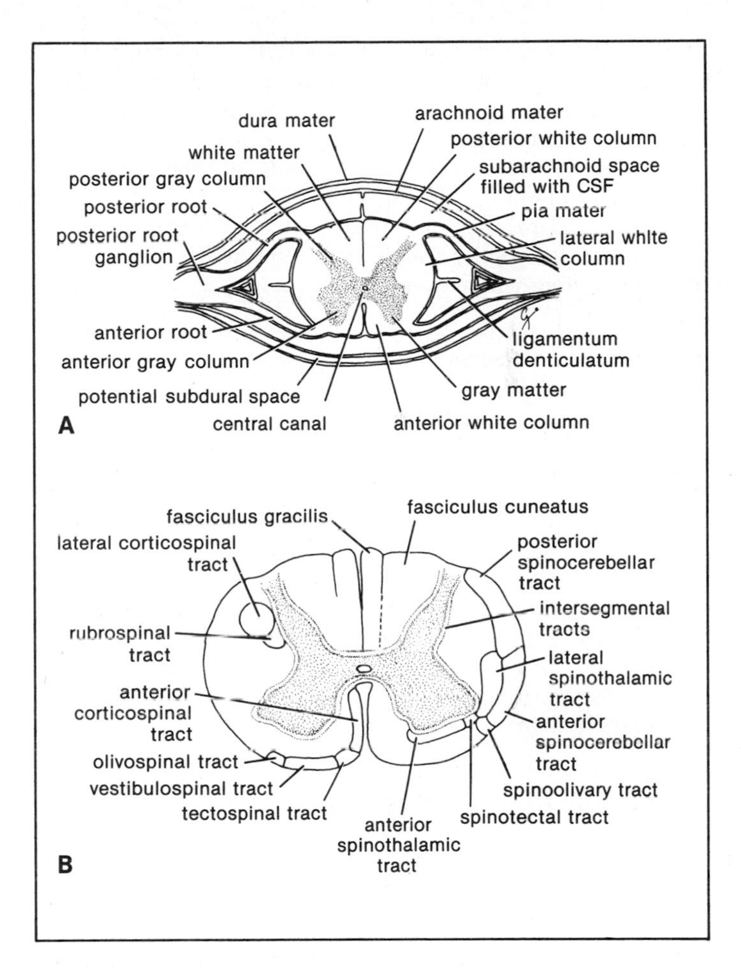

Figure 6–1. **A.** Transverse section of spinal cord, showing its meninges. **B.** Transverse section of spinal cord, showing the arrangement of the ascending tracts on the right side and the descending tracts on the left side.

The spinal nerve roots pass laterally and the anterior and posterior roots for each spinal cord segment unite in an intervertebral foramen, to form a **spinal nerve.** After emerging from the intervertebral foramen, each spinal nerve immediately divides into a large **anterior ramus** and a smaller **posterior ramus,** each containing both motor and sensory fibers.

The cervical roots of the **spinal part of the accessory nerve** emerge from the upper five cervical segments and run upward through the foramen magnum.

Structure of the Spinal Cord

The spinal cord is composed of an inner core of **gray matter,** which is surrounded by an outer covering of **white matter** (*Fig. 6–1*). The gray matter is seen on cross section as an H-shaped pillar with **anterior and posterior gray columns** or **horns,** united by a thin **gray commissure** containing the **central canal.** The central canal is lined with ciliated low columnar epithelium called the **ependyma** and is filled with **cerebrospinal fluid.** The canal opens above into the **fourth ventricle,** the cavity of the hindbrain.

The white matter, for purposes of description, may be divided into **anterior, lateral,** and **posterior white columns.** The arrangement of the tracts (bundles of nerve fibers) in the white columns is shown in *Figure 6–1.*

Meninges of the Spinal Cord

The spinal cord, like the brain, is covered by three meninges, the dura mater, the arachnoid mater, and the pia mater (*Fig. 6–1*).

Dura Mater. This is the most external membrane and is a dense fibrous sheet that encloses the spinal cord and the cauda equina. It is continuous above with the meningeal layer of dura covering the brain. Below it ends on the filum terminale at the level of the lower border of the second sacral vertebra. The dura gives sheaths to all the spinal nerve roots.

Arachnoid Mater. This is a delicate impermeable membrane that lies within the dura and outside the pia. It is separated from the pia mater by a wide space, the **subarachnoid space,** which is filled with **cerebrospinal fluid.** The arachnoid is continuous above through the foramen magnum with the arachnoid covering the brain. Inferiorly, it ends on the filum terminale at the level of the lower border of the second sacral vertebra. The arachnoid continues along the spinal nerve roots, forming small lateral extensions of the subarachnoid space.

Pia Mater. This is a vascular membrane that closely covers the spinal cord. It is thickened on either side between the nerve roots to form the **ligamentum denticulatum,** which passes laterally to adhere to the arachnoid and dura (*Fig. 6–1*). The pia mater extends along each nerve root as far as the spinal nerve. Inferiorly it becomes the **filum terminale.**

Blood Supply of the Spinal Cord

The **posterior spinal arteries** arise directly or indirectly from the vertebral arteries. There are two posterior spinal arteries on each side of the spinal cord. The **anterior spinal arteries** arise from the vertebral arteries and unite to form a single artery, which descends in the anterior medial fissure. The posterior and anterior spinal arteries are reinforced by **radicular arteries,** which are branches of local arteries.

Branches of the anterior spinal artery supply approximately the anterior two-thirds of the spinal cord, whereas the posterior third is supplied by the posterior spinal arteries.

The **veins** of the spinal cord drain into the internal vertebral venous plexus.

Anatomy of a Lumbar Puncture

With the patient lying on the side with the vertebral column well flexed, the space between adjoining laminae of the vertebrae in the lumbar region is opened to a maximum. An imaginary line joining the highest points on the iliac crests passes over the fourth lumbar spine. The physician passes the lumbar puncture needle into the vertebral canal above or below the fourth lumbar spine. The needle passes through the following anatomical structures before it enters the subarachnoid space: (1) skin, (2) fascia, (3) supraspinous ligament, (4) interspinous ligament, (5) ligamentum flavum, (6) fatty areolar tissue containing the internal vertebral venous plexus, (7) dura mater, and (8) arachnoid mater.

Remember the lower end of the spinal cord in the adult extends inferiorly as far as the lower border of the first lumbar vertebra and that the subarachnoid space extends down as far as the lower border of the second sacral vertebra.

BRAIN

The brain lies in the cranial cavity and is continuous with the spinal cord through the foramen magnum. The brain is conventionally divided into three major divisions. These are, in ascending order from the spinal cord, the **rhombencephalon** or hindbrain, the **mesencephalon** or midbrain, and the **prosencephalon** or forebrain. The rhombencephalon may be subdivided into the **myelencephalon** or medulla oblongata, the **metencephalon** or pons, and the cerebellum. The prosencephalon may also be subdivided into the **diencephalon** (between brain), which is the central part of the forebrain, and the **telencephalon** or cerebrum. The brainstem is a collective term used for the medulla oblongata, the pons, and the midbrain.

Hindbrain

Medulla Oblongata. The medulla oblongata is conical in shape and connects the pons superiorly to the spinal cord inferiorly. A median fissure is present on the anterior surface of the medulla, and on each side of this is a swelling, called the **pyramid** (*Fig. 6–2*). The pyramids are composed of descending nerve fibers that originate in large nerve cells in the precentral gyrus of the cerebral cortex. The pyramids taper below, and here the majority of the

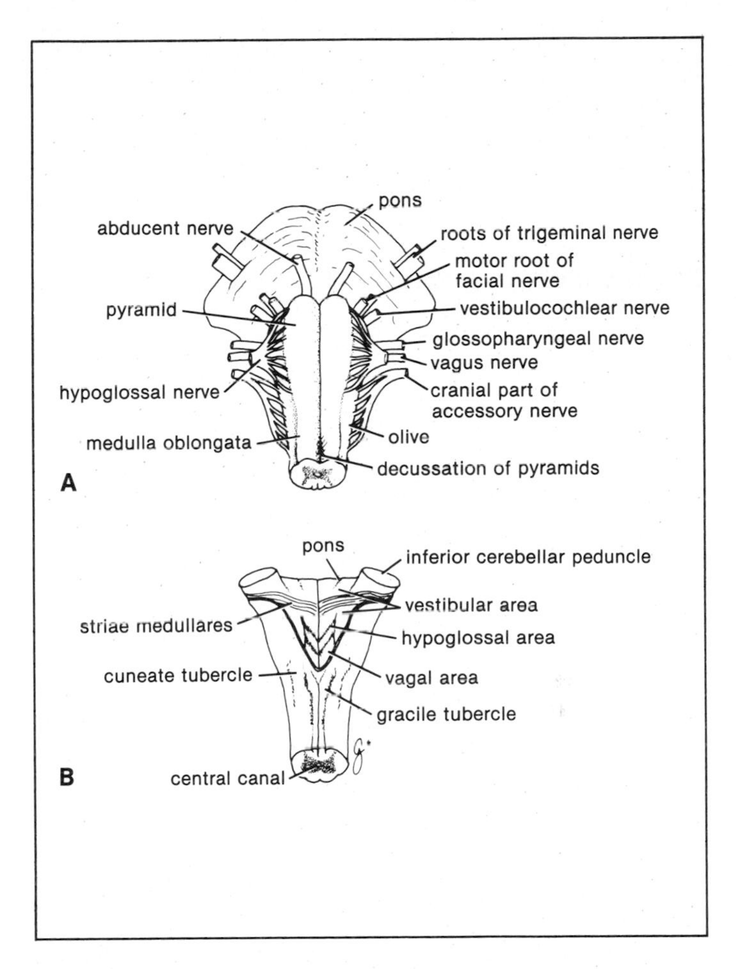

Figure 6–2. A. Anterior view of pons and medulla oblongata. **B.** Posterior view of pons and medulla oblongata. Note that the roof of the fourth ventricle and the cerebellum have been removed.

descending fibers cross over to the opposite side, forming the **decussation of the pyramids.**

Posterior to the pyramids are the **olives,** which are oval elevations produced by the underlying **olivary nuclei** (*Fig. 6–2*). Posterior to the olives are the **inferior cerebellar peduncles,** which connect the medulla to the cerebellum. On the posterior surface of the inferior part of the medulla oblongata are the **gracile** and **cuneate tubercles,** produced by the medially placed underlying **nucleus gracilis** and the laterally placed underlying **nucleus cuneatus** (*Fig. 6–2*).

Pons. The pons is situated on the anterior surface of the cerebellum, inferior to the midbrain, and superior to the medulla oblongata (*Figs. 6–2 and 6–3*). It is composed mainly of nerve fibers, which connect the two halves of the cerebellum. It also contains ascending and descending fibers connecting the forebrain, the midbrain, and the spinal cord. Some of the nerve cells within the pons serve as relay stations, whereas others form cranial nerve nuclei.

Cerebellum. The cerebellum lies within the posterior cranial fossa beneath the tentorium cerebelli (*Fig. 6–3*). It is situated posterior to the pons and the medulla oblongata. The cerebellum consists of two hemispheres connected by a median portion, the **vermis.** The cerebellum is connected to the midbrain by the **superior cerebellar peduncles,** to the pons by the **middle cerebellar peduncles,** and to the medulla by the **inferior cerebellar peduncles.**

The surface layer of each cerebellar hemisphere is called the **cortex** and is composed of gray matter. The cerebellar cortex is thrown into folds or **folia,** separated by transverse fissures. Certain masses of gray matter are found within the cerebellum in amongst the white matter. These masses are called the cerebellar nuclei and the largest is known as the **dentate nucleus.**

The cavity of the hindbrain is called the **fourth ventricle** (*Fig. 6–3*).

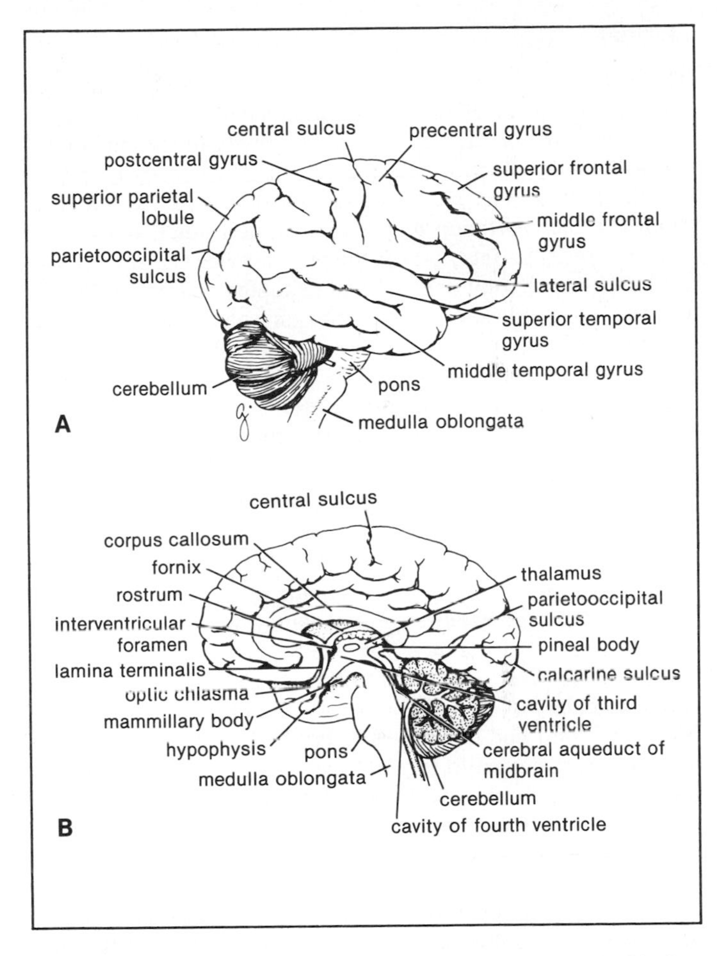

Figure 6–3. A. Brain viewed from its right lateral aspect. **B.** Median sagittal section of the brain showing the third and fourth ventricles.

Midbrain

The midbrain is the narrow part of the brain that passes through the tentorial notch of the tentorium cerebelli and connects the forebrain to the hindbrain. The narrow cavity of the midbrain is the **cerebral aqueduct,** which connects the third ventricle and the fourth ventricle (*Fig. 6–3*). The **tectum** is that part of the midbrain that lies posterior to the cerebral aqueduct; it has four swellings on its surface, called, the two **superior** and two **inferior colliculi.** The **cerebral peduncles** are situated anterior to the aqueduct. Each pedicle is divided into an anterior part, the **crus cerebri,** and a posterior part, the **tegmentum,** by a pigmented band of gray matter, the **substantia nigra.** A large ovoid mass of gray matter, called the **red nucleus** is situated in the tegmentum on each side at the level of the superior colliculus.

Forebrain

Diencephalon. The diencephalon is almost completely hidden from the surface of the brain. It is made up of a dorsal **thalamus** and a ventral **hypothalamus.** The thalamus is a large ovoid mass of gray matter that lies on either side of the third ventricle (*Figs. 6–3 and 6–4*). The anterior end of the thalamus forms the posterior boundary of the **interventricular foramen,** the opening between the third and lateral ventricles. The posterior end of the thalamus is expanded to form a large swelling, the **pulvinar.** The hypothalamus forms the lower part of the lateral wall and the floor of the third ventricle.

The following structures are found in the floor of the third ventricle, from anterior to posterior: the **optic chiasma,** the **tuber cinerium,** the **infundibulum,** the **mammillary bodies,** and the **posterior perforated substance.** The rostral (anterior) end of the third ventricle is bounded by a thin sheet, the **lamina terminalis** (*Fig. 6–3*).

Pineal Body. The pineal body is a small glandular structure that is attached by a stalk to the region of the posterior wall of the

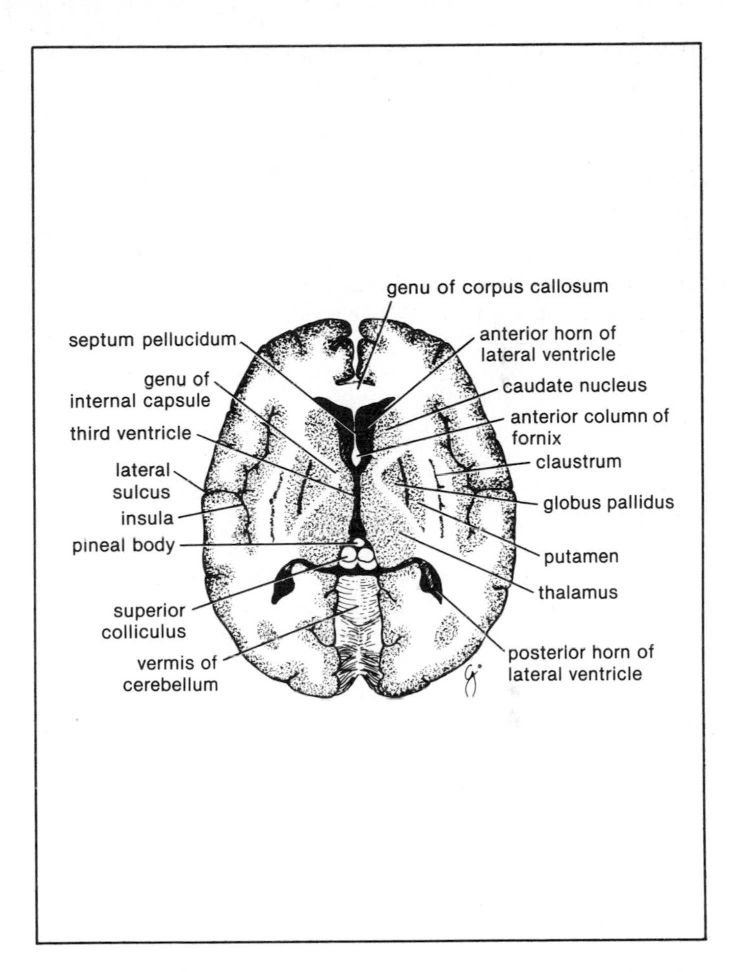

Figure 6–4. Horizontal section of the cerebrum showing the relationship between the globus pallidus, the caudate nucleus, the thalamus, and the internal capsule.

third ventricle (*Fig. 6–3*). The pineal commonly calcifies in middle age and thus it may be seen on radiographs of the skull.

Cerebrum. The cerebrum is the largest part of the brain and consists of two **cerebral hemispheres** connected by a mass of white matter called the **corpus callosum** (*Fig. 6–3*). Each hemisphere extends from the frontal to the occipital bones, superior to the anterior and middle cranial fossae; posteriorly, the cerebrum lies above the tentorium cerebelli. The hemispheres are separated by a deep cleft, the **longitudinal fissure,** into which projects the **falx cerebri.** The surface of each hemisphere is called the **cortex** and is composed of gray matter. The cerebral cortex is thrown into folds, or **gyri,** separated by fissures, or **sulci.** By this means the surface area of the cortex is greatly increased. A number of large sulci are used to subdivide the surface of each hemisphere into **lobes.** The lobes are named from the bones of the cranium under which they lie.

The **frontal lobe** is situated anterior to the **central sulcus** and superior to the **lateral sulcus** (*Fig. 6–3*). The **parietal lobe** is located posterior to the central sulcus and superior to the lateral sulcus. The occipital lobe lies inferior to the **parietooccipital sulcus.** Inferior to the lateral sulcus is located the **temporal lobe.**

The **precentral gyrus** lies immediately anterior to the central sulcus and is known as the **motor area** (*Fig. 6–3*). The large motor nerve cells in this area control voluntary movements on the opposite side of the body. The majority of the nerve fibers cross over to the opposite side of the brain at the **decussation of the pyramids** in the medulla oblongata. In the motor area the body is represented in an inverted position, with the movements of the foot being controlled by nerve cells located in the upper part, and those controlling movements of the face and the hands in the lower part.

The **postcentral gyrus** is located posterior to the central sulcus and is known as the **sensory area** (*Fig. 6–3*). The small nerve cells in this area receive and interpret common sensations from the opposite side of the body.

The **superior temporal gyrus** is located immediately below the lateral sulcus (*Fig. 6–3*). The middle of this gyrus is concerned with the reception and interpretation of sound and is known as the **auditory area.**

Broca's area, the **motor speech area,** lies just above the lateral sulcus (*Fig. 6–3*). It controls the movements of the larynx and mouth used in speech. It is dominant in the left hemisphere in right-handed persons and dominant in the right hemisphere in left-handed persons.

The **visual area** is situated on the posterior pole of the occipital lobe and on the medial aspect in the region of the calcarine sulcus (*Fig. 6–3*). It is the receiving area for visual impressions.

Inside the hemisphere is a central core of white matter, containing several large masses of gray matter, the **basal nuclei** or **ganglia** (*Fig. 6–4*). A fan-shaped collection of nerve fibers, termed the **corona radiata,** passes in the white matter to and from the cerebral cortex to the brainstem. The corona radiata fibers pass between the basal nuclei as the **internal capsule** (*Fig. 6–4*). The tailed nucleus situated on the medial side of the internal capsule is called the **caudate nucleus** and the lens-shaped nucleus on the lateral side of the internal capsule is called the **lentiform nucleus.**

The cavity present within each cerebral hemisphere is called the **lateral ventricle.** The lateral ventricle communicates with the third ventricle through the interventricular foramen.

Ventricular System and Cerebrospinal Fluid

Ventricular System. The ventricles of the brain are the **lateral ventricles** (one in each cerebral hemisphere), the **third ventricle** (in the midline of the diencephalon), and the **fourth ventricle** (cavity of the hindbrain). The two lateral ventricles communicate through the **interventricular foramina** with the third ventricle. The **cerebral aqueduct** connects the third ventricle to the fourth ventricle. The fourth ventricle is continuous with the **central canal** of the spinal cord and through the three foramina in its roof (the **single foramen of Magendie** and the two lateral **foramina of Luschka**), with the subarachnoid space. The ventricles are lined throughout with **ependyma** and are filled with cerebrospinal fluid.

Cerebrospinal Fluid. This is a clear, colorless, fluid formed mainly by the **choroid plexus** of the lateral, third, and fourth ventricles (some may originate as tissue fluid in the brain). The fluid circulates through the ventricular system and enters the subarachnoid space through the three foramina in the roof of the fourth ventricle. It now ascends in the subarachnoid space over the outer surface of the cerebrum. Some of the fluid moves inferiorly in the subarachnoid space around the spinal cord and cauda equina. The cerebrospinal fluid is mainly absorbed through the **arachnoid villi** that project into the dural venous sinuses, in particular the **superior sagittal sinus.** Groups of arachnoid villi are known as **arachnoid granulations.** Some of the cerebrospinal fluid is probably absorbed directly into the veins in the subarachnoid space and some into the perineural lymphatic vessels of cranial and spinal nerves.

The functions of the cerebrospinal fluid are (1) protection of the central nervous system from trauma, (2) regulation of the intracraniac pressure (volume adjusts in the presence of a space occupying tumor), (3) nourishment of the nervous tissue, and (4) removal of waste products from nervous tissue.

Meninges of the Brain

The brain, like the spinal cord, is surrounded by three membranes, the dura mater, the arachnoid mater, and the pia mater.

Dura Mater. The dura mater of the brain is formed of two layers, the endosteal layer and the meningeal layer. These are closely united except along certain lines, where they separate to form **venous sinuses.** The **endosteal layer** is the periosteum covering the inner surface of the skull. The **meningeal layer** is continuous through the foramen magnum with the dura mater of the spinal cord. The meningeal layer gives rise to four septa that divide the cranial cavity into spaces that lodge the subdivisions of the brain.

The **falx cerebri** is a sickle-shaped fold of dura that lies in the midline between the two cerebral hemispheres. It is attached in front to the crista galli and posteriorly to the upper surface of the tentorium cerebelli and the internal surface of the skull. The **superior sagittal sinus** runs in its upper fixed border and the **inferior sagittal sinus,** in its lower free margin; the **straight sinus** runs along its attachment to the tentorium cerebelli.

The **tentorium cerebelli** is a crescent-shaped fold of dura placed between the cerebrum and the cerebellum. It is suspended by the falx cerebri. In the anterior edge is a gap, the **tentorial notch,** for the passage of the midbrain. The fixed border of the tentorium is attached to the posterior clinoid processes, the petrous part of the temporal bone and the inner surface of the occipital bone. The free border is attached in front to the anterior clinoid processes. The **straight sinus** runs along its attachment to the falx cerebri, the **superior petrosal sinus,** along its attachment to the petrous bone, and the **transverse sinus** along its attachment to the occipital bone.

The **falx cerebelli** is a small vertical fold that projects forward from the occipital bone between the two cerebellar hemispheres.

The **diaphragma sellae** is a small circular fold that forms the roof of the sella turcica. A small hole in its center is for the passage of the stalk of the hypophysis cerebri.

Arachnoid Mater. The arachnoid mater is a delicate, impermeable membrane covering the brain and lying between the pia mater internally and the dura mater externally. The arachnoid bridges over the sulci on the surface of the brain and in certain locations is widely separated from the pia mater to form the **subarachnoid cisternae.** The **cisterna cerebellomedullaris** and the **cisterna interpeduncularis** are two examples.

The arachnoid projects into the venous sinuses (especially the superior sagittal) to form **arachnoid villi** and **arachnoid granulations.** These are sites where the cerebrospinal fluid diffuses into the bloodstream.

Pia Mater. The pia mater is a vascular membrane that closely covers the surface of the brain and descends into the sulci. As a two-layered fold, called the **tela choroidea,** it projects into the ventricles to form the **choroid plexuses.**

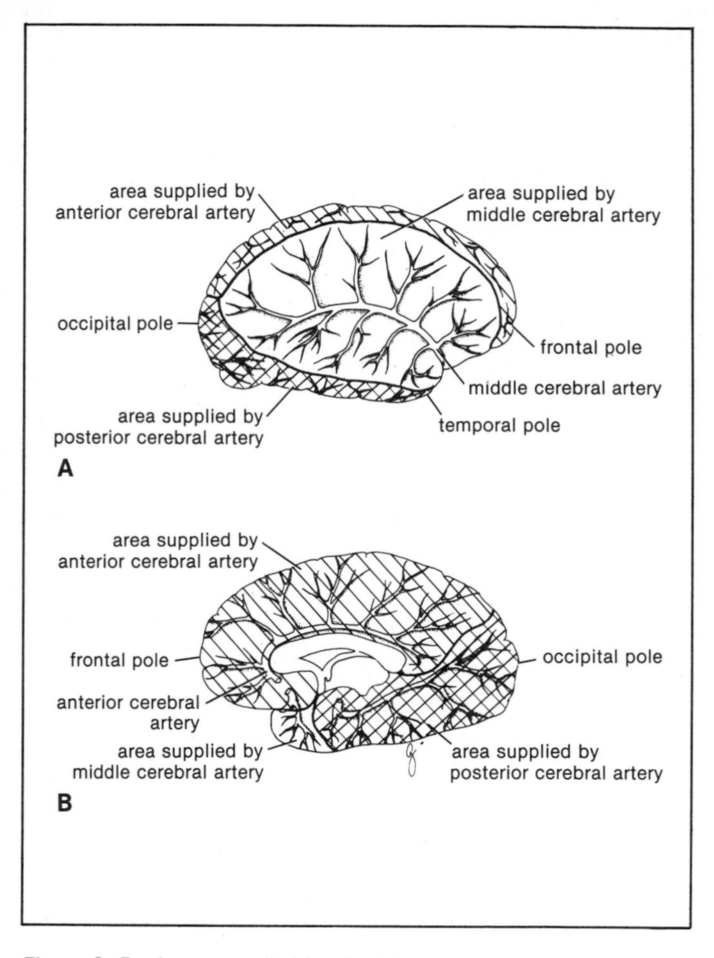

Figure 6–5. Areas supplied by the cerebral arteries. **A.** The lateral surface of the right cerebral hemisphere. **B.** The medial surface of the right cerebral hemisphere.

Blood Supply of the Brain

The brain is supplied by the two **internal carotid** and the two **vertebral arteries.** The four arteries anastomose on the inferior surface of the brain and form the **circulus arteriosus** (*see p. 186*). The cerebral arteries are situated in the subarachnoid space. The distribution of the cortical branches of the cerebral arteries is summarized in *Figure 6-5*.

The veins of the brain drain into the cranial venous sinuses.

7

Peripheral Nervous System

The peripheral nervous system consists of the cranial and spinal nerves and their associated ganglia.

CRANIAL NERVES

There are 12 pairs of cranial nerves that leave the brain and pass through foramina in the skull. All the nerves are distributed in the head and the neck except the tenth, which also supplies structures in the thorax and the abdomen. The cranial nerves are named as follows:

1. Olfactory
2. Optic
3. Oculomotor
4. Trochlear
5. Trigeminal
6. Abducent
7. Facial
8. Vestibulocochlear
9. Glossopharyngeal
10. Vagus
11. Accessory
12. Hypoglossal

The olfactory, the optic, and the vestibulocochlear nerves are entirely sensory. They carry sensations to the central nervous system. The oculomotor, trochlear, abducent, accessory, and hypoglossal nerves are entirely motor. They carry nervous impulses from the central nervous system to muscles or glands. The remaining cranial nerves are mixed nerves containing both sensory and motor fibers.

Olfactory Nerves

The olfactory nerves, or nerves of smell, arise from olfactory receptor nerve cells in the olfactory mucous membrane. The olfactory mucous membrane is located in the upper part of the nasal cavity above the level of the superior concha. Bundles of these nerve fibers pass through the openings of the cribriform plate of the ethmoid bone to enter the **olfactory bulb** inside the skull. The olfactory bulb is connected to the **olfactory area of the cerebral cortex** by the olfactory tract.

Optic Nerve

The optic nerve, or nerve of sight, is composed of the axons of the cells of the **ganglionic layer** of the retina. The **optic nerve** emerges from the back of the eyeball and leaves the orbital cavity of the skull through the optic canal. It unites with the optic nerve of the opposite side to form the **optic chiasma.**

In the chiasma, the fibers from the medial (nasal) half of each retina cross the midline and enter the **optic tract** of the opposite side, whereas the fibers from the lateral (temporal) half of each retina pass posteriorly in the optic tract of the same side. Most of the fibers of the optic tract terminate by synapsing with nerve cells in the **lateral geniculate body,** which is a small projection from the posterior part of the thalamus. A few fibers pass to the pretectal nucleus and superior colliculus of the midbrain and are concerned with light reflexes.

The axons of the nerve cells of the lateral geniculate body pass posteriorly as the **optic radiation** and terminate in the **visual cortex** of the cerebral hemisphere.

Oculomotor Nerve

The oculomotor nerve emerges on the anterior surface of the midbrain. It passes forward in the middle cranial fossa in the lateral wall of the cavernous sinus. Here, it divides into a superior and an inferior ramus that enter the orbital cavity through the superior orbital fissure.

The oculomotor nerve supplies the following extrinsic muscles of the eye, i.e., those that lie outside the eyeball: levator palpebrae superioris, superior rectus, medial rectus, inferior rectus, and inferior oblique. It also supplies through its branch to the ciliary ganglion and the short ciliary nerves parasympathetic nerve fibers to the following intrinsic muscles, i.e., those that lie inside the eyeball, namely, the constrictor pupillae of the iris and the ciliary muscles.

The oculomotor nerve is therefore entirely motor and is responsible for lifting the upper eyelid; turning the eye upward, downward, and medially; constricting the pupil; and accommodating the eye.

Trochlear Nerve

The trochlear nerve, the most slender of the cranial nerves, leaves the posterior surface of the midbrain and immediately decussates with (crosses) the nerve of the opposite side. The trochlear nerve passes forward through the middle cranial fossa in the lateral wall of the cavernous sinus and enters the orbit through the superior orbital fissure. The nerve supplies the superior oblique muscle of the eyeball. The trochlear nerve is entirely motor and assists in turning the eye downward and laterally.

Trigeminal Nerve

The trigeminal nerve is the largest cranial nerve and leaves the anterior aspect of the pons as a small motor root and a large sensory root. The nerve passes forward out of the posterior cranial fossa to reach the apex of the petrous part of the temporal bone in the middle cranial fossa. The large sensory root now expands to form the **trigeminal ganglion.** The trigeminal ganglion lies within a pouch of dura mater, called the **trigeminal cave.** The ophthalmic, maxillary, and mandibular nerves arise from the anterior border of the ganglion.

Ophthalmic Nerve. The **ophthalmic nerve** is entirely sensory. It runs forward in the lateral wall of the cavernous sinus in the middle cranial fossa and divides into three branches, the lacrimal, frontal, and nasociliary nerves, which enter the orbital cavity through the superior orbital fissure.

Lacrimal Nerve. The **lacrimal nerve** runs forward along the upper border of the lateral rectus muscle. It receives the zygomaticotemporal branch of the maxillary nerve, which contains the parasympathetic secretomotor branches to the lacrimal gland. The lacrimal nerve enters the lacrimal gland and is distributed to the conjuctiva and the skin of the upper eyelid.

Frontal Nerve. The **frontal nerve** runs forward on the upper surface of the levator palpebrae superioris and divides into the **supraorbital** and **supratrochlear nerves.** These nerves leave the orbital cavity and are distributed to the skin of the forehead, the anterior part of the scalp, and the frontal sinus.

Nasociliary Nerve. The **nasociliary nerve** crosses the optic nerve and runs forward on the upper border of the medial rectus muscle. It continues as the anterior **ethmoidal nerve** through the anterior ethmoidal foramen to enter the cranial cavity. It then descends through a slit at the side of the crista galli to enter the nasal cavity. It gives off two **internal nasal branches** and then supplies the skin of the tip of the nose with the **external nasal nerve.**

BRANCHES

1. **Sensory fibers** to the ciliary ganglion.
2. **Long ciliary nerves** that contain sympathetic fibers to the dilator pupillae and sensory fibers to the cornea.
3. **Infratrochlear nerve** to supply the skin of the eyelids.
4. **Posterior ethmoidal nerve** to the ethmoid and sphenoid sinuses.

Maxillary Nerve. This nerve is purely sensory. It leaves the skull through the foramen rotundum and crosses the pterygopalatine fossa to enter the orbit through the inferior orbital fissure. It continues as the **infraorbital nerve** in the infraorbital groove and emerges onto the face through the infraorbital foramen. It gives sensory fibers to the skin of the face and side of the nose.

Branches

1. **Meningeal branches.**
2. **Zygomatic.** It divides into the **zygomaticotemporal and zygomaticofacial branches.** These supply the skin of the face. The zygomaticotemporal branch gives parasympathetic secretomotor fibers to the lacrimal nerve for the lacrimal gland.
3. **Ganglionic branches.** Two short nerves that hold up the pterygopalatine ganglion in the pterygopalatine fossa. They contain sensory fibers that have passed through the ganglion from the nose, the palate, and the pharynx. They also contain postganglionic parasympathetic fibers that are going to the lacrimal gland.
4. **Posterior superior alveolar nerve.** This nerve supplies the maxillary sinus and the upper molar teeth and adjoining parts of the gum and the cheek.
5. **Middle superior alveolar nerve.** This nerve supplies the maxillary sinus, the upper premolar teeth, the gums, and the cheek.

6. **Anterior superior alveolar nerve.** This nerve supplies the maxillary sinus and the upper canine and incisor teeth.

Pterygopalatine Ganglion. This is a parasympathetic ganglion that is suspended from the maxillary nerve in the pterygopalatine fossa. It is secretomotor to the lacrimal and nasal glands (*see p. 320*).

Branches

1. **Orbital branches** enter the orbit through the inferior orbital fissure.
2. **Greater and lesser palatine nerves** supply the mucous membrane of the palate, the tonsil, and the nasal cavity.
3. **Nasal branches.**
4. **Pharyngeal branch** supplies the roof of the nasopharynx.

Mandibular Nerve. This nerve is motor and sensory. The sensory root leaves the trigeminal ganglion and passes out of the skull through the foramen ovale. The motor root of the trigeminal nerve also leaves the skull through the same foramen and joins the sensory root to form the trunk of the mandibular nerve. The nerve trunk then divides into a small anterior and a large posterior division.

Branches from the Main Trunk of the Mandibular Nerve

1. **Meningeal branch.**
2. **Nerve to medial pterygoid.** This branch supplies the medial pterygoid, the tensor tympani, and the tensor veli palatini muscles.

Branches from the Anterior Division of the Mandibular Nerve

1. **Masseteric nerve** to the masseter muscle.
2. **Deep temporal nerves** to the temporalis muscle.
3. **Nerve to the lateral pterygoid muscle.**

4. **Buccal nerve** to the skin of the cheek and the mucous membrane lining the cheek.

Branches from the Posterior Division of the Mandibular Nerve

1. **Auriculotemporal nerve** supplies the skin of the auricle, the external auditory meatus, the temporomandibular joint, and the scalp. Also conveys postganglionic parasympathetic secretomotor fibers from the otic ganglion to the parotid salivary gland.
2. **Lingual nerve.** This nerve descends in front of the inferior alveolar nerve and enters the mouth by passing under the superior constrictor muscle. It runs forward on the side of the tongue to reach its tip. In its course it crosses the submandibular duct. At the lower border of the lateral pterygoid muscle it is joined by the chorda tympani nerve. It ends by giving branches to the mucous membrane of the anterior two-thirds of the tongue and the floor of the mouth. It also gives off important preganglionic parasympathetic secretomotor fibers to the submandibular ganglion.
3. **Inferior alveolar nerve.** The nerve enters the mandibular canal to supply the teeth of the lower jaw. It emerges through the mental foramen to supply the skin of the face over the chin. It gives off the **mylohyoid nerve** that supplies the mylohyoid muscle and the anterior belly of the digastric muscle.
4. **Communicating branch.** This frequently runs from the inferior alveolar nerve to the lingual nerve.

Otic Ganglion. This is a small parasympathetic ganglion that is located medial to the mandibular nerve just below the foramen ovale. The ganglion is adherent to the nerve to the medial pterygoid muscle. The preganglionic fibers originate in the glosssopharyngeal nerve and reach the ganglion via the lesser petrosal nerve (*see p. 274*). The postganglionic secretomotor fibers reach the parotid salivary gland via the auriculotemporal nerve.

Submandibular Ganglion. This small parasympathetic ganglion lies between the submandibular salivary gland and the hyoglossus muscle. The ganglion is connected to the lingual nerve and the salivary gland by several branches. Preganglionic parasympathetic fibers reach the ganglion via the chorda tympani and the lingual nerves. Postganglionic secretomotor fibers pass to the submandibular and sublingual salivary glands.

Abducent Nerve

The abducent nerve is a small nerve that emerges from the anterior surface of the brain between the pons and the medulla oblongata. It passes forward through the cavernous sinus in the middle cranial fossa lying below and lateral to the internal carotid artery. The nerve then enters the orbit through the superior orbital fissure. The abducent nerve is entirely a motor nerve and supplies the lateral rectus muscle and is, therefore, responsible for turning the eye laterally.

Facial Nerve

The facial nerve has a medial motor root and a lateral sensory root, the **nervus intermedius.** The two roots emerge from the anterior surface of the brain between the pons and the medulla oblongata. They pass laterally in the posterior cranial fossa with the vestibulocochlear nerve and enter the internal acoustic meatus in the petrous part of the temporal bone. At the bottom of the meatus, the nerve enters the facial canal and runs laterally through the inner ear. On reaching the medial wall of the tympanic cavity, the nerve expands to form the sensory **geniculate ganglion** and turns sharply backward above the promontory. At the posterior wall of the tympanic cavity the facial nerve turns downward on the medial side of the aditus of the mastoid antrum. It descends behind the pyramid and emerges from the stylomastoid foramen.

Important Branches of Facial Nerve

1. The **greater petrosal nerve** arises from the facial nerve at the geniculate ganglion. It contains preganglionic parasym-

pathetic fibers that synapse in the pterygopalatine ganglion. The postganglionic fibers are secretomotor to the lacrimal gland and glands of the nose and the palate. The greater petrosal nerve also contains taste fibers from the mucous membrane of the palate.

2. **The nerve to stapedius** arises from the facial nerve in the facial canal. It supplies the stapedius muscle within the pyramid.
3. The **chorda tympani** arises from the facial nerve in the facial canal. It runs forward over the medial surface of the upper part of the tympanic membrane and crosses the root of the handle of the malleus. The nerve leaves the tympanic cavity through the **petrotympanic fissure** and enters the infratemporal fossa to join the lingual nerve. The nerve contains preganglionic parasympathetic secretomotor fibers to the submandibular and sublingual salivary glands. It also contains taste fibers from the anterior two-thirds of the tongue and the floor of the mouth.
4. **Posterior auricular, posterior belly of the digastric, and the stylohyoid nerves.** These muscular branches are given off from the main trunk as it emerges from the stylomastoid foramen.
5. **Five terminal branches to the muscles of facial expression.** As the facial nerve leaves the stylomastoid foramen it enters the substance of the parotid salivary gland lying between the superficial and deep parts. Here it gives off the **temporal, zygomatic, buccal, mandibular,** and **cervical** branches that emerge from the parotid gland and are distributed to the scalp and the facial muscles. Note that the buccal branch supplies the buccinator muscle. The cervical branch supplies the platysma and the depressor anguli oris muscles.

Vestibulocochlear Nerve

The vestibulocochlear nerve consists of two sets of sensory fibers, vestibular and cochlear. The two parts of the nerve leave the anterior surface of the brain between the lower border of the pons and the medulla oblongata. They run laterally in the posterior cranial fossa and enter the internal acoustic meatus with the facial nerve.

The **vestibular fibers** represent the central processes of nerve cells of the vestibular ganglion situated in the internal acoustic meatus. The vestibular fibers originate from the vestibule and the semicircular canals and are therefore concerned with the sense of position and movement of the head.

The **cochlear fibers** represent the central processes of nerve cells of the **spiral ganglion of the cochlea.** The cochlear fibers originate in the **spiral organ of Corti** and are therefore concerned with the sense of hearing.

Glossopharyngeal Nerve

The glossopharyngeal nerve is a motor and a sensory nerve. It emerges from the anterior surface of the medulla oblongata between the olive and the inferior cerebellar peduncle. It passes laterally in the posterior canal fossa and leaves the skull by passing through the jugular foramen. The superior and inferior glossopharyngeal ganglia are situated on the nerve as it passes through the jugular foramen. The glossopharyngeal nerve then descends through the upper part of the neck in company with the internal jugular vein and the internal carotid artery to reach the posterior border of the stylopharyngeus muscle. It then passes forward between the superior and middle constrictor muscles to be distributed to the pharynx and the back of the tongue.

Important Branches of the Glossopharyngeal Nerve

1. The **tympanic branch** passes through the floor of the tympanic cavity and contributes to the **tympanic plexus.** The secretomotor preganglionic parasympathetic fibers for the parotid salivary gland that originated in the glossopharyngeal nerve now leave the plexus as the **lesser petrosal nerve** and they synapse in the otic ganglion.
2. The **carotid branch** contains sensory fibers from the carotid sinus (pressoreceptor mechanism for the regulation of blood

pressure) and the carotid body (chemoreceptor mechanism sensitive to oxygen lack and carbon dioxide excess to regulate heart rate and respiration).

3. The **nerve to the stylopharyngeus muscle.**
4. The **pharyngeal branches** join the pharyngeal branch of the vagus and the pharyngeal branch of the sympathetic trunk to form the **pharyngeal plexus.** This gives sensory fibers to the mucous membrane of the pharynx.
5. The **lingual branch** supplies general sensation and special taste to the mucous membrane of the posterior third of the tongue and the vallate papillae.

Vagus Nerve

The vagus nerve is composed of motor and sensory fibers. It emerges from the anterior surface of the medulla oblongata between the olive and the inferior cerebellar peduncle. The nerve passes laterally through the posterior cranial fossa and leaves the skull through the jugular foramen. The vagus nerve possesses two sensory ganglia, a rounded **superior ganglion,** situated on the nerve within the jugular foramen, and a cylindrical **inferior ganglion,** which lies on the nerve just below the foramen. Below the inferior ganglion the cranial root of the accessory nerve joins the vagus nerve and is distributed mainly in its pharyngeal and recurrent laryngeal branches.

The vagus nerve descends vertically in the neck within the carotid sheath with the internal jugular vein and the internal and common carotid arteries.

The **right vagus nerve** crosses the anterior surface of the subclavian artery and enters the thorax posterolateral to the brachiocephalic artery, lateral to the trachea and medial to the azygos vein. It passes posteriorly to the root of the right lung contributing to the pulmonary plexus. It then passes onto the posterior surface of the esophagus and contributes to the esophageal plexus. It enters the abdomen behind the esophagus through the esophageal opening of the diaphragm. The posterior vagal trunk (which is the name now given to the right vagus) is distributed to the posterior surface of the stomach and by a large celiac branch to the duodenum, the liver, the kidneys, and the small and large intestines as far as the distal third of the transverse colon. This wide distribution is accomplished through the celiac, superior mesenteric, and renal plexuses.

The **left vagus** nerve enters the thorax between the left common carotid and the left subclavian arteries. It crosses the left side of the aortic arch and descends behind the root of the left lung contributing to the pulmonary plexus. The left vagus then descends on the anterior surface of the esophagus contributing to the esophageal plexus. It enters the abdomen through the esophageal opening of the diaphragm lying in front of the esophagus. The anterior vagal trunk (which is the name now given to the left vagus) divides into several branches that are distributed to the stomach, the liver, the upper part of the duodenum, and the head of the pancreas.

Important Branches of the Vagus Nerve

1. **Meningeal and auricular branches.**
2. The **pharyngeal branch** contains nerve fibers from the cranial root of the accessory nerve. It joins the pharyngeal branches of the glossopharyngeal nerves and the sympathetic trunk to form the **pharyngeal plexus.** The pharyngeal nerve supplies all the muscles of the pharynx except the stylopharyngeus and all the muscles of the soft palate except the tensor veli palatini.
3. The **superior laryngeal nerve** divides into internal and external laryngeal nerves. The **internal laryngeal nerve** is sensory to the mucous membrane of the piriform fossa and the larynx down as far as the vocal folds. The **external laryngeal nerve** is motor and located close to the superior thyroid artery; it supplies the cricothyroid muscle.
4. The **recurrent laryngeal nerve.** On the right side the nerve arises from the vagus as it crosses the first part of the subclavian artery. The recurrent laryngeal nerve then hooks backward and upward behind the artery and ascends in the

groove between the trachea and the esophagus. The nerve crosses either in front of or behind the inferior thyroid artery or may pass between its branches. It supplies all the muscles of the larynx except the cricothyroid and the mucous membrane of the larynx below the vocal folds and the mucous membrane of the upper part of the trachea.

On the left side, the recurrent laryngeal nerve arises from the vagus as the latter crosses the arch of the aorta. It hooks beneath the arch behind the ligamentum arteriosum and ascends into the neck in the groove between the trachea and the esophagus.

5. **Cardiac branches.** Two or three branches arise from the vagus in the neck. They descend into the thorax and end in the cardiac plexus. The vagus nerve thus innervates the heart and the great vessels within the thorax, the larynx, the trachea, the bronchi, and the lungs and much of the alimentary tract from the pharynx to the distal part of the transverse colon. It also supplies glands associated with the alimentary tract such as the liver and the pancreas.

The vagus has the most extensive distribution of all the cranial nerves and supplies the structures named above with afferent and efferent fibers.

Accessory Nerve

The accessory nerve is a motor nerve. It consists of a cranial root and a spinal root. The **cranial root** emerges from the anterior surface of the medulla oblongata between the olive and the inferior cerebellar peduncle. The nerve runs laterally in the posterior cranial fossa and joins the spinal root. The spinal root arises from nerve cells in the anterior gray column of the upper five segments of the cervical part of the spinal cord. The nerve ascends alongside the spinal cord and enters the skull through the foramen magnum. It then turns laterally to join the cranial root.

The two roots unite and leave the skull through the jugular foramen. The roots then separate and the cranial root joins the vagus nerve and is distributed in its pharyngeal and recurrent laryngeal branches to the muscles of the soft palate, pharynx, and larynx. The spinal root runs downward and laterally and enters the deep surface of the sternocleidomastoid muscle, which it supplies. The nerve then crosses the posterior triangle of the neck lying on the levator scapulae muscle. It then passes beneath the trapezius muscle, which it supplies.

The accessory nerve thus brings about movements of the soft palate, the pharynx, and the larynx and controls the movements of two large muscles in the neck.

Hypoglossal Nerve

The hypoglossal nerve is a motor nerve. The nerve emerges on the anterior surface of the medulla oblongata between the pyramid and the olive. It crosses the posterior cranial fossa and leaves the skull through the hypoglossal canal. The nerve then passes downward and forward in the neck between the internal carotid artery and the internal jugular vein until it reaches the lower border of the posterior belly of the digastric muscle. Here, it turns forward and crosses the internal and external carotid arteries and the loop of the lingual artery. It passes deep to the posterior margin of the mylohyoid muscle lying on the lateral surface of the hyoglossus muscle. It lies inferior to the submandibular duct and the lingual nerve. In the upper part of its course, the hypoglossal nerve is joined by C1 fibers from the cervical plexus.

Important Branches of the Hypoglossal Nerve

1. **Meningeal branch.**
2. The **descending branch,** consisting of C1 fibers descends in front of the internal and common carotid arteries. It is joined by the descending cervical nerve (C2 and C3) from the cervical plexus to form a loop called the **ansa cervicalis.** Branches from the loop supply the omohyoid, the sternohyoid, and the sternothyroid muscles.

3. The **nerve to the thyrohyoid,** consisting of C1 fibers, supplies the thyrohyoid muscle.
4. The **muscular branches** to all the muscles of the tongue except the palatoglossus. It thus supplies the styloglossus, the hyoglossus, the genioglossus, and the intrinsic muscles.
5. The **nerve to the geniohyoid,** consists of C1 fibers.

The hypoglossal nerve proper thus innervates the majority of the muscles of the tongue and controls the shape and movements of the tongue.

The different components of the cranial nerves, their functions, and the openings in the skull through which the nerves leave the cranial cavity are summarized in *Table 7–1.*

TABLE 7–1. CRANIAL NERVES

Name	Components	Function	Opening in Skull
I. Olfactory	Sensory	Smell	Openings in cribriform plate of ethmoid
II. Optic	Sensory	Vision	Optic canal
III. Oculomotor	Motor	Lifts upper eyelid, turns eyeball upward, downward, and medially; constricts pupil; accommodates eye	Superior orbital fissure
IV. Trochlear	Motor	Assists in turning eyeball downward and laterally	Superior orbital fissure

(continued)

TABLE 7–1. (cont.)

Name	Components	Function	Opening in Skull
V. Trigeminal			
Ophthalmic division	Sensory	Cornea, skin of forehead, scalp, eyelids; also mucous membrane of paranasal sinuses and nasal cavity	Superior orbital fissure
Maxillary division	Sensory	Skin of face over maxilla; teeth of upper jaw; mucous membrane of nose, the maxillary sinus, and palate	Foramen rotundum
Mandibular division	Motor	Muscles of mastication, mylohyoid, anterior belly of digastric, tensor veli palatini, and tensor tympani	Foramen ovale
	Sensory	Skin of cheek, skin over mandible and side of head, teeth of lower jaw and temporomandibular joint; mucous membrane of mouth and anterior part of tongue	
VI. Abducent	Motor	Lateral rectus muscle turns eyeball laterally	Superior orbital fissure

TABLE 7–1. (cont.)

Name	Components	Function	Opening in Skull
VII. Facial	Motor	Muscles of face and scalp, stapedius muscle, posterior belly of digastric and stylohyoid muscles	Internal acoustic meatus, facial canal, stylomastoid foramen
	Sensory	Taste from anterior part of tongue, floor of mouth and palate	
	Secretomotor parasympathetic	Submandibular and sublingual salivary glands, the lacrimal gland and glands of nose and palate	
VIII. Vestibulocochlear			
Vestibular	Sensory	Position and movement of head	Internal acoustic meatus
Cochlear	Sensory	Hearing	
IX. Glossopharyngeal	Motor	Stylopharyngeus muscle-assists in swallowing	
	Secretomotor parasympathetic	Parotid salivary gland	Jugular foramen
	Sensory	Posterior part of tongue and pharynx, carotid sinus and carotid body	

(continued)

TABLE 7–1. (cont.)

Name	Components	Function	Opening in Skull
X. Vagus	Motor and Sensory	Heart and great thoracic vessels; larynx, trachea, bronchi and lungs; alimentary tract from pharynx to splenic flexure of colon; liver, kidneys, and pancreas	Jugular foramen
XI. Accessory			
Cranial root	Motor	Muscles of soft palate, pharynx, and larynx	Jugular foramen
Spinal root	Motor	Sternocleidomastoid and trapezius muscles	
XII. Hypoglossal	Motor	Muscles of tongue (except palatoglossus) controlling its shape and movement	Hypoglossal canal

SPINAL NERVES

Composition

There are 31 pairs of spinal nerves that leave the spinal cord and pass through intervertebral foramina in the vertebral column. Each spinal nerve is connected to the spinal cord by two roots, the anterior root and the posterior root. The **anterior root** consists of bundles of nerve fibers carrying nerve impulses away from the central nervous system; these nerve fibers are called efferent or motor fibers. The **posterior root** consists of bundles of nerve fibers carrying nerve impulses to the central nervous system; these nerve fibers are called afferent fibers. Because these fibers are concerned with carrying information to the central nervous system, they are called sensory fibers. The cell bodies of these nerve fibers are situated in a swelling on the posterior root called the **posterior root ganglion.** Because a spinal nerve is formed by the union of an anterior root and a posterior root, it possesses both motor and sensory nerve fibers and is therefore a **mixed nerve.**

Because during development the spinal cord grows in length more slowly than the vertebral column, when growth ceases in the adult the lower end of the spinal cord only reaches inferiorly as far as the lower border of the first lumbar vertebra. To accommodate for this disproportionate growth in length the length of the roots increases progressively from above downward. In the upper cervical region the spinal nerve roots are short and run almost horizontally, but the roots of the lumbar and sacral nerves below the level of the termination of the cord form a vertical bundle of nerves that resembles a horse's tail and is called the **cauda equina.**

Names of Spinal Nerves

The 31 pairs of spinal nerves are named according to the regions of the vertebral column with which they are associated:

- 8 cervical
- 12 thoracic
- 5 lumbar
- 5 sacral
- 1 coccygeal

Note that there are eight cervical nerves and only seven cervical vertebrae and that there is one coccygeal nerve and four coccygeal vertebrae.

Branches of Spinal Nerves

After emerging from the intervertebral foramen, each spinal nerve immediately divides into a large **anterior ramus** and a smaller

posterior ramus, each containing both motor and sensory nerve fibers. The posterior ramus passes posteriorly around the vertebral column to supply the muscles and the skin of the back of the neck and the trunk. The anterior ramus supplies the muscles and the skin over the anterolateral body wall and the muscles and the skin of the limbs.

In addition to the anterior and posterior rami the spinal nerves give off a small **meningeal branch** that supplies the vertebrae and the meninges. The spinal nerves also have branches called the **rami communicantes** that are associated with the sympathetic part of the autonomic nervous system (*see p. 310*).

Plexuses

At the root of the limbs the anterior rami join one another to form complicated nerve plexuses. The **cervical** and **brachial plexuses** are found at the root of the upper limbs and the **lumbar** and **sacral plexuses** are found at the root of the lower limbs.

Cervical Plexus

The cervical plexus is formed by the anterior rami of the first four cervical nerves. The rami are joined by connecting branches, which form loops that lie in front of the origins of the levator scapulae and the scalenus medius muscles. The cervical plexus supplies the skin and the muscles of the head, the neck, and the shoulders.

Phrenic Nerve. The phrenic nerve is a large and very important branch of the cervical plexus; it is the only motor nerve supply to the diaphragm. The phrenic nerve runs vertically downward through the neck on the scalenus anterior muscle. Because of the obliquity of the scalenus anterior muscle, the nerve crosses the muscle from its lateral to its medial border. The nerve enters the thorax by passing anterior to the subclavian artery and it crosses the internal thoracic artery from lateral to medial.

The **right phrenic nerve** descends along the right side of the superior vena cava and in front of the root of the right lung. It then passes along the right side of the pericardium to the diaphragm. The **left phrenic nerve** descends along the left side of the left subclavian artery and crosses the left side of the aortic arch and the left vagus nerve. It passes in front of the root of the left lung and descends on the pericardium to the diaphragm.

In addition to the motor fibers to the diaphragm, the phrenic nerve contains sensory fibers from the pericardium and mediastinal parietal pleura and the pleura and peritoneum covering the upper and lower surfaces of the central part of the diaphragm.

Table 7–2 summarizes the branches of the cervical plexus and their distribution.

TABLE 7–2. BRANCHES OF THE CERVICAL PLEXUS AND THEIR DISTRIBUTION

Branches	Distribution
Cutaneous	
Lesser occipital	Skin of scalp behind ear
Greater auricular	Skin over parotid gland, auricle, and angle of jaw
Transverse cutaneous	Skin over side and front of neck
Supraclavicular	Skin over upper part of chest and shoulder
Musclar	
Segmental	Prevertebral muscles, levator scapulae
Ansa cervicalis (C1, 2, 3)	Omohyoid, sternohyoid, sternothyroid
Cl fibers via hypoglossal nerve	Thyrohyoid, geniohyoid
Phrenic nerve (C3, 4, 5)	Diaphragm (most important muscle of respiration)
Sensory to Pericardium, Pleura and Peritoneum	Pericardium, mediastinal parietal pleura, pleura, and peritoneum covering central diaphragm
Phrenic nerve (C3, 4, 5)	

Brachial Plexus

The brachial plexus is formed in the posterior triangle of the neck by the union of the anterior rami of the fifth, sixth, seventh, and eighth cervical and the first thoracic spinal nerves (*Fig. 7–1*). The plexus may be divided up into **roots, trunks, divisions, and cords.** The roots of C5 and C6 unite to form the upper trunk, the root of C7 continues as the **middle trunk,** the roots of C8 and T1 unite to form the **lower trunk.** Each trunk now divides into **anterior and posterior divisions.** The anterior divisions of the upper and middle trunks unite to form the **lateral cord,** the anterior division of the lower trunk continues as the **medial cord,** the posterior divisions of all three trunks join to form the **posterior cord.**

The roots of the brachial plexus enter the posterior triangle at the base of the neck between the scalenus anterior and scalenus medius muscles and behind the sternocleidomastoid muscle. The trunks and divisions cross the posterior triangle of the neck and the cords become arranged around the axillary artery in the axilla.

Branches of the brachial plexus arise from the roots, the trunks, and the cords. The courses of the more important nerves and their branches are as follows:

Musculocutaneous Nerve. The musculocutaneous nerve arises from the lateral cord of the brachial plexus (C5, 6, and 7). It pierces the coracobrachialis muscle and descends between the biceps and the brachialis muscles. It pierces the deep fascia in the region of the elbow and is distributed along the lateral side of the forearm as the **lateral cutaneous nerve of the forearm.** The musculocutaneous nerve supplies the coracobrachialis, both heads of biceps, and the greater part of the brachialis muscles.

Median Nerve. The median nerve arises from the medial and lateral cords of the brachial plexus (C5, 6, 7, 8, and T1). The nerve descends on the lateral side of the axillary artery and at first on the lateral side of the brachial artery. At the middle of the arm it crosses the brachial artery to reach its medial side. The

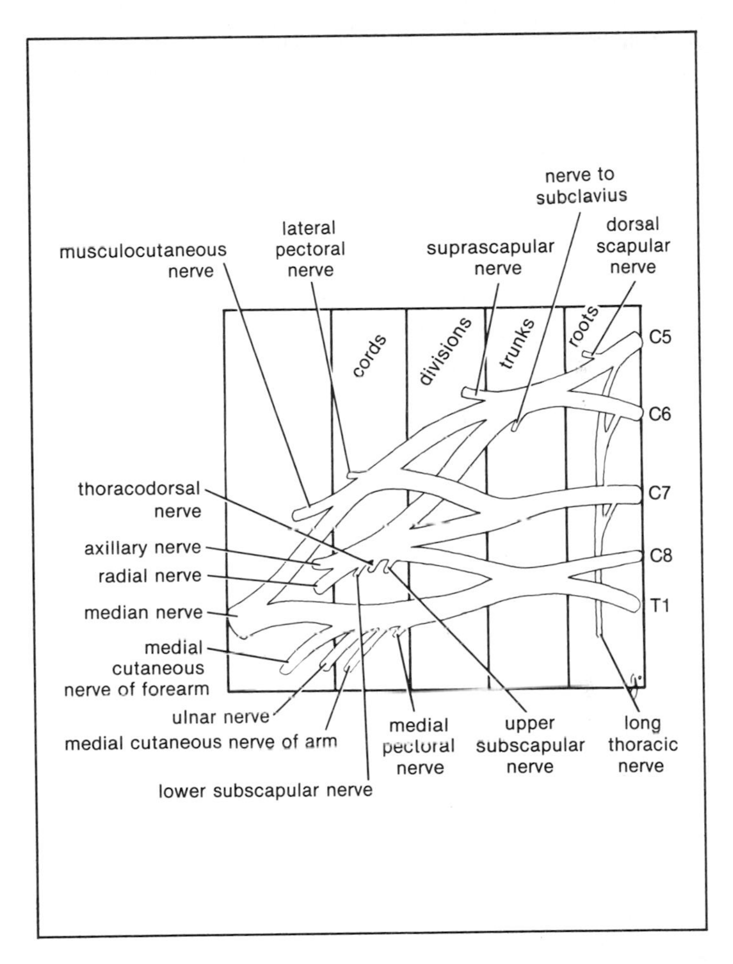

Figure 7–1. Formation of brachial plexus and its branches.

nerve enters the forearm between the two heads of pronator teres and descends on the posterior surface of the flexor digitorum superficialis lying upon the flexor digitorum profundus. At the wrist it lies posterior to the tendon of palmaris longus and between the tendons of flexor carpi radialis and flexor digitorum superficialis. The median nerve enters the palm by passing **behind** the flexor retinaculum and through the carpal tunnel.

Branches of the Median Nerve in the Forearm

1. **Muscular branches.** Pronator teres, flexor carpi radialis, palmaris longus, flexor digitorum superficialis.
2. **Articular branches.** Elbow joint.
3. **Anterior interosseous nerve.**
   a. **Muscular branches.** Flexor pollicis longus, pronator quadratus, lateral half of flexor digitorum profundus.
   b. **Articular branches.** Wrist and carpal joints.
4. **Palmar branch** is distributed to the skin over the lateral part of the palm.

Branches of the Median Nerve in the Palm

1. **Muscular branches.** Abductor pollicis brevis, flexor pollicis brevis, opponens pollicis, and the first and second lumbrical muscles.
2. **Cutaneous branches** to the palmar aspect of the lateral three and one-half fingers and the distal half of the dorsal aspect of each finger as well.

Ulnar Nerve. The ulnar nerve arises from the medial cord of the brachial plexus (C8, T1). It descends along the medial side of the axillary and brachial arteries in the anterior compartment of the arm. At the middle of the arm it pierces the medial intermuscular septum and passes down **behind** the medial epicondyle of the humerus. It then enters the anterior compartment of the forearm by passing between the two heads of flexor carpi ulnaris. It descends behind the flexor carpi ulnaris medial to the ulnar artery. At the wrist it passes **anterior** to the flexor retinaculum, lateral to the pisiform bone. It then divides into superficial and deep terminal branches.

Branches of the Ulnar Nerve in the Forearm

1. **Muscular branches.** Flexor carpi ulnaris, medial half of flexor digitorum profundus.
2. **Articular branches.** Elbow joint.
3. **Palmar branch.** Supplies skin over hypothenar eminence.
4. **Dorsal cutaneous branch.** Supplies skin over medial side of back of hand and back of medial one and one-half fingers over the proximal phalanges.

Branches of the Ulnar Nerve in the Hand. The **superficial terminal branch** descends into the palm and gives off the following branches:

1. **Muscular branch.** Palmaris brevis.
2. **Cutaneous branches.** Supplies skin over palmar aspect of medial one and one-half fingers including their nail beds.

The **deep terminal branch** runs backward between the abductor digiti minimi and the flexor digiti minimi and pierces the opponeus digiti minimi. It gives off the following branches:

1. **Muscular branches.** Abductor digiti minimi, flexor digiti minimi, opponens digiti minimi, all palmar and all dorsal interossei, third and fourth lumbricals, adductor pollicis.
2. **Articular branches.** Carpal joints.

Radial Nerve. The radial nerve arises from the posterior cord of the brachial plexus (C5, 6, 7, 8, and T1). It descends behind the axillary and brachial arteries and passes posteriorly between the long and medial heads of the triceps muscle to enter the posterior compartment of the arm. The radial nerve then winds round in the spiral groove on the back of the humerus with the profunda brachii vessels. Piercing the lateral intermuscular septum just above the elbow, it continues downward into the cubital fossa between the brachialis and the brachioradialis muscles. At the level of the lateral epicondyle it divides into superficial and deep branches.

Branches of the Radial Nerve in the Axilla

1. **Muscular branches.** Long and medial heads of the triceps.
2. **Cutaneous branch.** Posterior cutaneous nerve of the arm.

Branches of the Radial Nerve in the Spiral Groove

1. **Muscular branches.** Lateral and medial heads of the triceps, anconeus.
2. **Cutaneous branches.** Lower lateral cutaneous nerve of the arm. Posterior cutaneous nerve of the forearm.

Branches of the Radial Nerve in the Anterior Compartment of the Arm

1. **Muscular branches.** Brachialis, brachioradialis, extensor carpi radialis longus.
2. **Articular branches.** Elbow joint.

Superficial Branch of Radial Nerve. This is the direct continuation of the radial nerve. It descends under cover of the brachioradialis muscle on the lateral side of the radial artery. In the lower part of the forearm it passes backward under the tendon of brachioradialis and descends onto the back of the hand.

BRANCHES. **Cutaneous branches** to the lateral two-thirds of the posterior surface of the hand (variable area) and the posterior surface of the lateral two and one-half fingers over the proximal phalanges.

Deep Branch of the Radial Nerve. This branch winds around the lateral side of the neck of the radius in the supinator muscle to enter the posterior compartment of the forearm. It descends between the superficial and deep layers of muscle and reaches the back of the interosseous membrane.

BRANCHES

1. **Muscular branches.** Extensor carpi radialis brevis, supinator, extensor digitorum, extensor digiti minimi, extensor carpi ulnaris, abductor pollicis longus, extensor pollicis brevis, extensor pollicis longus, extensor indicis.
2. **Articular branches.** Wrist and carpal joints.

Axillary Nerve. The axillary nerve arises from the posterior cord of the brachial plexus (C5 and C6). It passes backward through the quadrilateral space below the shoulder joint in company with the posterior circumflex humeral vessels. The main trunk supplies the shoulder joint and its **anterior terminal branch** winds around the surgical neck of the humerus beneath the deltoid muscle; the anterior branch supplies the deltoid and the skin that covers its lower half. A **posterior terminal branch** supplies the teres minor muscle, the deltoid muscle, and then becomes the **upper lateral cutaneous nerve of the arm,** which supplies the skin over the lower part of the deltoid muscle.

A summary of the branches of the brachial plexus and their distribution is shown in *Table 7–3.*

Thoracic Spinal Nerves

There are 12 thoracic spinal nerves. The posterior rami supply the deep back muscles and the skin over the posterior surface of the back. The anterior rami do not join together to form a plexus. Instead the anterior rami of the first 11 thoracic spinal nerves run forward between ribs in intercostal spaces and are known as **intercostal nerves.** The anterior ramus of the twelfth thoracic nerve lies below the twelfth rib in the abdomen and runs forward in the abdominal wall.

TABLE 7–3. BRANCHES OF THE BRACHIAL PLEXUS AND THEIR DISTRIBUTION

Branches	Distribution
Roots	
Dorsal scapular nerve (C5)	Rhomboid minor, rhomboid major, and the levator scapulae muscles
Long thoracic nerve (C5, 6, 7)	Serratus anterior muscle
Upper Trunk	
Suprascapular nerve	Supraspinatus and infraspinatus muscles
Nerve to subclavius	Subclavius muscle
Lateral Cord	
Lateral pectoral	Pectoralis major muscle
Musculocutaneous nerve (C5, 6, 7)	Coracobrachialis, biceps brachii, brachialis muscles and supplies skin along lateral border of forearm when it becomes the lateral cutaneous nerve of forearm
Lateral root of median nerve	(See median nerve medial root)
Posterior Cord	
Upper subscapular nerve	Subscapularis muscle
Thoracodorsal nerve	Latissmus dorsi muscle
Lower subscapular nerve	Subscapularis and teres major muscles
Axillary nerve (C5, 6)	Deltoid and teres minor muscles; upper lateral cutaneous nerve of arm supplies skin over lower half of deltoid muscle
Radial nerve (C5, 6, 7, 8, T1)	Triceps, anconeus, part of brachialis, extensor carpi radialis longus; via deep radial nerve branch supplies extensor muscles of forearm: supinator, extensor carpi radialis brevis, extensor carpi

TABLE 7–3. (cont.)

Branches	Distribution
	ulnaris, extensor digitorum, extensor digit minimi, extensor indicis, abductor pollicis longus, extensor pollicis longus, extensor pollicis brevis; skin, lower lateral cutaneous nerve of arm, posterior cutaneous nerve of arm and posterior cutaneous nerve of forearm; skin on lateral side of dorsum of hand and dorsal surface of lateral 3 and a half fingers. Articular branches to elbow, wrist, and carpal joints
Medial Cord	
Medial pectoral	Pectoralis major and minor muscles
Medial cutaneous nerve of arm joined by intercostal brachial nerve from second intercostal nerve	Skin of medial side of arm
Medial cutaneous nerve of forearm	Skin of medial side of forearm
Ulnar nerve (C8 and T1)	Flexor carpi ulnaris and medial half of flexor digitorum profundus, flexor digiti minimi, abductor digiti minimi, opponens digiti minimi, adductor pollicis, third and fourth lumbricals, interossei, palmaris brevis, skin of medial half of dorsum of hand and palm, skin of palmar and dorsal surfaces of medial 1 and a half fingers

(continued)

TABLE 7–3. (cont.)

Branches	Distribution
Medial root of median nerve (with lateral root) forms median nerve (C5, 6, 7, 8, T1)	Pronator teres, flexor carpi radialis, palmaris longus, flexor digitorum superficialis, abductor pollicis brevis, flexor pollicis brevis, opponens pollicis, first two lumbricals, by way of anterior interosseous branch flexor pollicis longus, flexor digitorum profundus (lateral half) pronator quadratus. Palmar cutaneous branch to lateral half of palm and digital branches to palmar surface lateral 3 and a half fingers. Articular branches to elbow, wrist and carpal joints

Intercostal Nerves. The intercostal nerves are the anterior rami of the first 11 thoracic spinal nerves. Each intercostal nerve enters an intercostal space (*Fig. 7–2*) and then runs forward inferiorly to the intercostal vessels in the subcostal groove of the corresponding rib, between the transversus thoracis and internal intercostal muscle. The first six nerves are distributed within their intercostal spaces. The seventh to ninth intercostal nerves leave the anterior ends of their intercostal spaces by passing deep to the costal cartilages, to enter the abdominal wall. In the case of the tenth and eleventh nerves, because the corresponding ribs are floating, these nerves pass directly into the abdominal wall.

Branches of the Intercostal Nerves

1. **Collateral branch.** Runs forward below main nerve in lower part of intercostal space.
2. **Lateral cutaneous branch.** Arises in midaxillary line and di-

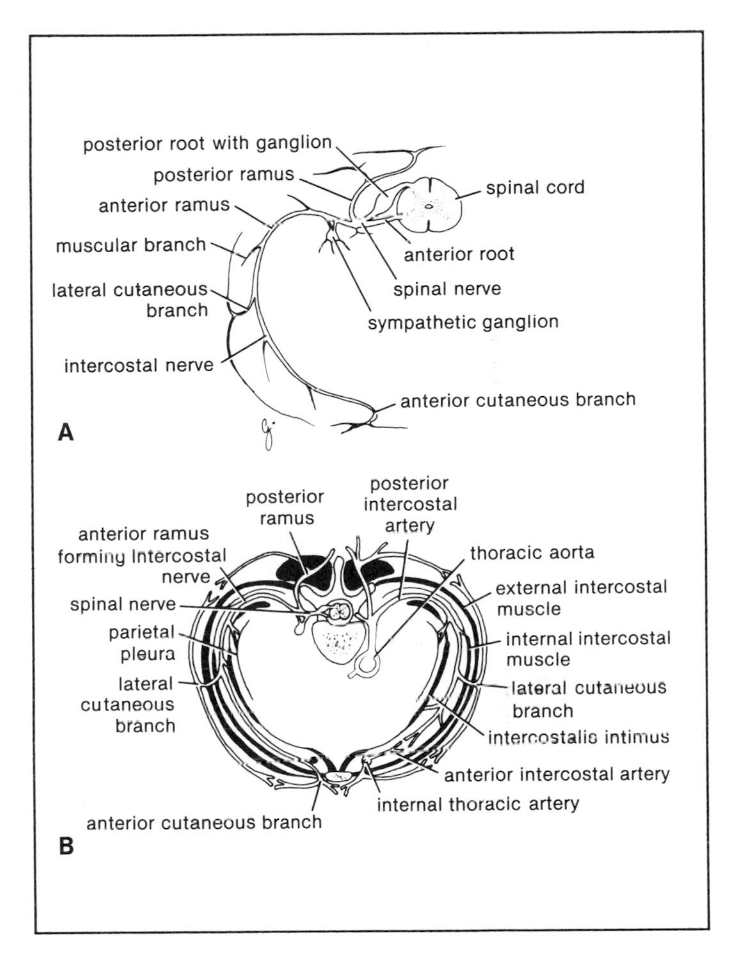

Figure 7–2. A. Cross section of the thoracic region of the spinal cord, showing spinal nerve roots, spinal nerve, and anterior and posterior rami and their branches. **B.** Cross section of thorax, showing distribution of a typical intercostal nerve and a posterior and an anterior intercostal artery.

vides into anterior and posterior branches that supply the skin.

3. **Anterior cutaneous branch** is the terminal portion of main trunk. It reaches the skin near the midline and divides into a medial and a lateral branch.
4. **Muscular branches.** Numerous muscular branches are given off by the main nerve and its collateral branch.
5. **Pleural branches** to parietal pleura and **peritoneal branches** (7 to 11 intercostal nerves only) to parietal peritoneum. These are sensory nerves. It is thus seen that the first six intercostal nerves give off numerous branches that supply: (1) the skin and the parietal pleura covering the outer and inner surface of each intercostal space, respectively, and (2) the intercostal muscles of each intercostal space.

In addition, the seventh to eleventh intercostal nerves supply (1) the skin and the parietal peritoneum covering the outer and inner surfaces of the abdominal wall, respectively, and (2) the anterior abdominal muscles, which include the external oblique, internal oblique, transversus abdominis, and rectus abdominis muscles. It should be noted that the first and second intercostal nerves are an exception.

The **first intercostal nerve** is joined to the brachial plexus by a large branch that is equivalent to the lateral cutaneous branch of typical intercostal nerves. The remainder of the first intercostal nerve is small.

The **second intercostal nerve** is joined to the medial cutaneous nerve of the arm by the **intercostobrachial nerve.** The second intercostal nerve therefore supplies the skin of the armpit and the upper medial side of the arm. In coronary artery heart disease, pain is referred along this nerve to the medial side of the arm.

Lumbar Plexus

The lumbar plexus, which is one of the main nervous pathways supplying the lower limb, is formed inside the psoas muscle from the anterior rami of the upper four lumbar nerves (*Fig. 7-3*). The branches of the plexus emerge from the lateral and medial bor-

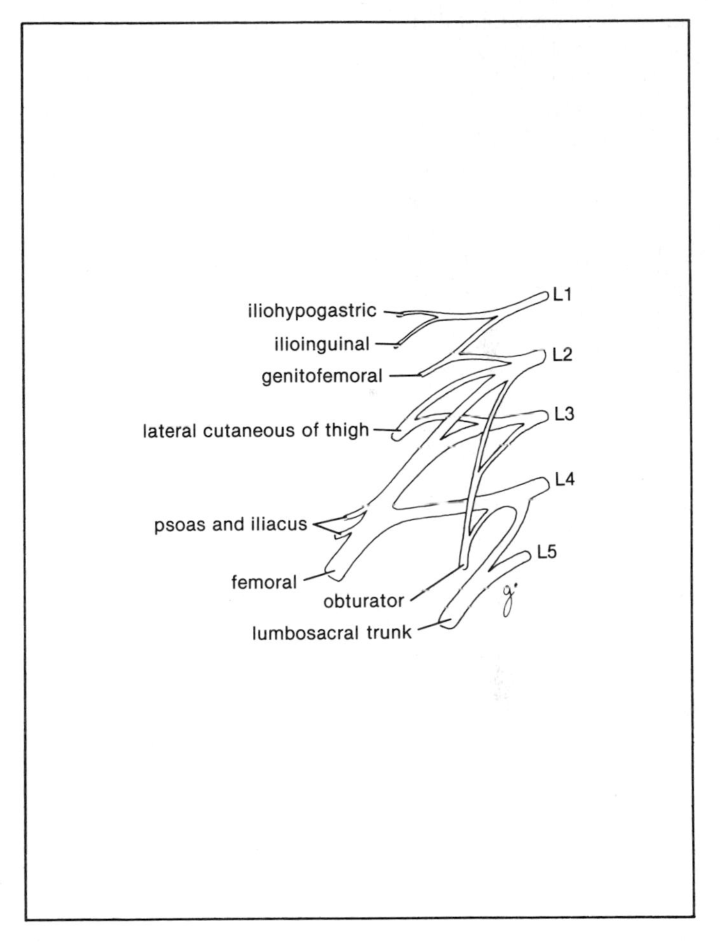

Figure 7–3. Lumbar plexus of nerves.

ders of the muscle and from its anterior surface. The courses of the more important nerves and their branches are as follows:

Femoral Nerve. This is the largest branch of the lumbar plexus and arises from L2, 3, and 4 lumbar nerves. It emerges from the lateral border of the psoas muscle in the abdomen and descends between the psoas and the iliacus muscles. It enters the thigh behind the inguinal ligament and lies lateral to the femoral vessels and the femoral sheath. About 4 cm (1½ inches) below the inguinal ligament it terminates by dividing into anterior and posterior divisions.

Branches of the Femoral Nerve in the Abdomen. Muscular branches to iliacus.

Branches of the Femoral Nerve in the Thigh

1. **Cutaneous branches.** The **medial cutaneous nerve of the thigh** supplies the skin on the medial side of the thigh. The **intermediate cutaneous nerve of the thigh** supplies the skin on the anterior surface of the thigh. The **saphenous nerve** descends into the adductor canal crossing the femoral artery on its anterior surface. The nerve emerges on the medial side of the knee joint between the tendons of the sartorius and gracilis. It descends along the medial side of the leg in company with the great saphenous vein. It passes **anterior** to the medial malleous and along the medial border of the foot as far as the ball of the big toe.
2. **Muscular branches.** Sartorius, pectineus, quadriceps femoris. The muscular branch to the rectus femoris also supplies the hip joint; the branches to the three vasti muscles also supply the knee joint.

Obturator Nerve. This nerve arises from the lumbar plexus (L2, 3, and 4). It emerges on the medial border of the psoas muscle within the abdomen. The nerve descends and crosses the pelvic brim anterior to the sacroiliac joint and posterior to the common iliac vessels. It then runs downward and forward on the lateral wall of the pelvis in the angle between the internal and external iliac vessels. Here it is accompanied by the obturator vessels. On reaching the obturator canal (upper part of the obturator foramen) it divides into anterior and posterior divisions.

Branches of the Obturator Nerve

1. **Parietal peritoneum.** Sensory fibers to parietal peritoneum on lateral wall of pelvis.
2. **Anterior division.** Descends into thigh anterior to obturator externus and adductor brevis.
   a. **Muscular branches.** Gracilis, adductor brevis, adductor longus and sometimes pectineus.
   b. **Sensory branch** to skin on medial side of thigh.
   c. **Articular branch.** Hip joint.
3. **Posterior division.** Descends through obturator externus and passes behind adductor brevis and in front of adductor magnus.
   a. **Muscular branches.** Obturator externus, adductor magnus (adductor part), and sometimes adductor brevis.
   b. **Articular branch.** Knee joint.

A summary of the branches of the lumbar plexus and their distribution is shown in *Table 7–4.*

Sacral Plexus

The sacral plexus is formed from the anterior rami of the fourth and fifth lumbar nerves and the anterior rami of the first, second, third, and fourth sacral nerves (*Fig. 7–4*). Note that the contribution from the fourth lumbar nerve joins the fifth lumbar nerve to form the **lumbosacral trunk.** The lumbosacral trunk passes down into the pelvis and joins the sacral nerves as they emerge from the anterior sacral foramina. Parasympathetic branches arise from the second, third, and fourth sacral nerves to form the **pelvic splanchnic nerves.**

TABLE 7–4. BRANCHES OF THE LUMBAR PLEXUS AND THEIR DISTRIBUTION

Branches	Distribution
Iliohypogastric nerve	External oblique, internal oblique, transversus abdominis muscles of anterolateral abdominal wall; skin over lower anterolateral abdominal wall and buttock
Ilioinguinal nerve	External oblique, internal oblique, transversus abdominis muscles of anterolateral abdominal wall; skin of upper medial aspect of thigh, root of penis and scrotum in the male, mons pubis and labia majora in the female.
Lateral cutaneous nerve of thigh	Skin of anterior and lateral surfaces of the thigh
Genitofemoral nerve (L1, 2)	Cremaster muscle in scrotum of the male; skin over anterior surface of thigh. Nervous pathway for cremasteric reflex
Femoral nerve (L2, 3, 4)	Iliacus muscle, pectineus, sartorius, quadriceps femoris, by medial and intermediate cutaneous branches to the skin of the anterior surface of the thigh and by saphenous branch to skin of medial side of leg and medial side of foot. Articular branches. Hip and knee joints
Obturator nerve (L2, 3, 4)	Gracilis, adductor brevis, adductor longus, obturator externus, ? pectineus, adductor magnus (adductor portion), and skin on medial surface of thigh. Articular branches. Hip and knee joints
Segmental branches	Quadratus lumborum and psoas muscles

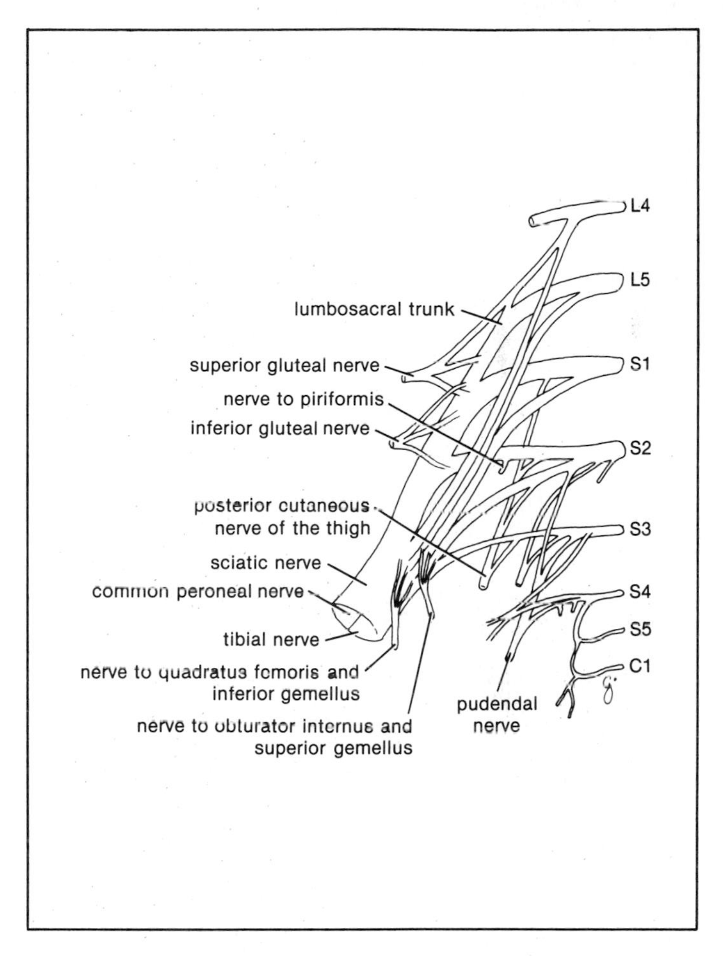

Figure 7–4. Sacral plexus of nerves.

The courses of the most important nerves and their branches are described below.

Sciatic Nerve. The sciatic nerve (*Fig. 7–4*) is the largest nerve in the body and arises from L4, 5, S1, 2, and 3. It passes out of the pelvis into the gluteal region through the greater sciatic foramen. The nerve appears below the piriformis muscle and curves downward and laterally lying on the superior gemellus, the obturator internus, the inferior gemellus, and the quadratus femoris to reach the back of the adductor magnus muscle. It is covered by the gluteus maximus and the long head of the biceps femoris. In the lower third of the thigh (occasionally at a higher level) it ends by dividing into the tibial and common peroneal nerves.

Branches of the Sciatic Nerve

MUSCULAR BRANCHES. Biceps femoris (long head), semitendinosus, semimembranosus and the hamstring part of adductor magnus.

ARTICULAR BRANCHES. Hip joint.

TIBIAL NERVE. The tibial nerve runs downward through the popliteal fossa crossing the popliteal artery from the lateral to the medial side. It then passes deep to the gastrocnemius and soleus muscles and descends to the interval between the medial malleolus and the heel. It is covered here by the flexor retinaculum and divides into the medial and lateral plantar nerves.

BRANCHES OF THE TIBIAL NERVE

1. **Cutaneous branches.** The **sural nerve** is usually joined by the sural communicating branch of the common peroneal nerve. It supplies the skin of the calf and the back of the leg and accompanies the small saphenous vein behind the lateral malleolus to supply the skin of the lateral border of the foot and the lateral side of the little toe. The **medial**

calcaneal branch supplies the skin over the medial surface of the heel.

2. **Muscular branches.** Gastrocnemius, plantaris, soleus, popliteus, flexor digitorum longus, flexor hallucis longus, tibialis posterior.
3. **Articular branches.** Knee joint, ankle joint.
4. **Medial plantar nerve.** Runs forward deep to the abductor hallucis with the medial plantar artery. Branches:
   a. **Cutaneous.** Medial part of sole and plantar digital nerves supply medial three and one half toes and the nail beds.
   b. **Muscular.** Abductor hallucis, flexor digitorum brevis, flexor hallucis brevis, first lumbrical.
5. **Lateral plantar nerve.** Runs forward deep to the abductor hallucis and flexor digitorum brevis in company with the lateral plantar artery. Branches:
   a. **Cutaneous.** Skin of lateral part of sole and plantar digital branches to lateral one and one-half toes and the nail beds.
   b. **Muscular.** Flexor digitorum accessorius, abductor digiti minimi, flexor digiti minimii brevis, adductor hallucis, interosseous muscles, second, third, and fourth lumbricals.

COMMON PERONEAL NERVE. This nerve runs downward through the popliteal fossa following the tendon of the biceps femoris muscle. It then passes laterally around the neck of the fibula and pierces the peroneus longus muscle and divides into the superficial peroneal nerve and the deep peroneal nerve.

BRANCHES OF COMMON PERONEAL NERVE

1. **Cutaneous branches. Sural communicating branch** joins sural nerve. **Lateral cutaneous nerve of the calf** supplies the skin on the lateral side of the back of the leg.
2. **Muscular branch.** Short head of the biceps femoris.
3. **Articular branches.** Knee joint.

4. **Superficial peroneal nerve.** It descends between the peroneus longus and brevis and then between the peroneus brevis and the extensor digitorum longus. Branches:
   a. **Cutaneous.** Medial and lateral branches are distributed to skin on front of leg and dorsum of foot except cleft between big and second toes.
   b. **Muscular.** Peroneus longus and brevis.
5. **Deep peroneal nerve.** It leaves the peroneus longus and enters the anterior compartment of the leg. It descends deep to the extensor digitorum longus muscle and is accompanied by the anterior tibial vessels. The nerve enters the dorsum of the foot on the lateral side of the dorsalis pedis artery and divides into medial and lateral terminal branches. Branches:
   a. **Cutaneous.** Adjacent side of big and second toes.
   b. **Muscular.** Tibialis anterior, extensor digitorum longus, peroneus tertius, extensor hallucis longus, extensor digitorum brevis.
   c. **Articular.** Ankle joint, tarsal joints.

Pudendal Nerve. The pudendal nerve arises from S2, 3, and 4 spinal nerves. It leaves the pelvis through the greater sciatic foramen and after a brief course in the gluteal region, enters the perineum through the lesser sciatic foramen. The nerve passes forward in the **pudendal canal** with the internal pudendal vessels on the lateral wall of the ischiorectal fossa. It ends by dividing into the perineal nerve and the dorsal nerve of the penis (clitoris).

Branches of the Pudendal Nerve

1. **Inferior rectal nerve.** Crosses ischiorectal fossa to anal canal. Branches:
   a. **Sensory** to mucous membrane of lower half of anal canal and perianal skin.
   b. **Muscular.** External anal sphincter.
2. **Perineal nerve.** Branches:
   a. **Cutaneous. Posterior scrotal (labial) nerves** to posterior surface of scrotum or labia majora.

b. **Muscular.** Superficial and deep transverse perineal muscles, bulbospongiosus, ischiocavernosus, sphincter urethrae, levator ani.

3. **Dorsal nerve of the penis (clitoris)** supplies the skin and deeper structures of the penis (clitoris).

The branches of the sacral plexus and their distribution are shown in *Table 7–5*.

Coccygeal Nerve

The anterior ramus of the coccygeal nerve joins with branches of the anterior rami of the fourth and fifth sacral nerves to form the **coccygeal plexus.** Branches from this plexus supply the skin over the coccyx.

TABLE 7–5. BRANCHES OF THE SACRAL PLEXUS AND THEIR DISTRIBUTION

Branches	Distribution
Superior gluteal nerve	Gluteus medius, gluteus minimus and tensor fasciae latae muscles
Inferior gluteal nerve	Gluteus maximus muscle
Nerve to piriformis	Piriformis muscle
Nerve to obturator internus	Obturator internus and superior gemellus muscles
Nerve to quadratus femoris	Quadratus femoris and inferior gemellus muscles
Perforating cutaneous nerve	Skin over medial aspect of buttock
Posterior cutaneous nerve of thigh	Skin over posterior surface of thigh and popliteal fossa, also skin over lower part of buttock, scrotum or labium majus

(continued)

TABLE 7–5. (cont.)

Branches	Distribution
Sciatic nerve (L4, 5, S1, 2, 3)	
Tibial portion	Hamstring muscles (semitendinosus, semimembranosus, biceps femoris—long head, adductor magnus—hamstring part), gastrocnemius, soleus, plantaris, popliteus, tibialis posterior, flexor digitorum longus, flexor hallucis longus and via medial and lateral plantar branches to muscles of sole of foot. Sural branch supplies skin on lateral side of leg and foot
Common peroneal portion	Biceps femoris muscle (short head) and via deep peroneal branch–tibialis anterior, extensor hallucis longus, extensor digitorum longus, peroneus tertius and extensor digitorum brevis muscles. Skin over cleft between first and second toes. The superficial peroneal branch supplies the peroneus longus and brevis muscles and skin over lower third of anterior surface of leg and dorsum of foot
Pudendal nerve	Muscles of perineum including the external anal sphincter, mucous membrane of lower half of anal canal, perianal skin, skin of penis, scrotum, clitoris, labia majora and minora

DERMATOMES

The skin covering the whole body is supplied segmentally by spinal nerves. The area of skin supplied by a single spinal nerve, and therefore a single segment of the spinal cord, is called a **dermatome.** On the trunk, adjacent dermatomes overlap considerably; therefore, to produce a region of complete anesthesia at least three adjacent spinal nerves have to be sectioned. In the upper and lower limbs the arrangement of the dermatomes is more complicated, and this is due to the embryological changes that take place as the limbs grow out from the body wall. The skin of the face is largely supplied by the three divisions of the trigeminal cranial nerve. Here there is little or no overlap of the areas of skin supplied by each of the divisions.

8

Autonomic Part of the Nervous System

The autonomic part of the nervous system is concerned with the innervation of involuntary structures such as the heart, the smooth muscles, and the glands. It is distributed throughout the central and peripheral nervous systems. The autonomic system is divided into two parts, the **sympathetic** and the **parasympathetic**, and in both parts there are afferent and efferent nerve fibers.

SYMPATHETIC PART OF THE AUTONOMIC NERVOUS SYSTEM

The activities of the sympathetic part of the autonomic system prepare the body for an emergency. The heart rate is increased, the peripheral blood vessels are constricted, and the blood pressure is raised. There is a redistribution of blood so that it leaves the skin and the gastrointestinal tract and passes to the brain, the heart, and the skeletal muscles. The sympathetic nerves inhibit peristalsis of the gastrointestinal tract and close the sphincters.

Efferent Nerve Fibers

The lateral gray columns (horns) of the spinal cord from the first thoracic segment to the second lumbar segment (sometimes third lumbar segment) possess the cell bodies of the sympathetic connector neurons (*Fig. 8–1*). The myelinated axons of these cells leave the cord in the anterior nerve roots and pass via the **white rami communicantes** to the **paravertebral ganglia** of the **sympathetic trunk**. Once these fibers (preganglionic) reach the ganglia in the sympathetic trunk, they are distributed as follows:

1. They synapse with an excitor neuron in the ganglion (*Figs. 8–1 and 8–2*). The gap between the two neurons is bridged by a neurotransmitter called **acetylcholine**. The postganglionic nonmyelinated axons leave the ganglion and pass to the thoracic spinal nerves as **gray rami communicantes**. These axons are distributed in the branches of the spinal nerves to smooth muscle in the blood vessel walls, the sweat glands, and the arrector pili muscles of the skin.
2. They may travel up in the sympathetic trunk to synapse in ganglia in the cervical region (*Fig. 8–2*). Here again, the postganglionic nerve fibers pass via gray rami communicantes to join the cervical spinal nerves. Many of the preganglionic fibers entering the lower part of the sympathetic trunk from the lower thoracic and upper two lumbar segments of the spinal cord travel down to synapse in ganglia in the lower lumbar and sacral regions. Here again, the postganglionic nerve fibers pass via gray rami communicantes to join the lumbar, sacral, and coccygeal spinal nerves (*Fig. 8–2*).
3. They may pass through the ganglia of the sympathetic trunk without synapsing. These myelinated fibers leave the sympathetic trunk as the **greater splanchnic** (T5 to T9), **lesser splanchnic** (T10 to T11), and **lowest or least splanchnic nerves** (T12). The greater and lesser splanchnic nerves pierce the diaphragm and synapse with excitor cells in ganglia of the **celiac plexus** (*Fig. 8–2*). The lowest splanchnic nerve

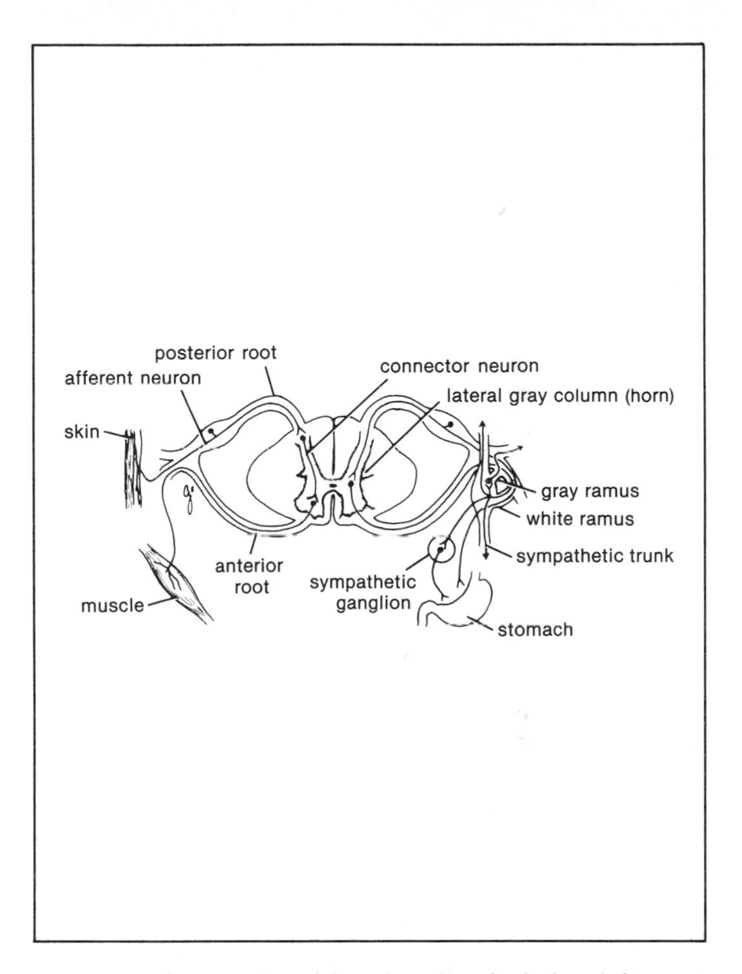

Figure 8–1. Cross section of thoracic region of spinal cord. Arrangement of somatic part of nervous system (**left**); Arrangement of autonomic part of nervous system (**right**).

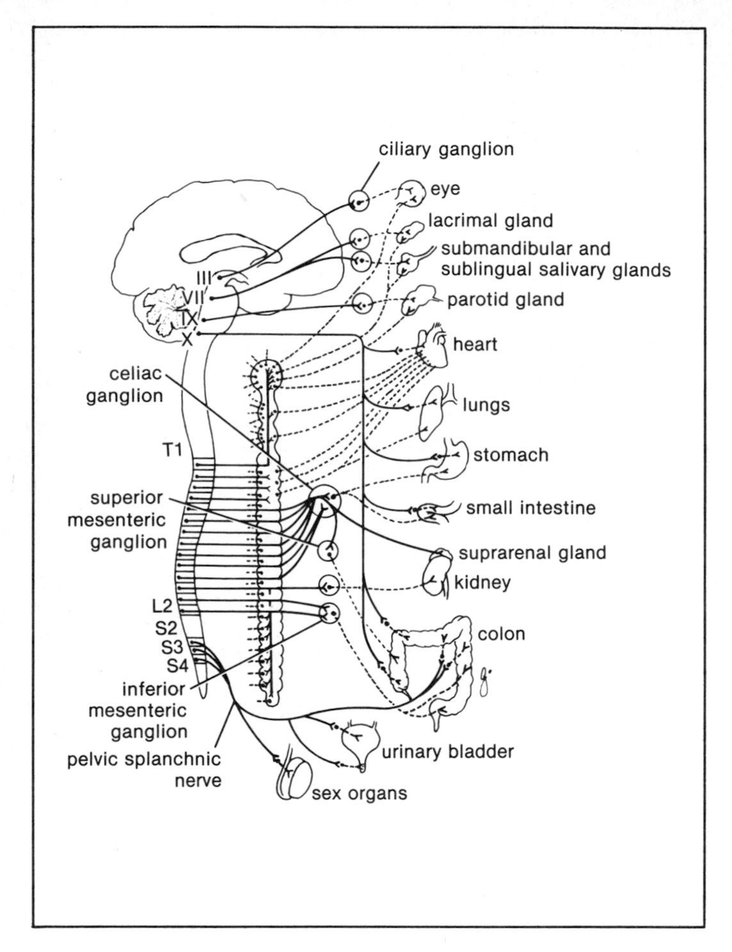

Figure 8–2. Efferent part of the autonomic nervous system. Preganglionic fibers are shown as solid lines; postganglionic fibers are shown as interrupted lines.

pierces the diaphragm and synapses with excitor cells in the **renal plexus**. The postganglionic fibers arising from the cells in the peripheral plexus are distributed to the smooth muscle and glands of the viscera. A few preganglionic axons, traveling in the greater splanchnic nerve, end directly on the cells of the suprarenal medulla.

Afferent Nerve Fibers

The afferent myelinated nerve fibers travel from the viscera through the sympathetic ganglia without synapsing. They pass to the spinal nerve via white rami communicantes and reach their cell bodies in the posterior root ganglion of the corresponding spinal nerve (*Fig. 8–1*). The central axons then enter the spinal cord and may form the afferent component of a local reflex arc or ascend to higher centers.

Sympathetic Trunks

The sympathetic trunks are two ganglionated nerve trunks that extend the whole length of the vertebral column (*Fig. 8–2*). In the neck each trunk has three ganglia, in the thorax there are eleven or twelve ganglia, in the lumbar region four or five ganglia, and in the pelvis four or five ganglia. In the neck the trunks lie anterior to the transverse processes of the cervical vertebrae, in the thorax they are anterior to the heads of the ribs, in the abdomen they are anterolateral to the bodies of the lumbar vertebrae, and in the pelvis they are anterior to the sacrum. Below the two trunks end by joining together to form a single ganglion, the **ganglion impar**.

Cervical Part of the Sympathetic Trunk. The cervical part of the sympathetic trunk extends upward to the base of the skull and below to the first rib, where it becomes continuous with the thoracic part of the sympathetic trunk. The trunk lies behind the carotid sheath. The cervical part of the trunk possesses three ganglia, the superior, middle, and inferior cervical ganglia (*Fig. 8–2*).

Superior Cervical Ganglion. Lies immediately below the skull.

BRANCHES

1. The **internal carotid nerve**, consisting of postganglionic nerve fibers, accompanies the internal carotid artery into the carotid canal. It divides into branches that form the **internal carotid plexus** around the artery.
2. **Gray rami communicantes**, consisting of postganglionic nerve fibers, pass to the upper four cervical spinal nerves.
3. **Arterial branches** to the common and external carotid arteries.
4. **Cranial nerve branches** to the ninth, tenth, and twelfth cranial nerves.
5. **Pharyngeal branches,** which join the pharyngeal branches of the ninth and tenth cranial nerves to form the **pharyngeal plexus**.
6. **Superior cardiac branch** descends in the neck to end in the cardiac plexus in the thorax.

Middle Cervical Ganglion. Lies at the level of the cricoid cartilage.

BRANCHES

1. **Gray rami communicantes** to the fifth and sixth cervical spinal nerves.
2. **Thyroid branches** that reach the thyroid gland along the inferior thyroid artery.
3. **Middle cardiac branch** descends in the neck to join the cardiac plexus in the thorax.

Inferior Cervical Ganglion. In the majority of individuals this is fused with the first thoracic ganglion to form the **stellate ganglion**. It is located between the transverse process of the seventh cervical vertebra and the neck of the first rib.

BRANCHES

1. **Gray rami communicantes** to the seventh and eighth cervical nerves.
2. **Arterial branches** to the subclavian and vertebral arteries.
3. **Inferior cardiac branch** that descends to join the cardiac plexus.

The portion of the sympathetic trunk that connects the middle cervical ganglion to the inferior or stellate ganglion is in the form of two or more nerve bundles. The anterior bundle crosses anterior to the first part of the subclavian artery and is known as the **ansa subclavia**.

Thoracic Part of the Sympathetic Trunk. The thoracic part of the sympathetic trunk is the most laterally placed structure in the mediastinum and runs downward on the heads of the ribs. It leaves the thorax by passing behind the medial arcuate ligament to become continuous with the lumbar part of the sympathetic trunk. The trunk has 11 or 12 segmentally arranged ganglia (*Fig. 8-2*).

Branches

1. **White rami communicantes** join each ganglion to a corresponding thoracic spinal nerve. A white ramus contains preganglionic nerve fibers and afferent sensory nerve fibers.
2. **Gray rami communicantes** join each ganglion to a corresponding thoracic spinal nerve. A gray ramus contains postganglionic nerve fibers.
3. The first five ganglia give postganglionic fibers to the heart, aorta, lungs, and esophagus.
4. The lower seven ganglia give preganglionic fibers and receive afferent sensory fibers that are grouped together to form splanchnic nerves and supply abdominal viscera. They enter the abdomen by piercing the crura of the diaphragm. The **greater splanchnic nerve** arises from ganglia 5 to 9, the

lesser splanchnic nerve arises from ganglia 10 and 11, and the **lowest splanchnic nerve** arises from the last thoracic ganglion.

Lumbar Part of the Sympathetic Trunk. The lumbar part of the sympathetic trunk runs downward on the bodies of the lumbar vertebrae along the medial border of the psoas muscle. It is continuous above with the thoracic part of the trunk by passing behind the medial arcuate ligament. It becomes continuous below with the pelvic part of the trunk by passing behind the common iliac vessels. The right trunk lies posterior to the right margin of the inferior vena cava and the left trunk lies close to the left margin of the aorta. The trunk has four or five segmentally arranged ganglia (*Fig. 8–2*).

Branches

1. **White rami communicantes** join the first two ganglia to the first two lumbar spinal nerves. A white ramus contains preganglionic nerve fibers and afferent sensory nerve fibers.
2. **Gray rami communicantes** join each ganglion to a corresponding lumbar spinal nerve. A gray ramus contains postganglionic nerve fibers.
3. Fibers pass medially to the sympathetic plexuses on the abdominal aorta and its branches. (These plexuses also receive fibers from splanchnic nerves and the vagus.)
4. Fibers pass downward and medially anterior to the commom iliac vessels to enter the pelvis, where, together with branches from sympathetic nerves in front of the aorta, they form a large bundle of nerve fibers called the **superior hypogastric plexus.**

Pelvic Part of the Sympathetic Trunk. The pelvic part of the sympathetic trunk is continuous above, behind the common iliac vessels, and with the lumbar part of the trunk. Below, the two trunks come together in front of the coccyx to form the **ganglion impar**. Each trunk descends behind the rectum in front of the sacrum, medial to the anterior sacral foramina. The sympathetic trunk has four or five segmentally arranged ganglia (*Fig. 8–2*).

Branches

1. Gray rami communicantes to the sacral and coccygeal spinal nerves.
2. Fibers that join the hypogastric plexuses.

LARGE AUTONOMIC PLEXUSES

Large collections of sympathetic and parasympathetic efferent nerve fibers and their associated ganglia, together with visceral afferent fibers, form autonomic nerve plexuses in the thorax, the abdomen, and the pelvis. Branches from these plexuses innervate the viscera. In the abdomen the plexuses are associated with the aorta and its branches and the subdivisions of these autonomic plexuses are named according to the branch of the aorta along which they are lying. The following plexuses are summarized here: cardiac, pulmonary, celiac, superior mesenteric, inferior mesenteric, aortic, and superior and inferior hypogastric plexus.

Cardiac Plexuses

Superficial and deep cardiac plexuses are found at the base of the heart. The sympathetic fibers cause cardiac acceleration and dilatation of the coronary arteries. The parasympathetic fibers from the vagus cause slowing of the heart and constriction of the coronary arteries.

Pulmonary Plexuses

Anterior and posterior pulmonary plexuses are present at the hilus of each lung. The sympathetic fibers cause bronchodilatation and vasoconstriction. The parasympathetic (vagal) fibers cause bronchoconstriction, vasodilatation, and bronchial secretion.

Celiac Plexus

A network of fibers and ganglion cells form two celiac ganglia around the celiac artery (*Fig. 8–2*). The network receives sympathetic fibers from the greater and lesser splanchnic nerves and parasympathetic fibers from the vagus. Branches are distributed along branches of the celiac artery.

Superior Mesenteric Plexus

The superior mesenteric plexus is continuous above with the celiac plexus and consists of a network of fibers and ganglion cells (*Fig. 8–2*). It also receives a branch from the right vagus. Branches are distributed along the branches of superior mesenteric artery.

Inferior Mesenteric Plexus

The inferior mesenteric plexus is continuous with the aortic plexus and receives branches from the lumbar part of the sympathetic trunk. It also receives parasympathetic branches from the pelvic splanchnic nerve.

Aortic Plexus

The aortic plexus is a continuous plexus around the abdominal part of the aorta. Regional concentrations around the origins of the celiac, renal, superior mesenteric, and inferior mesenteric arteries form the celiac plexus, renal plexus, superior mesenteric plexus, and inferior mesenteric plexus, respectively.

Superior Hypogastric Plexus

The superior hypogastric plexus is located at the bifurcation of the abdominal part of the aorta and extends down in front of the promontary of the sacrum. It is formed as a continuation of the aortic plexus and from branches of the third and fourth lumbar ganglia. It divides inferiorly into right and left hypogastric nerves. It contains sympathetic and sacral parasympathetic nerve fibers.

Inferior Hypogastric Plexus

The inferior hypogastric plexus lies in the extraperitoneal connective tissue on each side of the rectum. Each plexus is formed from a hypogastric nerve and from the pelvic splanchnic nerve. Numerous branches pass to the pelvic viscera. Note that parasympathetic fibers from the pelvic splanchnic nerve ascend through the hypogastric nerve to the superior hypogastric plexus and the inferior mesenteric plexus to be distributed to the splenic flexure, and descending and sigmoid parts of the colon.

PARASYMPATHETIC PART OF THE AUTONOMIC NERVOUS SYSTEM

The activities of the parasympathetic part of the autonomic system are directed toward conserving and restoring energy. The heart rate is slowed, the peristalsis of the gastrointestinal tract is increased, and the glandular activity is augmented; the sphincters are opened.

Efferent Nerve Fibers

The craniosacral outflow of the efferent part of the autonomic nervous system is located in the nuclei of the oculomotor (third), the facial (seventh), the glossopharyngeal (ninth), and the vagus (tenth) cranial nerves and the second, third, and fourth sacral segments of the spinal cord (*Fig. 8–2*). The parasympathetic nucleus of the oculomotor nerve is often called the **Edinger-Westphal nucleus**, those of the facial, the **lacrimatory and superior salivary nuclei**, that of the glossopharyngeal, the **inferior salivatory nucleus**, and that of the vagus, the **dorsal nucleus of the vagus**. The axons of these connector nerve cells are myelinated and emerge from the brain within the cranial nerves.

The sacral connector nerve cells give rise to myelinated axons that leave the spinal cord in the anterior nerve roots of the corresponding spinal nerves. They then leave the sacral nerves and form the **pelvic splanchnic nerves** (*Fig. 8–2*).

The efferent fibers of the craniosacral outflow are preganglionic and synapse in peripheral ganglia located close to the viscera they innervate. Here again, acetylcholine is the neurotransmitter. The cranial parasympathetic ganglia are the **ciliary**, the **pterygopalatine**, the **submandibular**, and the **otic**. In certain locations, the ganglion cells are placed in nerve plexuses, such as the **cardiac plexus**, the **pulmonary plexus**, the **myenteric plexus (Auerbach's plexus)** and the **mucosal plexus (Meissner's plexus)**; the last two plexuses are associated with the gastrointestinal tract. The pelvic splanchnic nerves synapse in ganglia in the hypogastric plexuses. The postganglionic parasympathetic fibers are nonmyelinated and short in length.

Afferent Nerve Fibers

The afferent myelinated fibers leave the viscera and reach their cell bodies in the sensory ganglia of cranial nerves or in posterior root ganglia of the sacral spinal nerves. The central axons then enter the central nervous system and form local reflex arcs or ascend to higher centers.

HIGHER CONTROL OF THE AUTONOMIC NERVOUS SYSTEM

The sympathetic outflow in the spinal cord (T1 to L2(3)) and the parasympathetic craniosacral outflow (cranial nuclei 3, 7, 9, and 10, spinal cord S2, 3, and 4) are controlled by the hypothalamus. The hypothalamus appears to integrate the autonomic and neuroendocrine systems, thus preserving body homeostasis. The hypothalamus receives information from all parts of the nervous system and afferent input from the viscera. At the same time it is receiving information concerning the hormonal levels of the blood. All this input is integrated within the hypothalamus and appropriate information is transmitted to the lower autonomic centers in the brainstem and spinal cord by descending tracts. In a similar manner, **releasing factors** or **release inhibiting factors** are liberated into the circulation to exert direct hormonal control of tissues, or to control endocrine glands, so that they in turn will influence the activities of tissues.

For important anatomic, physiologic, and pharmacologic differences between the sympathetic and parasympathetic parts of the autonomic system, *see Table 8–1.*

TABLE 8–1. ANATOMIC, PHYSIOLOGIC, AND PHARMACOLOGIC DIFFERENCES AND SIMILARITIES BETWEEN THE SYMPATHETIC AND PARASYMPATHETIC PARTS OF THE AUTONOMIC NERVOUS SYSTEM

Feature	Sympathetic	Parasympathetic
Action	Prepares body for emergency	Conserves and restores energy
Outflow	T1–L2 (3)	CN 3, 7, 9, 10; S2, 3, 4
Preganglionic fibers	Myelinated	Myelinated
Ganglia	Paravertebral (sympathetic trunks); prevertebral (e.g., celiac, superior mesenteric)	Small ganglia close to viscera (e.g., otic, ciliary) or ganglion cells in plexuses (e.g., cardiac, pulmonary)
Neurotransmitter within ganglia	Acetylcholine	Acetylcholine
Postganglionic fibers	Long—nonmyelinated	Short—nonmyelinated
Characteristic activity	Widespread due to many postganglionic fibers and liberation of epinephrine and norepinephrine from suprarenal medulla	Discrete action with few postganglionic fibers
Neurotransmitter at post-ganglionic endings	Norepinephrine at most endings and acetylcholine at few endings (sweat glands)	Acetylcholine at all endings
Higher control	Hypothalamus	Hypothalamus

9

Sense Organs

EYE

The eyeball is recessed into a bony socket called the **orbital cavity**. Only about one-sixth of the eye is exposed, the remainder is embedded in orbital fat. The eyeball is spherical in shape, with the segment of a smaller sphere, the cornea, superimposed anteriorly (*Fig. 9–1*). The eyeball is composed of three coats that, from without inward, are: (1) fibrous, (2) vascular pigmented, and (3) nervous.

Coats of the Eyeball

Fibrous Coat. The fibrous coat is made up of a posterior opaque part, the sclera, and an anterior transparent part, the cornea (*Fig. 9–1*).

The **sclera** is white in color and composed of dense fibrous tissue. Posteriorly, it is pierced by the optic nerve and is fused with the dural sheath of that nerve. The **lamina cribrosa** is the area of the sclera that is pierced by the nerve fibers of the optic nerve. The sclera is also pierced by the ciliary arteries, veins, and nerves. The sclera is directly continuous in front with the cornea at the **corneoscleral junction**, or **limbus**.

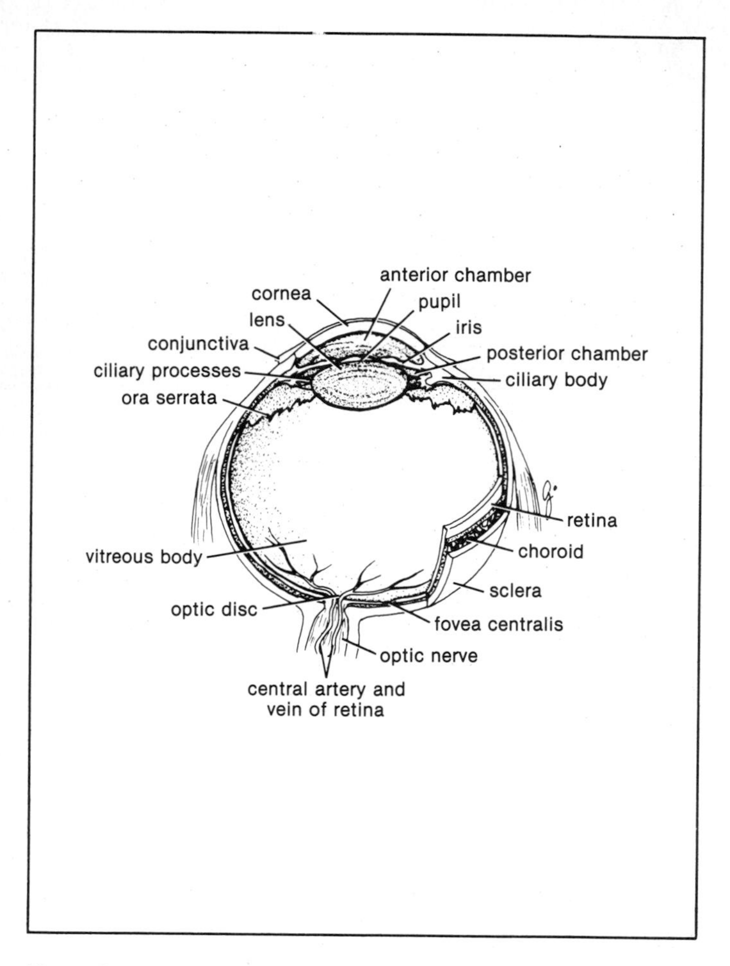

Figure 9–1. Horizontal section through the eyeball and the optic nerve.

The **cornea** forms the anterior part of the fibrous coat. It is transparent and largely responsible for the refraction (bending) of the light entering the eye.

Vascular Pigmented Coat. The vascular pigmented coat consists, from back to front, of the choroid, the ciliary body, and the iris (*Fig. 9–1*).

Choroid. The **choroid** is composed of an outer pigmented layer and an inner vascular layer.

Ciliary Body. The **ciliary body** lies behind the peripheral margin of the iris and is continuous posteriorly with the choroid (*Fig. 9–2*). It is a complete ring that runs around inside the sclera. It is composed of the ciliary processes and the ciliary muscle. The **ciliary processes** are radially arranged folds, or ridges, that are connected to the suspensory ligaments of the lens. The **ciliary muscle**, which is responsible for changing the shape of the lens, is composed of meridional and circular fibers of smooth muscle. The meridional fibers run backward from the region of the corneoscleral junction to the ciliary processes. The circular fibers run around the eyeball within the ciliary body.

NERVE SUPPLY OF THE CILIARY MUSCLE. Parasympathetic fibers within the oculomotor nerve synapse in the ciliary ganglion. Postganglionic fibers reach the eyeball in the short ciliary nerves.

ACTION OF THE CILIARY MUSCLE. The ciliary muscle pulls the ciliary body forward. This slackens the suspensory ligaments and the elastic lens becomes more convex and the refractive power is increased.

Iris. The **iris** is the colored part of the eye and consists of a thin, contractile, pigmented sheet with a central hole, the **pupil** (*Fig. 9–2*). The iris is suspended in the aqueous humor between the cornea and the lens. The periphery of the iris joins the wall of the eyeball at the **iridocorneal** angle and is attached to the anterior surface of the ciliary body. It divides the space between the

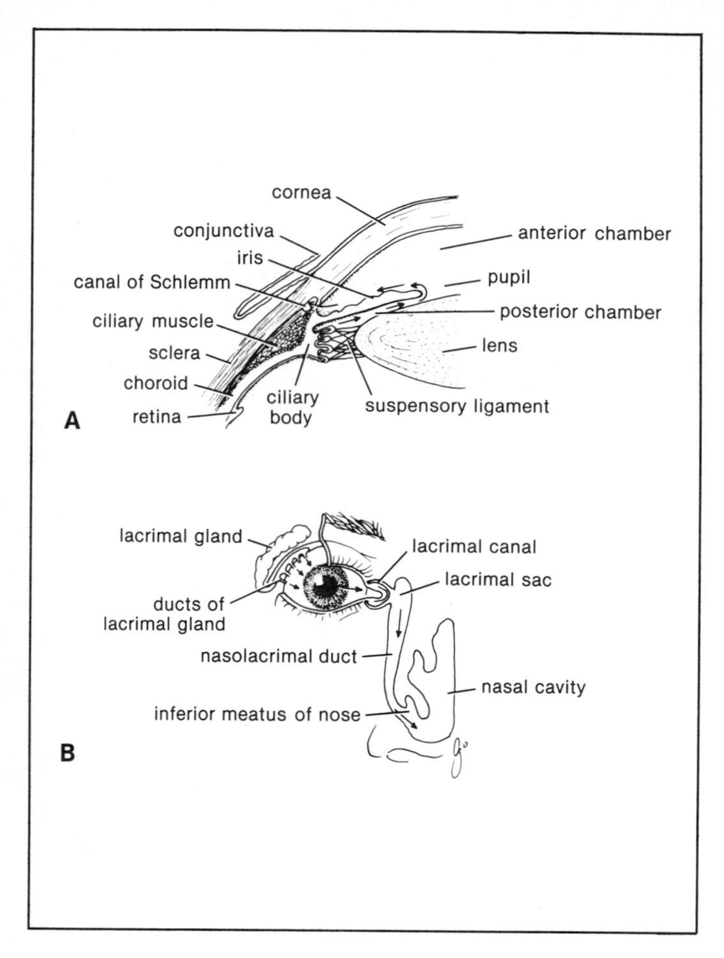

Figure 9–2. A. Structures seen in a section through the anterior portion of the eye; the arrows indicate the pathway taken by the aqueous humor during its circulation. **B.** The lacrimal gland, its ducts, and the lacrimal passages. The arrows indicate the path taken by the tears as they pass across the front of the eye to drain into the nose.

cornea and the lens into an **anterior and a posterior chamber** (*Fig. 9–2*). The function of the iris is to control the amount of light entering the eyeball.

The smooth muscle fibers of the iris consist of circular and radiating fibers. The circular fibers form the **sphincter pupillae** and are arranged around the margin of the pupil. The radial fibers form the **dilator pupillae.**

NERVE SUPPLY OF THE SMOOTH MUSCLE OF THE IRIS. The sphincter pupillae is supplied by parasympathetic fibers from the oculomotor nerve. After synapsing in the ciliary ganglion, the postganglionic fibers reach the eyeball in the short ciliary nerves. The dilator pupillae is supplied by sympathetic nerve fibers that reach the eyeball in the long ciliary nerves.

ACTION OF THE IRIS. The sphincter pupillae constricts the pupil in the presence of bright light and during accommodation. The dilator pupillae dilates the pupil in the presence of light of low intensity or in the presence of excessive sympathetic activity.

Nervous Coat: The Retina. The **retina** is the internal layer of the eyeball and consists of an anterior one-quarter that is insensitive and a posterior three-quarters that is the photoreceptor organ. The retina consists of an outer **pigmented layer** and an inner **nervous layer**. Its outer surface is in contact with the choroid, and its inner surface is in contact with the vitreous body (*Fig. 9–1*).

The posterior receptive part of the retina extends forward from the optic nerve to a point just posterior to the ciliary body. Here the nervous tissues of the retina end and its anterior edge forms a wavy line called the **ora serrata** (*Fig. 9–1*). The anterior nonreceptive part of the retina consists merely of pigment cells, with a deeper layer of columnar epithelium. This anterior part of the retina covers the ciliary processes and the back of the iris.

At the center of the posterior part of the retina is an oval yellowish area, the **macula lutea**. It has a central depression, the **fovea centralis** (*Fig. 9–1*), which is the area of the retina for the most distinct vision. The reason for this is that here the photoreceptors are present in greater numbers than elsewhere in the retina, and the inner layers of the retina are displaced laterally (hence the depression) allowing an unobstructed passage of the light rays to the layer of photoreceptors.

The optic nerve leaves the retina about 3 mm to the medial side of the macula lutea by the optic disc. The **optic disc** is slightly depressed at its center, and here it is pierced by the **central artery of the retina** (*Fig. 9–1*). At the optic disc there is a complete absence of **rods** and **cones**, so that it is insensitive to light and is referred to as the "**blind spot.**" On ophthalmoscopic examination, the optic disc is seen to be pale pink in color, much paler than the surrounding retina.

Contents of the Eyeball

The contents of the eyeball consist of the refractive media: the aqueous humor, the vitreous body, and the lens.

Aqueous Humor. The aqueous humor is a clear fluid that fills the anterior and posterior chambers of the eyeball (*Fig. 9–2*) and nourishes the cornea and the lens. The aqueous humor is kept under constant pressure and it assists in the maintenance of the normal shape of the eyeball. The aqueous humor is believed to be a secretion or transudate from the ciliary processes, from which it enters the posterior chamber. It then flows into the anterior chamber through the pupil and is drained away through the **canal of Schlemm** at the iridocorneal angle into a venous sinus. Obstruction to the draining of the aqueous humor results in a rise in intraocular pressure called **glaucoma**.

Vitreous Body The vitreous body fills the eyeball behind the lens (*Fig. 9–1*). Its function is to maintain the shape of the eyeball and assist in the support of the lens and the retina; in this latter function it keeps the retina against the wall of the eyeball. It differs from aqueous humor in that it cannot be replaced if lost as the result of injury because it is not continuously being formed. The vitreous body is a transparent gel enclosed by the **vitreous mem-**

brane. The **hyaloid canal** is a narrow channel that runs through the vitreous body from the optic disc to the posterior surface of the lens; in the fetus, it is filled by the hyaloid artery, which disappears before birth.

In front, in the region of the margin of the lens, the vitreous membrane is thickened and consists of two layers. The posterior layer covers the vitreous body; the anterior layer consists of a series of delicate, radially arranged fibers. Collectively, the fibers form the **suspensory ligament of the lens**; they are attached laterally to the ciliary processes and centrally to the capsule of the lens in the region of the equator (*Fig. 9–2*).

Lens. The lens (*Fig. 9–1*) is a transparent, biconvex structure enclosed in a transparent capsule. It is situated behind the iris and in front of the vitreous body and is encircled by the ciliary processes. The lens has considerable flexibility. The lens consists of (1) an elastic **capsule**, (2) a **cuboidal epithelium**, which is confined to the anterior surface of the lens, and (3) **lens fibers**.

The elastic capsule envelops the entire lens and is under tension, so that the lens is constantly endeavoring to assume a globular rather than a disc shape. The equatorial region, or circumference, of the lens is attached to the ciliary processes of the ciliary body by the suspensory ligament. The pull of the radiating fibers of the suspensory ligament tends to keep the elastic lens flattened, so that the eye may be focused on distant objects.

Accommodation of the Lens. To accommodate the eye for close objects, the ciliary muscle contracts and pulls the ciliary body forward and inward, so that the radiating fibers of the suspensory ligament are relaxed. This process allows the elastic lens to assume nearly globular shape. With advancing age the lens becomes denser and less elastic, and, as a result, the ability to accommodate is lessened (presbyopia).

Constriction of the Pupil During Accommodation of the Lens. To insure that the light rays pass through the thickest central part of the lens during accommodation for near objects, the sphincter pupillae muscle contracts so that the pupil becomes smaller.

Convergence of the Eyes During Accommodation of the Lens. In the human eye the retinae of both eyes focus on only one set of objects (single binocular vision). When an object moves from a distance toward an individual the eyes converge so that one continues to see a single object and not two objects. Convergence of the eyes is brought about by the coordinated contraction of the medial rectus muscles (see below).

Eye Muscles. The eye muscles may be divided into two groups, extrinsic and intrinsic. The extrinsic muscles are six in number and run from the posterior wall of the orbital cavity to the eyeball. They are the superior rectus, the inferior rectus, the medial rectus, the lateral rectus, and the superior and inferior oblique muscles. The intrinsic muscles are the ciliary muscle and the constrictor and dilator pupillae muscles of the iris (*see pp. 325, 327*).

Because the superior and inferior recti are inserted on the medial side of the vertical axis of the eyeball, they not only raise and depress the cornea respectively, but **rotate it medially**. For the superior rectus muscle to raise the cornea directly upward, the inferior oblique must assist, and for the inferior rectus to depress the cornea directly downward, the superior oblique must assist. Note that the tendon of the superior oblique muscle passes through a fibrocartilaginous pulley (trochlea) attached to the frontal bone. The tendon now turns backward and laterally, and is inserted into the sclera beneath the superior rectus.

The origins, insertions, nerve supply, and actions of the muscles of the eyeball are summarized in *Table 9–1.*

Fascial Sheath of the Eyeball

The fascial sheath is a thin membrane that surrounds the eyeball from the optic nerve to the corneoscleral junction. It separates the eyeball from the orbital fat and is perforated by the tendons of insertion of the orbital muscles.

TABLE 9–1. MUSCLES OF THE EYEBALL AND EYELIDS

Name of Muscle	Origin	Insertion	Nerve Supply	Action
Extrinsic Muscles of Eyeball (Striated skeletal muscle)				
Superior rectus	Tendinous ring on posterior wall of orbital cavity	Superior surface of eyeball just posterior to corneoscleral junction	Oculomotor nerve (third cranial nerve)	Raises cornea upward and medially
Inferior rectus	Tendinous ring on posterior wall of orbital cavity	Inferior surface of eyeball just posterior to corneoscleral junction	Oculomotor nerve (third cranial nerve)	Depresses cornea downward and medially
Medial rectus	Tendinous ring on posterior wall of orbital cavity	Medial surface of eyeball just posterior to corneoscleral junction	Oculomotor nerve (third cranial nerve)	Rotates eyeball so that cornea looks medially

(continued)

TABLE 9–1. (cont.)

Name of Muscle	Origin	Insertion	Nerve Supply	Action
Extrinsic Muscles of Eyeball (cont.)				
Lateral rectus	Tendinous ring on posterior wall of orbital cavity	Lateral surface of eyeball just posterior to corneoscleral junction	Abducent nerve (sixth cranial nerve)	Rotates eyeball so that cornea looks laterally
Superior oblique	Posterior wall of orbital cavity	Having passed through pulley it is attached to superior surface of eyeball beneath superior rectus	Trochlear nerve (fourth cranial nerve)	Rotates eyeball so that cornea looks downward and laterally
Inferior oblique	Floor of orbital cavity	Lateral surface of eyeball deep to lateral rectus	Oculomotor nerve (third cranial nerve)	Rotates eyeball so that cornea looks upward and laterally

Intrinsic Muscles of Eyeball (Smooth muscle)				
Sphincter pupillae of iris			Parasympathetic via oculomotor nerve (third cranial nerve)	Constricts pupil
Dilator pupillae of iris			Sympathetic	Dilates pupil
Ciliary muscle			Parasympathetic via oculomotor nerve (third cranial nerve)	Controls shape of lens in accommodation—makes it more globular
Muscles of Eyelids				
Orbicularis oculi (sphincter)	*(See Table 3–2)*			
Levator palpebrae superioris	Back of orbital cavity	Anterior surface of superior tarsal plate	Striated muscle oculomotor nerve; smooth muscle sympathetic	Raises upper lid

Eyelids (Palpebrae)

The eyelids are thin, movable folds placed in front of the eye and by their closure protect it from injury and excessive light. The upper eyelid is larger and more mobile than the lower, and they meet each other at the **medial and lateral angles** (canthi). The **palpebral fissure** is the elliptical opening between the eyelids and is the entrance into the conjunctival sac. The superficial surface of the eyelids is covered by skin, and the deep surface is covered by the **conjunctiva**. The **eyelashes**, which are short, curved hairs, are present on the free edges of the eyelids. The lateral angle of the palpebral fissure is more acute than the medial and lies directly in contact with the eyeball. The more rounded medial angle is separated from the eyeball by a small space, the **lacus lacrimalis**, in the center of which is a small, reddish-yellow elevation, the **caruncula lacrimalis**. A reddish semilunar fold, called the **plica semilunaris**, lies on the lateral side of the caruncle.

Near the medial angle of the eye, there is a small elevation, the **papilla lacrimalis**. On the summit of the papilla is a small hole, the **punctum lacrimale**, which leads into the **canaliculus lacrimalis**. The papilla lacrimalis projects into the lacus, and the punctum and canaliculus serve to carry tears down into the nose (*Fig. 9–2*).

The **conjunctiva** is a thin mucous membrane that lines the eyelids and is reflected at the **superior and inferior fornices** onto the anterior surface of the eyeball. The upper lateral part of the superior fornix is pierced by the ducts of the lacrimal gland (*Fig. 9–2*).

The fibrous framework of the eyelids is formed by a membranous sheet, the **orbital septum**. This is attached to the periosteum of the orbital margin. The orbital septum is thickened at the margins of the lids to form the **tarsal plates**. These are crescent-shaped laminae of dense fibrous tissue that give support to the eyelids. The ends of the plates are attached to the orbital margins by the **lateral and medial palpebral ligaments**. The superficial surface of the tarsal plates and the orbital septum are covered by the **orbicularis oculi muscle**. The aponeurosis of insertion of the **levator**

palpebrae superioris muscle pierces the orbital septum, to reach the anterior surface of the superior tarsal plate and the skin (for the orbicularis oculi muscle, *see Table 3–2*). The origin, insertion, nerve supply, and action of the levator palpebrae superioris muscle are summarized in *Table 9–1*.

Movements of the Eyelids. The position of the eyelids at rest depends on the tone of the orbicularis oculi and the levator palpebrae superioris muscles and the position of the eyeball. The eyelids are closed by the contraction of the orbicularis oculi and the relaxation of the levator palpebrae superioris muscles. The eye is opened by the levator palpebrae superioris raising the upper lid.

Lacrimal Apparatus

The lacrimal apparatus is concerned with the formation of fluid, the **tears**, and its drainage into the nose. Tears bathe the cornea and keep it moist and prevent this transparent structure from becoming opaque due to exposure to air. Tears also cleanse the anterior surface of the eye by washing away foreign particles and killing bacteria.

The **lacrimal gland** secretes the tears and is situated above the eyeball in the anterior and upper part of the orbit (*Fig. 9–2*). Several ducts open from the gland into the superior part of the conjunctival sac.

Nerve Supply of the Lacrimal Gland. The parasympathetic secretomotor nerve supply from the facial nerve (seventh cranial nerve).

The **tears** circulate across the cornea and accumulate in the **lacus lacrimalis**. They clean, lubricate, and moisten the cornea. Tears enter the canaliculi that pass medially and open into the **lacrimal sac**. This is the upper blind end of the nasolacrimal duct.

The **nasolacrimal duct** emerges from the lower end of the sac and descends downward and laterally in a bony canal to open into the inferior meatus of the nose (*Fig. 9–2*). The opening is guarded by a fold of mucuous membrane that prevents air from being forced up the duct into the lacrimal sac on blowing the nose.

EAR

The ear consists of two functional entities that are anatomically closely related: (1) hearing and (2) balance. The ear may be divided into the external ear, the middle ear or tympanic cavity, and the internal ear or labyrinth.

External Ear

The external ear consists of the auricle and the external auditory meatus (*Fig. 9–3*). The **auricle** has a characteristic shape and serves to collect air vibrations. It consists of a thin plate of elastic cartilage covered by skin (for muscles of the auricle, *see Table 3–1*).

The **external auditory meatus** is a curved tube that leads from the auricle to the tympanic membrane (eardrum) and serves to conduct sound waves from the auricle to the tympanic membrane. In the adult the external auditory meatus measures about 2.5 cm (1 inch) long and may be straightened for the insertion of an otoscope by pulling the auricle upward and backward. The framework of the outer third of the meatus is elastic cartilage, whereas the inner two-thirds is bone. The meatus is lined by skin, and its outer third is provided with **hairs**, **sebaceous**, and **ceruminous glands**.

Tympanic Cavity (Middle Ear)

The tympanic cavity lies between the external and internal parts of the ear (*Fig. 9–3*). It is separated from the external part of the ear by the eardrum. The tympanic cavity is an air-filled cavity in the petrous part of the temporal bone and is lined with mucous membrane. It contains the auditory ossicles, whose function is to transmit the vibrations of the tympanic membrane (eardrum) to the perilymph of the internal ear. It is a narrow, slitlike cavity

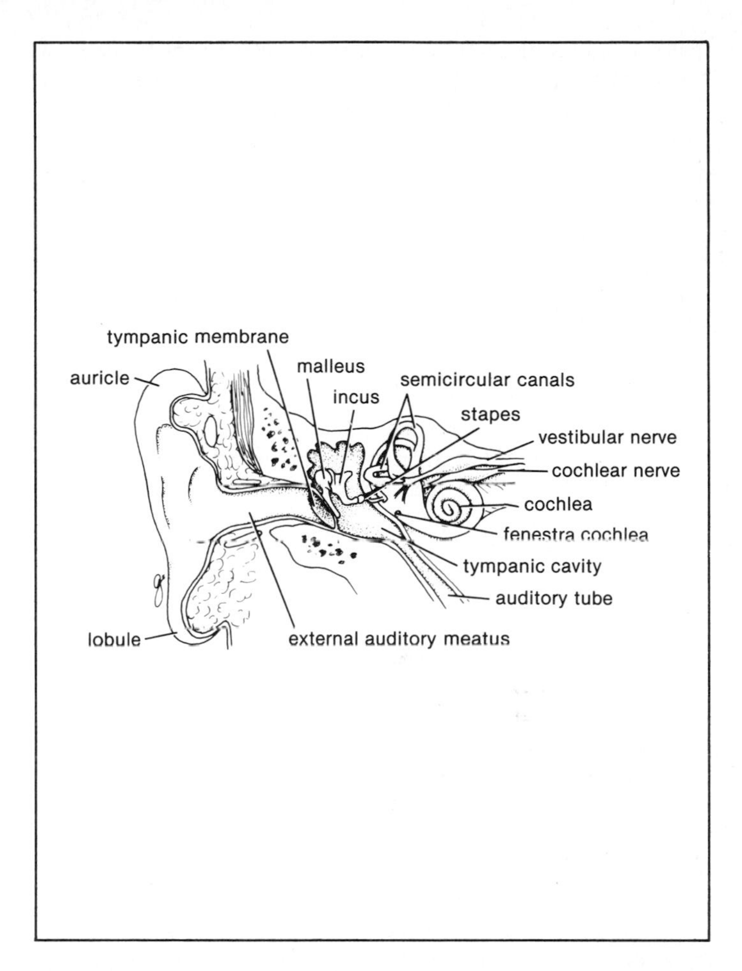

Figure 9–3. External, middle, and internal portions of the ear, viewed from the front.

that communicates in front through the **auditory tube** with the nasopharynx and behind with the mastoid antrum. The roof of the tympanic cavity is formed by a thin plate of bone, the **tegmen tympani,** which separates the cavity from the meninges and the temporal lobe of the brain in the middle cranial fossa. The floor is also formed by a thin plate of bone that separates the tympanic cavity from the upper part of the internal jugular vein. The middle ear is thus located very close to the brain and a large vein.

Tympanic Membrane. This membrane, also known as the eardrum, is a thin, fibrous membrane that is pearly-gray in color. It is covered on the outer surface with stratified squamous epithelium and on its inner surface with low columnar epithelium. The membrane is obliquely placed and is concave laterally (*Fig. 9–3*). At the depth of the concavity is a small depression, the **umbo,** produced by the tip of the handle of the malleus. When the membrane is illuminated through an otoscope, the concavity produces a "cone of light," which radiates anteriorly and inferiorly from the umbo. The tympanic membrane is circular and measures about 1 cm in diameter. It possesses two folds, termed the **anterior and posterior malleolar folds**, which pass to the lateral process of the malleus. The small triangular area on the tympanic membrane that is bounded by the folds is slack and is called the **pars flaccida**. The remainder of the membrane is tense and is called the **pars tensa**. The handle of the malleus is bound down to the inner surface of the tympanic membrane by the mucous membrane (*Fig. 9–4*).

The tympanic membrane is extremely sensitive to pain and is innervated on its outer surface by the trigeminal and vagus nerves.

The **medial wall** of the tympanic cavity is formed by the lateral wall of the inner ear. The greater part of the wall shows a rounded projection called the **promontory**, which is due to the underlying first turn of the cochlea. Above and behind the promontory is the **fenestra vestibuli** (oval window), which is oval in shape and closed by the footpiece, or base, of a small bone called the stapes. On the medial side of the window is the perilymph of the scala vestibuli of the internal ear. Below the posterior end of the promontory lies the **fenestra cochleae**, which is round in shape and closed by the **secondary tympanic membrane**. On the medial side of the window is the perilymph of the blind end of the scala tympani (*see p. 343*). The **prominence of the facial canal,** for the facial nerve, runs backward above the promontory and on reaching the posterior wall, it curves downward behind the pyramid; it forms a ridge on the medial wall of the aditus of the mastoid antrum.

The **posterior wall** has an opening, the **aditus to the mastoid antrum**. Below this is a hollow conical projection, the **pyramid**, from whose apex emerges the tendon of **stapedius muscle**. The **anterior wall** has the canal for the **tensor tympani** muscle and the **auditory tube.**

Auditory Ossicles. The auditory ossicles are small bones called the malleus, incus, and stapes (*Fig. 9–4*). The **malleus** (hammer) is the largest ossicle and possesses a head, a neck, a long process or handle, an anterior process, and a lateral process. The **incus** (anvil) possesses a large body and two processes. The **stapes** (stirrup) has a head, a neck, two limbs, and a base. The **two limbs** diverge from the neck and are attached to the oval **base**. The edge of the base is attached to the margin of the fenestra vestibuli by a ring of fibrous tissue, the **anular ligament**.

Movements of the Ossicles. The ossicles connect the tympanic membrane or eardrum to the perilymph of the internal ear. The bones articulate with one another by means of synovial joints and they are held in position by ligaments attached to the walls of the tympanic cavity. The movements of the tympanic membrane are conveyed from the malleus to the incus and then from the incus to the stapes (*Fig. 9–4*). When the base of the stapes is pushed medially in the fenestra vestibuli the motion is communicated to the perilymph in the scala vestibuli. Liquid being incompressible, the perilymph causes an outward bulging of the secondary tympanic membrane in the fenestra cochlea at the lower end of the scala tympani. Note that during the passage of the vibrations from the tympanic membrane to the perilymph via the small ossicles, the leverage increases at a rate of 1.3 to 1.

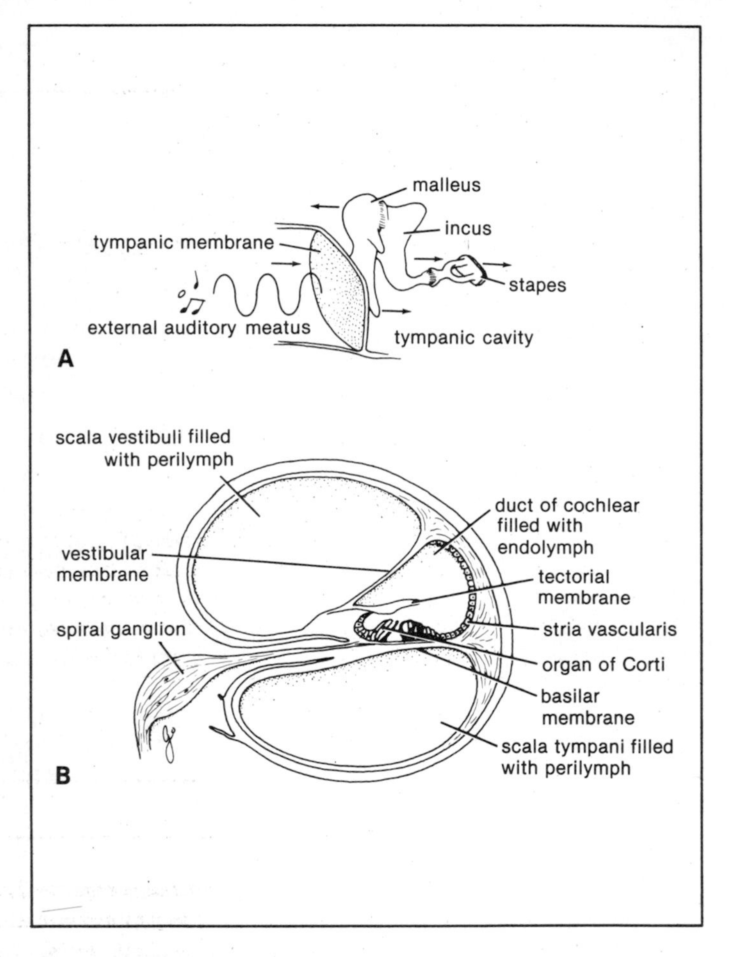

Figure 9–4. A. Air vibrations passing down the external auditory meatus causing the tympanic membrane to move medially; the head of the malleus and incus move laterally, and the long process of the incus with the stapes, moves medially. **B.** Section of the cochlea showing the different parts; note the position of the organ of Corti.

Muscles of the Ossicles. The two muscles of the tympanic cavity are the tensor tympani and the stapedius. The origins, insertions, nerve supply, and actions of these muscles are summarized in *Table 9–2*.

Auditory Tube

The auditory tube extends from the anterior wall of the tympanic cavity to the nasal pharynx (*Fig. 9–3*). Its posterior third is bony, and its anterior two-thirds is cartilaginous. It serves to equalize air pressures in the tympanic cavity and nasal pharynx. Swallowing or yawning causes the tube to dilate, which helps the passage of air. Unfortunately the tube provides a passageway for infection to travel from the pharynx to the middle ear.

Mastoid Antrum

The mastoid antrum lies behind the tympanic cavity in the petrous part of the temporal bone. It is rounded in shape and may be as large as a 1 cm in diameter. It communicates with the tympanic cavity through the posterior tympanic wall.

Mastoid Air Cells

The mastoid air cells are a series of communicating cavities within the mastoid process, which are continuous above with the antrum and the tympanic cavity. They are lined with mucous membrane.

Internal Ear, or Labyrinth

The labyrinth is situated in the petrous part of the temporal bone, medial to the middle ear (*Fig. 9–3*). It consists of (1) the bony labyrinth, comprising a series of cavities within the bone and (2) the membranous labyrinth, comprising a series of membranous sacs and ducts contained within the bony labyrinth.

Bony Labyrinth. The bony labyrinth consists of three parts: (1) the vestibule, (2) the semicircular canals, and (3) the cochlea. These are cavities situated in the substance of dense bone. They

TABLE 9–2. MUSCLES OF THE MIDDLE EAR

Name of Muscle	Origin	Insertion	Nerve Supply	Action
Tensor tympani	Cartilaginous wall of auditory tube	Handle of malleus	Mandibular division of trigeminal nerve (fifth cranial nerve)	Reflexly dampens down vibrations of malleus by making tympanic membrane more tense. Protects ear from excessively loud sounds
Stapedius	Pyramid (bony projection on posterior wall of middle ear)	Neck of stapes	Facial nerve (seventh cranial nerve)	Reflexly dampens down movements of stapes. Protects ear from excessively loud sounds

are lined by endosteum and contain a clear fluid, the **perilymph,** in which is suspended the membranous labyrinth.

The **vestibule** is the central part of the bony labyrinth and lies posterior to the cochlea and anterior to the semicircular canals. In its lateral wall is the **fenestra vestibuli** (oval window), which is closed by the base of the stapes and its anular ligament, and the **fenestra cochleae** (round window), which is closed by the **secondary tympanic membrane**. Lodged within the vestibule are the saccule and utricle of the membranous labyrinth.

The **semicircular canals** open into the posterior part of the vestibule and are three in number, **superior** (anterior), **posterior**, and **lateral** (*Fig. 9–3*). Each canal has a swelling at one end called the **ampulla**. The canals open into the vestibule by five orifices, one of which is common to two of the canals. Lodged within the canals are the **semicircular ducts**.

The superior (anterior) semicircular canal is vertical and anterior in position and lies at right angles to the posterior semicircular canal, which is vertical and posterior in position. The lateral canal is set in a horizontal position. The two lateral canals of the right and left ears are on the same plane, and the superior canal of one side is parallel to the posterior canal of the other.

The **cochlea** resembles a snail shell (*Fig. 9–3*). It opens into the anterior part of the vestibule. Basically, it consists of a central pillar, the **modiolus,** around which a hollow bony tube makes two-and-one-half spiral turns. The first basal turn of the cochlea is responsible for the promontory seen on the medial wall of the tympanic cavity.

The modiolus has a broad base, which is situated at the bottom of the internal acoustic meatus (see interior of skull, *p. 8*). It is perforated by branches of the cochlear nerve. A spiral ledge, the **spiral lamina,** winds around the modiolus and projects into the interior of the canal and partially divides it. A fibrous membrane called the **basilar membrane** stretches from the free edge of the spiral lamina to the outer bony wall (*Fig. 9–4*).

Membranous Labyrinth. The membranous labyrinth is located within the bony labyrinth. It is filled with endolymph and surrounded by perilymph. It consists of the utricle and saccule, which lie within the bony vestibule; the three semicircular ducts, which lie within the bony semicircular canals; and the duct of the cochlea, which lies within the bony cochlea. All these structures freely communicate with each other.

The **utricle** is the larger of the two vestibular sacs. It is indirectly connected to the saccule and the **ductus endolymphaticus** by the **ductus utriculosaccularis**.

The **saccule** is globular in shape and connected to the utricle, as described previously. The ductus endolymphaticus, after being joined by the ductus utriculosaccularis, passes on to end in a small blind pouch, the **saccus endolymphaticus**.

The **semicircular ducts** are smaller in diameter than the semicircular canals, but they have the same configuration. They open by five orifices into the utricle.

The **duct of the cochlea** is triangular in cross section and is connected to the saccule by the **ductus reuniens**. The highly specialized epithelium that lies on the **basilar membrane** forms the **spiral organ of Corti**, and contains the sensory receptors for hearing (*Fig. 9–4*). In addition to the basilar membrane there is a second membane, called the **vestibular membrane** sometimes called **Reissner's membrane**. This extends across the cochlea from the spiral lamina to the outer wall of the cochlea.

Examination of a cross section of the cochlea (*Fig. 9–4*) shows that the basilar and vestibular membranes divide it up into three distinct portions: a **scala vestibuli** above (filled with perilymph), a **scala tympani** (filled with perilymph), and the **cochlear duct or scala media** (filled with endolymph). The perilymph within the scala vestibuli is separated from the tympanic cavity by the base of the stapes and the anular ligament, at the fenestra vestibuli (oval window). The perilymph in the scala tympani is separated from the tympanic cavity by the secondary tympanic membrane at the fenestra cochleae (round window). The duct of the cochlea is joined to the saccule by the ductus reuniens and ends as a blind sac at the apex of the cochlea. The **helicotrema** is the name given to the point at the apex of the cochlea where the scala vestibuli and scala tympani become continuous with one another.

10

Respiratory System

NOSE

The nose consists of the external nose and the nasal cavity, both of which are divided by a septum into right and left halves.

External Nose

The external nose has a free **tip** and is attached to the forehead by the **root**. The external orifices of the nose are the two **nostrils**, or **nares**. Each nostril is bounded laterally by the **ala** and medially by the **nasal septum**.

The framework of the external nose is made up above by the nasal bones, the frontal processes of the maxillae, and the nasal part of the frontal bone. Below, the framework is formed of plates of hyaline cartilage, which include the **upper** and **lower nasal cartilages** and the **septal cartilage**.

Nasal Cavity

The **nasal cavity** is divided into right and left halves by the **nasal septum**. The cavity extends from the nostrils in front to the posterior nasal apertures or **choanae** behind where the nose opens into the nasopharynx. Each half has a floor, a roof, and a lateral and medial wall.

The **floor** consists of the palatine process of the maxilla and the horizontal plate of the palatine bone.

The **roof** is narrow and is formed by the body of the sphenoid, the cribriform plate of the ethmoid, the frontal bone, and the nasal bone from behind forward.

The **lateral** wall is irregular and has three projections called the **superior, middle,** and **inferior nasal conchae** (*Fig. 10–1*). The space below each concha is referred to as a **meatus.**

The **sphenoethmoidal recess** is a small area of the nose that lies above the superior concha. It receives the opening of the **sphenoidal air sinus.**

The **superior meatus** lies below the superior concha. It receives the openings of the **posterior ethmoidal sinuses** (*Fig. 10–1*).

The **middle meatus** lies below the middle concha. It has on its lateral wall a rounded prominence, the **bulla ethmoidalis,** caused by the bulging of the underlying middle ethmoidal sinuses, which open on its upper border. A curved opening, the **hiatus semilunaris,** lies just below the bulla (*Fig. 10–1*). The anterior end of the hiatus leads into a funnel-shaped channel called the **infundibulum.** The **maxillary sinus** opens into the middle meatus through the hiatus semilunaris. The **frontal sinus** opens into and is continuous with the infundibulum. The **anterior ethmoidal sinuses** also open into the infundibulum.

The middle meatus is continuous in front with a shallow depression called the **atrium.** The atrium is limited above by a ridge, the **agger nasi.** Below and in front of the atrium, and just within the nostril is an area bounded laterally by the ala, called the **vestibule.** This is lined by modified skin and possesses short, curved hairs.

The inferior meatus lies below the inferior concha and receives the opening of the **nasolacrimal duct** (*Fig. 10–1*). The opening is guarded by a fold of mucous membrane that forms an imperfect valve.

The **medial wall** is formed by the nasal septum, which is made up of the **septal cartilage,** the vertical plate of the ethmoid, and

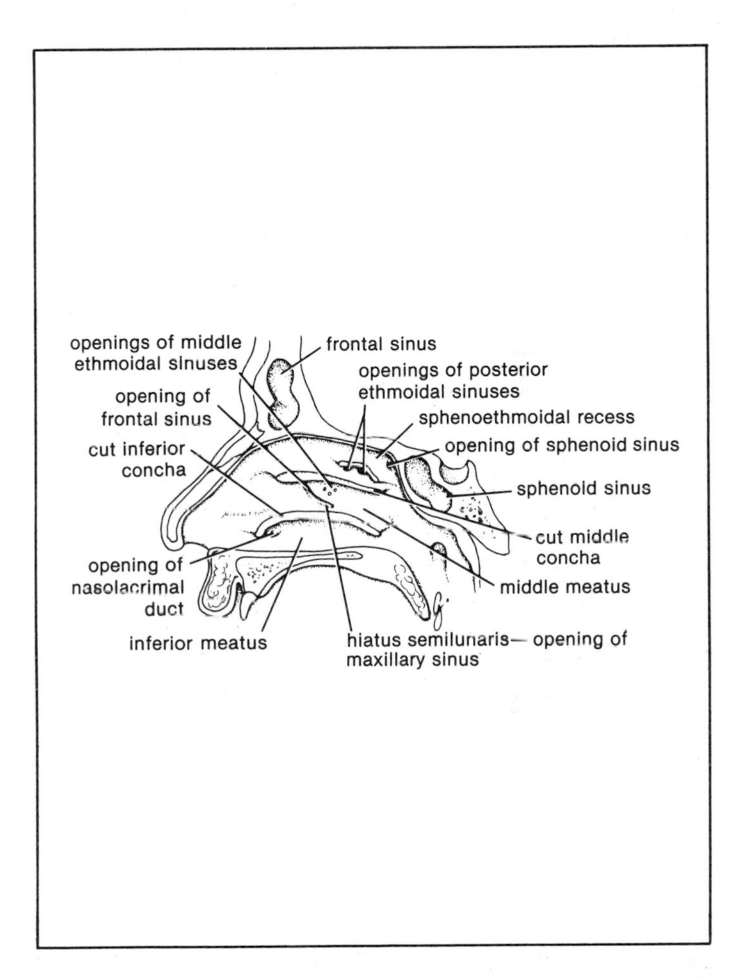

Figure 10–1. Sagittal section through the nose showing the lateral wall. Note the conchae, the meati, and the openings of the paranasal sinuses.

by the vomer. It rarely lies in the median plane, usually curving to one or the other side.

Mucous Membrane of the Nasal Cavity. The mucous membrane lines the nasal cavities and there are three regions (1) vestibular, (2) olfactory, and (3) respiratory.

The vestibule of the nose lies just within the nostrils and is lined with modified skin. The olfactory mucous membrane lines the upper surface of the superior concha and the sphenoethmoidal recess. It also lines the roof and a corresponding area on the nasal septum. The respiratory mucous membrane lines the lower part of the nasal cavities.

Nerve Supply of the Nasal Cavity. The **olfactory nerves** arise from the olfactory cells in the olfactory mucous membrane and pass through the openings in the cribriform plate of the ethmoid to end in the olfactory bulbs.

The remainder of the nasal cavity is supplied by branches of the ophthalmic and maxillary divisions of the trigeminal nerve.

Arterial Supply of the Nasal Cavity. The majority of the arteries are branches of the maxillary artery. The sphenopalatine branch anastomoses with the septal branch of the facial artery in the vestibule. This is a common site for nose bleeds.

Lymphatic Drainage of the Nasal Cavity. Lymph from the vestibule drains into the submandibular nodes; the remainder drains into the deep cervical nodes.

PARANASAL SINUSES

The paranasal sinuses are cavities found in the maxilla, frontal, sphenoid, and ethmoid bones. They are lined with mucoperiosteum and communicate with the nasal cavity through small openings (*Fig. 10–1*).

The **maxillary sinus** is pyramidal in shape and on each side lies within the body of the maxilla. The roof is formed by the floor of the orbit and the floor is formed by the alveolar process. The roots of the first and second premolars and the third molar project upward into the sinus. The maxillary sinus opens into the middle meatus of the nose through the hiatus semilunaris.

The **frontal sinuses**, two in number, are contained within the frontal bone. Rudimentary or absent at birth, they are well developed by the eighth year. They are situated above and behind the medial end of the eyebrow. Each frontal sinus opens into the middle meatus of the nose through the infundibulum.

The **sphenoidal sinuses**, two in number, lie within the body of the sphenoid bone. Each sinus opens into the sphenoethmoidal recess above the superior concha.

The **ethmoidal sinuses** are anterior, middle, and posterior and are contained within the ethmoid bone on each side. They are situated between the nose and the orbit. The anterior sinus opens into the infundibulum, the middle sinus opens into the middle meatus, on or above the bulla ethmoidalis, and the posterior sinus opens into the superior meatus.

LARYNX

The larynx is an organ that provides a protective sphincter at the inlet of the air passages and is responsible for voice production. It is situated below the tongue and hyoid bone and between the great blood vessels of the neck. It opens above into the laryngeal part of the pharynx, and below, it is continuous with the trachea.

The framework of the larynx is formed of cartilages that are held together by ligaments and membranes, and moved by muscles. The larynx is lined with mucous membrane.

Cartilages of the Larynx

Thyroid Cartilage. This is the largest cartilage of the larynx (*Fig. 10–2*). It consists of two laminae of hyaline cartilage that meet

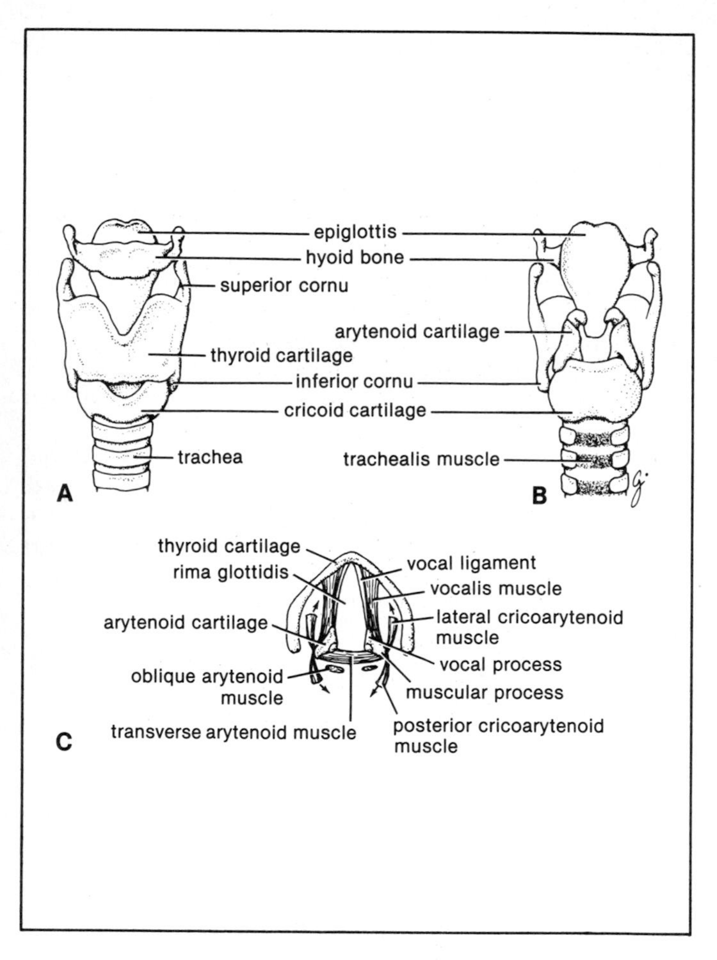

Figure 10–2. The larynx. **A.** From the front. **B.** From behind. **C.** Shows the muscles that move the vocal ligaments.

in the midline in the prominent V angle of the Adam's apple. The posterior border extends upward into a **superior cornu** and downward into an **inferior cornu**. On the outer surface of each lamina is an **oblique line** for the attachment of the sternothyroid, the thyrohyoid, and the inferior constrictor muscles.

Cricoid Cartilage. This is formed of hyaline cartilage and shaped like a signet ring having a broad lamina behind and a shallow arch in front. The cricoid cartilage lies below the thyroid cartilage and on each side of the lateral surface there is a facet for articulation with the inferior cornu of the thyroid cartilage. Posteriorly, the lamina has on its upper border on each side a facet for articulation with the arytenoid cartilage. All these joints are synovial joints.

Arytenoid Cartilages. These are two in number, small and pyramidal in shape. They are situated at the back of the larynx, on the lateral part of the upper border of the lamina of the cricoid cartilage. Each cartilage has an **apex** above that articulates with the corniculate cartilage, a **base** below that articulates with the lamina of the cricoid cartilage, and a **vocal process** that projects forward and gives attachment to the vocal ligament. There is a **muscular process** that projects laterally and gives attachment to the posterior and lateral cricoarytenoid muscles.

Corniculate Cartilages. These are two small conical shaped cartilages that articulate with the arytenoid cartilages. They give attachment to the aryepiglottic folds.

Cuneiform Cartilages. These are two small rod-shaped cartilages that are found in and serve to strengthen the aryepiglottic folds.

Epiglottis. This is a leaf-shaped lamina of elastic cartilage lying behind the root of the tongue. It is attached by its stalk to the back of the thyroid cartilage. The sides of the epiglottis are attached to the arytenoid cartilages by the aryepiglottic folds of mucous membrane. The upper edge of the epiglottis is free. The covering of mucous membrane passes forward onto the posterior surface of the tongue as the **median glossoepiglottic fold**; the depression on each side of the fold is called the **vallecula**. Laterally the mucous membrane passes on to the wall of the pharynx as the **lateral glossoepiglottic folds (pharyngoepiglottic folds)**.

Membranes and Ligaments of the Larynx

Thyrohyoid Membrane. This connects the upper margin of the thyroid cartilage to the upper margin of the hyoid bone. In the midline it is thickened to form the **median thyrohyoid ligament**. The membrane is pierced on each side by the superior laryngeal vessels and the internal laryngeal nerve.

Cricotracheal Ligament. This connects the cricoid cartilage to the first ring of the trachea.

Fibroelastic Membrane. This lies beneath the mucous membrane lining the larynx. The upper part is thin and called the **quadrangular membrane**. It extends between the epiglottis and the arytenoid cartilages, its inferior margin forming the **vestibular ligaments**. The vestibular ligaments form the interior of the vestibular folds. The lower part of the fibroelastic membrane is called the **cricothyroid ligament**. The anterior part of the cricothyroid ligament is thick and connects the cricoid cartilage to the lower margin of the thyroid cartilage. The lateral part of the ligament is thin and is attached below to the upper margin of the cricoid cartilage. The superior margin of the ligament, instead of being attached to the lower margin of the thyroid cartilage, ascends on the medial surface of the thyroid cartilage. The upper margin, composed almost entirely of elastic tissue, forms the important **vocal ligament** on each side. The vocal ligaments form the interior of the **vocal folds** (vocal cords). The anterior end of each vocal ligament is attached to the thyroid cartilage. The posterior end is attached to the vocal process of the arytenoid cartilage.

Cavity of the Larynx

The **inlet** of the larynx looks backward and upward into the laryngeal part of the pharynx. The opening is bounded in front by the epiglottis; laterally, by the aryepiglottic fold of mucous membrane; and posteriorly by the arytenoid cartilages with the corniculate cartilages. The cuneiform cartilage produces a small elevation on the upper border of each aryepiglottic fold.

The cavity of the larynx extends from the inlet to the lower border of the cricoid cartilage where it is continuous with the cavity of the trachea. It is divided into three regions by the presence of an upper and a lower pair of folds of mucous membrane that project into the cavity from the sides of the larynx. The upper pair of folds are fixed and called the **vestibular folds** and the gap between them is named the **rima vestibuli**. The lower pair are mobile and concerned with voice production and are called the **vocal folds**; the gap between them is named the **rima glottidis** or **glottis**.

The three regions of the cavity of the larynx are (1) the upper region called the **vestibule**, situated between the inlet and the vestibular folds, (2) the middle region, situated between the vestibular and the vocal folds, and (3) the lower region, situated between the vocal folds and the lower border of the cricoid cartilage.

The **vestibular fold** on each side is formed of mucous membrane and contains within it the vestibular ligament, which is the thickened lower edge of the quadrangular membrane. The fold is fixed (does not move with respiration) and is pink in color when viewed with a laryngoscope.

The **vocal fold** on each side is formed of mucous membrane and contains within it the vocal ligament that is the thickened upper edge of the cricothyroid ligament. The fold is mobile and moves with respiration and is white in color when viewed with a laryngoscope.

The **sinus of the larynx** is a small recess on each side of the larynx situated between the vestibular and the vocal folds. It is lined with mucous membrane.

The **saccule of the larynx** is a diverticulum that ascends from the sinus of the larynx between the vestibular fold and the thyroid cartilage.

Muscles of the Larynx

The muscles of the larynx may be divided into two groups: (1) extrinsic and (2) intrinsic.

Extrinsic Muscles. These muscles move the larynx up and down during the act of swallowing. Note that many of these muscles are attached to the hyoid bone and the hyoid bone is attached to the thyroid cartilage by the thyrohyoid membrane. It follows that movements of the hyoid bone are accompanied by movements of the larynx.

The **elevator muscles** of the larynx include the digastric, the stylohyoid, the mylohyoid, the geniohyoid, the stylopharyngeus, the salpingopharyngeus, and the palatopharyngeus muscles. The **depressor muscles** of the larynx include the sternothyroid, the sternohyoid, and the omohyoid muscles.

Intrinsic Muscles. These muscles modify the inlet of the larynx and move the vocal folds.

Muscles Controlling the Laryngeal Inlet

Oblique Arytenoid

ORIGIN. Muscular process of arytenoid cartilage.

INSERTION. Apex of opposite arytenoid cartilage. Some of the fibers continue beyond the apex of the arytenoid cartilage and reach the epiglottis via the aryepiglottic fold. The latter fibers are named the **aryepiglottic muscle**.

ACTION. The muscles of the two sides acting together bring the aryepiglottic folds together and narrow the inlet.

NERVE SUPPLY. Recurrent laryngeal nerve.

Thyroepiglottic

ORIGIN. Thyroid cartilage on medial surface.

INSERTION. Lateral margin of epiglottis and aryepiglottic fold.

ACTION. Widens the inlet by pulling the aryepiglottic folds apart.

NERVE SUPPLY. Recurrent laryngeal nerve.

Muscles Controlling the Movements of the Vocal Folds. The vocal folds may be tensed or relaxed; they may be adducted or abducted (*Fig. 10–2*).

Cricothyroid (Tensor)

ORIGIN. Side of cricoid cartilage.

INSERTION. Lower border of lamina and anterior border of inferior cornu of thyroid cartilage.

NERVE SUPPLY. External laryngeal nerve.

ACTION. Tenses vocal ligaments and folds.

Thyroarytenoid (Relaxor)

ORIGIN. Inner surface of angle between laminae of thyroid cartilage.

INSERTION. Anterolateral surface of arytenoid cartilage. Fibers that lie on lateral side of vocal ligament are called **vocalis** muscle.

NERVE SUPPLY. Recurrent laryngeal nerve.

ACTION. Pulls arytenoid cartilage forward toward thyroid cartilage thus relaxing vocal ligament.

Lateral Cricoarytenoid (Adductor)

ORIGIN. Upper border of arch of cricoid cartilage.

INSERTION. Muscular process of arytenoid cartilage.

NERVE SUPPLY. Recurrent laryngeal nerve.

ACTION. Rotates forward muscular process of arytenoid so that vocal process moves medially thus adducting the vocal fold.

Posterior Cricoarytenoid (Abductor)

ORIGIN. Back of cricoid cartilage.

INSERTION. Muscular process of arytenoid cartilage.

NERVE SUPPLY. Recurrent laryngeal nerve.

ACTION. Rotates backward muscular process of arytenoid so that vocal process moves laterally thus abducting the vocal fold.

Transverse Arytenoid

ORIGIN. Back and medial surface of arytenoid cartilage.

INSERTION. Back and medial surface of opposite arytenoid cartilage.

NERVE SUPPLY. Recurrent laryngeal nerve.

ACTION. Approximates arytenoid cartilages thus closing the posterior part of the rima glottidis.

Movements of the Vocal Folds with Respiration. In quiet respiration, the rima glottidis is triangular in shape, with the apex in front. With forced inspiration, the rima glottidis assumes a diamond shape due to the lateral rotation of the arytenoid cartilages.

Nerve Supply of the Larynx

The **sensory nerve supply** to the mucous membrane above the vocal folds is from the internal laryngeal nerve. Below the level of the vocal folds the mucous membrane is supplied by the recurrent laryngeal nerve. The **motor nerve supply** to the intrinsic muscles is the recurrent laryngeal nerve, except for the crico-thyroid muscle which is supplied by the external laryngeal nerve.

Blood Supply and Lymphatic Drainage of the Larynx

The **arterial supply of the larynx** is from branches of the superior and inferior thyroid arteries. The **lymphatic drainage** is into the deep cervical lymph nodes.

TRACHEA AND BRONCHI

Trachea

The trachea is a mobile cartilaginous and membranous tube (*Fig. 10–3*). It begins as a continuation of the larynx at the lower border of the cricoid cartilage at the level of the sixth cervical vertebra. It descends in the midline of the neck and in the thorax ends by dividing into right and left principal (main) bronchi at the level of the sternal angle (opposite the disc between the fourth and fifth thoracic vertebrae). During respiration the bifurcation rises by about one vertebral level during expiration, and during deep inspiration may be lowered as far as the sixth thoracic vertebra.

The trachea is about 11.25 cm (4½ inches) long and 2.5 cm (1 inch) in diameter in the adult. The fibroelastic tube is kept patent by the presence of U-shaped cartilaginous bars (rings) of hyaline cartilage embedded in its wall. The posterior free ends of the cartilage are connected by smooth muscle, the **trachealis muscle**.

Relations in the Neck. *Anteriorly.* Skin, fascia, isthmus of thyroid gland (in front of the second, third, and fourth rings), inferior thyroid veins, jugular arch, thyroidea ima artery (if present), and left brachiocephalic vein in the child, overlapped by the sternothyroid and sternohyoid muscles. *Posteriorly.* Right and left recurrent laryngeal nerves, esophagus. *Laterally.* Lobes of thyroid gland and the carotid sheath and contents.

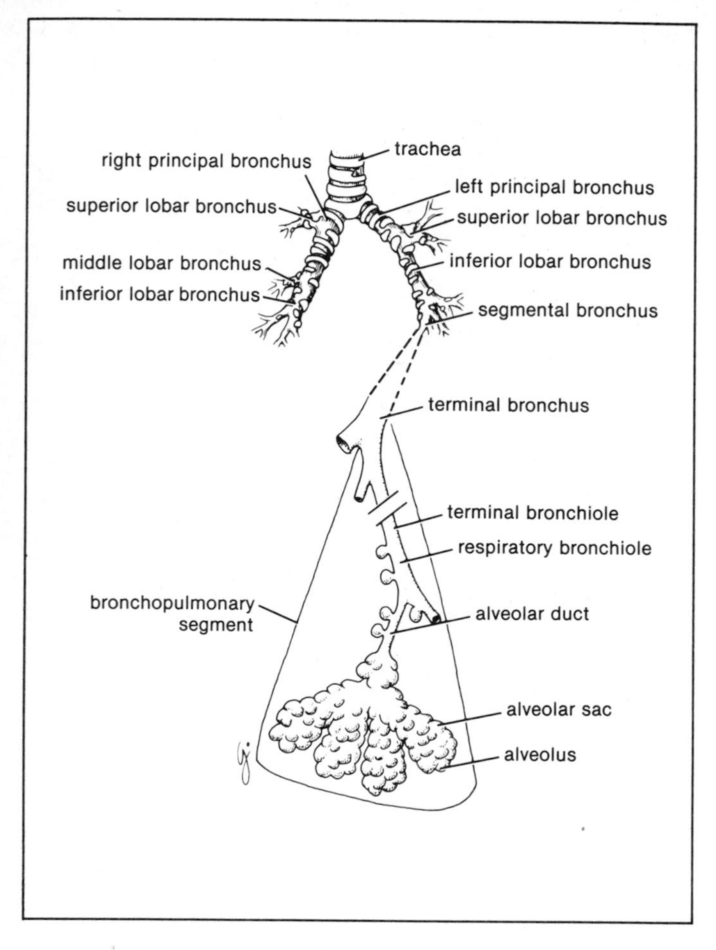

Figure 10–3. Shows the conducting and respiratory parts of the trachea and lungs.

Relations in Superior Mediastinum of Thorax. *Anteriorly.* Sternum, thymus, left brachiocephalic vein, origins of brachiocephalic and left common carotid arteries, arch of aorta. *Posteriorly.* Esophagus and left recurrent laryngeal nerve. *Right Side.* Azygos vein, right vagus nerve, and the pleura. *Left Side.* Arch of aorta, left common carotid, left subclavian arteries, left vagus nerve, left phrenic nerve, and the pleura.

Bronchi

The **right principal (main) bronchus** is wider, shorter, and more vertical than the left and is about 2.5 cm (1 inch) long (*Fig. 10–3*). Before entering the hilum of the right lung it gives off the **superior lobar bronchus**. On entering the hilum it divides into a **middle and inferior lobar bronchus**.

The **left principal (main) bronchus** is narrower, longer, and more horizontal than the right and is about 5 cm (2 inches) long. It passes to the left below the arch of the aorta and in front of the esophasgus. On entering the hilum of the left lung it divides into a **superior and an inferior lobar bronchus**.

PLEURAE AND LUNGS

The thoracic cavity can be divided into a median partition called the **mediastinum**, that lies between the sternum and the vertebral column, and the laterally placed pleurae and lungs.

Pleurae

The pleurae are two serous sacs surrounding and covering the lungs. Each pleura has two parts: (1) a **parietal layer**, which lines the thoracic wall, covers the thoracic surface of the diaphragm and the lateral surface of the mediastinum and (2) a **visceral layer**, which covers the outer surfaces of the lungs and extends into the interlobar fissures. The parietal layer of each pleura becomes continuous with the visceral layer at the **hilum** of each lung. Here, they form a cuff that surrounds the structures entering and leaving the lung at the **lung root**. The **pulmonary ligament** is a loose extension of the cuff below the lung root to allow for movement during respiration.

The **pleural cavity** is a slit-like space that separates the parietal and visceral layers of pleura. It normally contains a small amount of **pleural fluid** that lubricates the apposing pleural surfaces. The **costodiaphragmatic recess** is the lowest area of the pleural cavity into which the lung expands on inspiration.

Surface Marking of the Parietal Pleura. The lines, which indicate the limits of the parietal pleura and where they are reflected close to the body surface, are as follows: The pleura extends upward into the neck from the sternoclavicular joint along a curved line about 2.5 cm (1 inch) above the junction of the medial and intermediate thirds of the clavicle. The **anterior border of the right pleura** descends from the sternoclavicular joint to the midline behind the sternal angle. It then descends to the xiphisternal joint. The **anterior border of the left pleura** has a similar course, but at the level of the fourth costal cartilage it turns laterally from the midline to the lateral margin of the sternum, forming the **cardiac notch**. It then turns downward to the xiphisternal joint. The **lower border of the pleura** on both sides crosses the eighth rib in the midclavicular line and the tenth rib in the midaxillary line, and reaches the twelfth rib adjacent to the vertebral column (for surface marking of visceral pleura and lungs, *see p. 363*).

Lungs

The lungs, two in number, are situated one on each side of the mediastinum. They are therefore separated from each other by the heart and great vessels and other structures in the mediastinum. Each lung is conical in shape and is covered with visceral pleura. It is suspended free in its own pleural cavity, being attached to the mediastinum only by its root.

Each lung has a blunt **apex**, which projects upward for about 2.5 cm (1 inch) above the clavicle; a concave **base** that overlies the diaphragm; a convex **costal surface**, that corresponds to the chest wall; and a concave **mediastinal surface** that is molded to the pericardium and other mediastinal structures. About the middle of this surface is the **hilum**, a depression where the bronchi, vessels, and nerves enter the lung to form the **root**.

The **anterior border** is thin and overlaps the heart; it is here on the left lung that there is a notch, called the **cardiac notch**. The **posterior border** is thick and lies beside the vertebral column.

Lobes and Fissures. The right lung is slightly larger than the left and is divided by the oblique and horizontal fissures into three lobes, the **upper**, **middle**, and **lower lobes**. The **oblique fissure** runs from the inferior border upward and backward across the medial and costal surfaces until it cuts the posterior border about 6.25 cm (2½ inches) below the apex. The **horizontal fissure** runs horizontally across the costal surface at the level of the fourth costal cartilage to meet the oblique fissure in the midaxillary line. The middle lobe is thus a small triangular lobe bounded by the horizontal and oblique fissures.

The left lung is divided by a similar oblique fissure into two lobes, the **upper** and **lower lobes**. There is no horizontal fissure in the left lung.

Bronchopulmonary Segments. Each lobar (secondary) bronchus, which passes to a lobe of the lung, gives off branches called **segmental** (tertiary) **bronchi** (*Fig. 10–3*). Each segmental bronchus passes to a structurally and functionally independent unit of a lung lobe called a **bronchopulmonary segment**. A bronchopulmonary segment of lung tissue is pyramidal in shape, having its apex toward the root of the lung and its base toward the lung surface. Each bronchopulmonary segment is surrounded by connective tissue and in addition to its own bronchus it receives an artery, lymph vessels, and autonomic nerves. The vein from each segment travels in the connective tissue between adjacent bronchopulmonary segments.

The main bronchopulmonary segments are as follows:

Right Lung

Superior lobe	Apical, posterior, anterior
Middle lobe	Lateral, medial
Inferior lobe	Superior (apical), medial basal, anterior basal, lateral basal, posterior basal

Left Lung

Superior lobe	Apical, posterior, anterior, superior lingular, inferior lingular
Inferior lobe	Superior (apical), medial basal, anterior basal, lateral basal, posterior basal

Blood Supply of the Lungs. The bronchi, the connective tissue of the lung, and the visceral pleura receive their blood supply from the bronchial arteries, which are branches of the descending aorta. The bronchial veins, which communicate with the pulmonary veins, drain into the azygos and hemiazygos veins.

The alveoli receive deoxygenated blood from the terminal branches of the pulmonary arteries. The tributaries of the pulmonary veins join together and two pulmonary veins leave each lung root.

Lymph Drainage of the Lungs. The lymph vessels originate in superficial and deep plexuses. The superficial plexus lies beneath the visceral pleura, the deep plexus travels along the bronchi and the pulmonary vessels toward the root of the lung. The vessels from the deep plexus drain into **pulmonary nodes** that are located close to the hilum. All lymph from the pulmonary nodes and from the superficial plexus drain into the bronchopulmonary nodes in the hilum. The lymph then drains into the tracheobronchial nodes and from there into the bronchomediastinal lymph trunks.

Nerve Supply of the Lungs. Each lung is supplied by the **pulmonary plexus.** The plexus is formed from branches of the sympathetic trunk and receives parasympathetic fibers from the vagus nerve.

Nerve Supply of the Pleura

Parietal Pleura. The costal pleura is supplied by intercostal nerves; the mediastinal pleura is supplied by the phrenic nerve; and the diaphragmatic pleura is supplied over the domes by the phrenic nerve and around the periphery by the lower intercostal nerves.

Visceral Pleura. Receives an autonomic supply from the pulmonary plexus. It is sensitive to stretch but is insensitive to common sensations such as pain and touch.

Surface Marking of the Lungs. The **apex of each lung** projects up into the neck and extends from the sternoclavicular joint to a point 2.5 cm (1 inch) above the junction of the medial and intermediate thirds of the clavicle.

The **anterior border of the right lung** extends from behind the sternoclavicular joint downward and medially to almost reach the midline behind the sternal angle. It then descends to reach the xiphisternal joint.

The **anterior border of the left lung** has a similar course, but at the level of the fourth costal cartilage it deviates laterally and extends for a variable distance beyond the lateral margin of the sternum to form the **cardiac notch**. This notch is produced by the heart displacing the lung to the left. The anterior border then turns sharply downward to the level of the xiphisternal joint.

The **lower borders of each lung** at the end of quiet inspiration follow a curving line, which crosses the sixth rib in the midclavicular line and the eighth rib in the midaxillary line, and reaches the tenth rib adjacent to the vertebral column.

The **posterior border of each lung** extends downward from the spinous process of the seventh cervical vertebra to the level of the tenth thoracic vertebra, and lies about 4 cm (1½ inches) from the midline.

The **oblique fissure** of each lung can be indicated on the surface by a line that extends from the root of the spine of the scapula downward, laterally, and anteriorly, following the course of the sixth rib to the costochondral junction. In the left lung the upper lobe lies above and anterior to this line; the lower lobe lies below and posterior to it.

In the right lung the **horizontal fissure** may be represented by a line drawn horizontally along the fourth costal cartilage to meet the oblique fissure in the midaxillary line. Above the horizontal fissure lies the upper lobe and below it, the middle lobe; below and posterior to the oblique fissure lies the lower lobe.

MEDIASTINUM

The mediastinum is a movable partition that lies between the pleurae and the lungs (*Fig. 10–4*). It extends superiorly to the thoracic inlet and inferiorly to the diaphragm. It extends anteriorly to the sternum and posteriorly to the thoracic vertebrae.

For descriptive purposes the mediastinum is divided into superior and inferior mediastina by an imaginary plane passing from the sternal angle anteriorly to the lower border of the body of the fourth thoracic vertebra posteriorly. The **inferior mediastinum** is further subdivided into the middle mediastinum, which consists of the pericardium and heart; the **anterior mediastinum,** which is a space between the pericardium and the sternum; and the **posterior mediastinum**, which lies between the pericardium and the vertebral column.

Contents of the Superior Mediastinum

From Anterior to Posterior. Remains of thymus; brachiocephalic veins, upper part of superior vena cava; brachiocephalic artery,

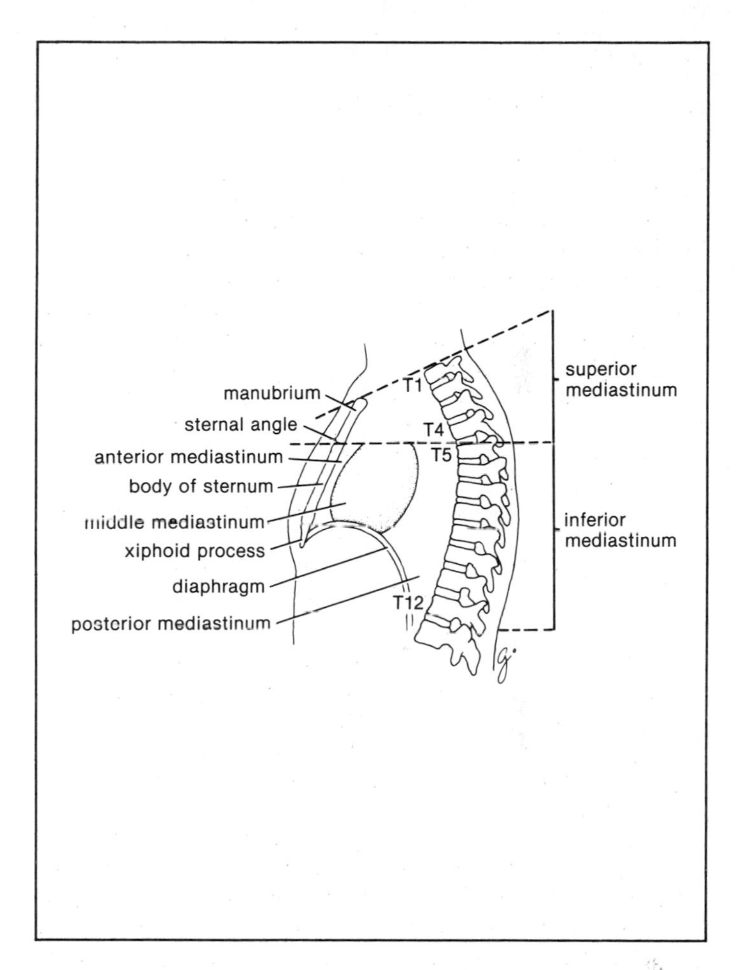

Figure 10–4. Subdivisions of the mediastinum.

left common carotid artery, left subclavian artery, arch of aorta, both phrenic and vagus nerves, left recurrent laryngeal and cardiac nerves, trachea and lymph nodes, esophagus and thoracic duct, and sympathetic trunks.

Contents of the Inferior Mediastinum

Contents of the Anterior Mediastinum. Sternopericardial ligaments, lymph nodes, remains of thymus.

Contents of the Middle Mediastinum. Pericardium, heart and roots of great vessels, phrenic nerves, bifurcation of trachea, lymph nodes.

Contents of the Posterior Mediastinum. Descending thoracic aorta, esophagus, thoracic duct, azygos, and hemiazygos veins, vagus nerves, splanchnic nerves, sympathetic trunks, lymph nodes.

Digestive System

The digestive system consists of the organs involved in digestion and includes the mouth, the pharynx, the esophagus, the stomach, and the small and large intestines (*Fig. 11–1*). Other organs involved in digestion, which lie outside the alimentary tract, are the salivary glands, the liver, the bile passages, and the pancreas.

MOUTH

The mouth extends from the lips to the entrance into the pharynx (*Fig. 11–1*). It is subdivided into the **vestibule,** which lies between the lips and the cheeks externally and the gums and the teeth internally, and the **mouth cavity proper,** which lies within the gums and the teeth.

The **vestibule** is a slitlike space that is bounded laterally by the cheek. The cheek is made up of the buccinator muscle and is covered on the outside by skin and is lined by mucuous membrane. Opposite the upper second molar, a small papilla is present on the mucous membrane for the opening of the **duct of the parotid salivary gland.**

The **mouth proper** has a roof, which is formed by the hard palate anteriorly and the soft palate posteriorly. The floor is

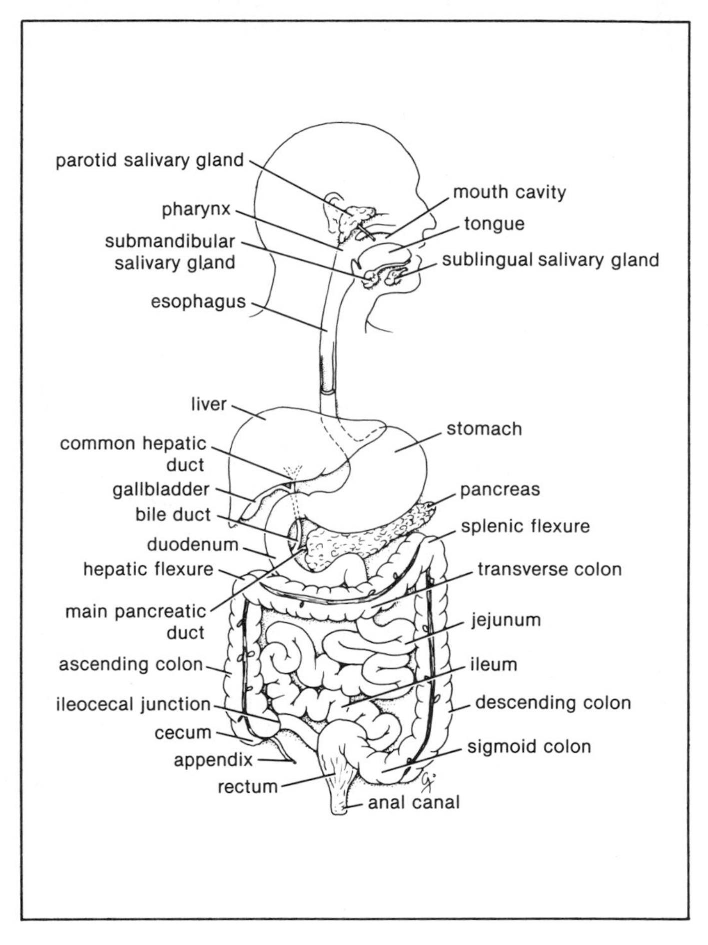

Figure 11–1. The digestive system.

formed largely by the anterior two-thirds of the tongue and by the reflection of the mucous membrane from the sides of the tongue to the gum on the mandible. In the midline, a fold of mucous membrane called the **frenulum of the tongue** connects the undersurface of the tongue to the floor of the mouth (*Fig. 11-2*). On each side of the frenulum there is a small papilla for the **orifice of the duct of the submandibular salivary gland.** The mucuous membrane lateral to the papilla forms a rounded fold, the **sublingual fold,** produced by the underlying sublingual salivary gland. The ducts of the gland (18 to 20 in number) open on the summit of the fold.

Sensory Innervation of the Mouth. The roof is supplied by the greater palatine and nasopalatine nerves. The sensory fibers travel in the maxillary nerve. The floor is supplied by the lingual nerve (common sensation) and the chorda tympani (taste). The cheek is supplied by the buccal nerve, a branch of the mandibular nerve.

Teeth

There are two sets of teeth, deciduous and permanent.

The **deciduous teeth** are 20 in number; four incisors, two canines, and four molars in each jaw. They begin to erupt at about the sixth month and have all erupted by the end of the second year. The teeth of the lower jaw usually appear before those of the upper jaw. The **permanent teeth** are 32 in number, including four incisors, two canines, four premolars, and six molars in each jaw. They begin to erupt at the sixth year. The last tooth to erupt is the third molar, and this may take place between the seventeenth and thirtieth year. The teeth of the lower jaw usually appear before those of the upper jaw.

The teeth are set in sockets on the alveolar processes of the mandible and the maxillae. The alveolar processes are covered by a mucoperiosteum, known as the **gums** or **gingivae.** The part of the tooth that projects above the gums is the **crown,** the portion that lies within the socket is the root. In the center of each tooth is the **pulp cavity**, which is filled with connective tissue containing

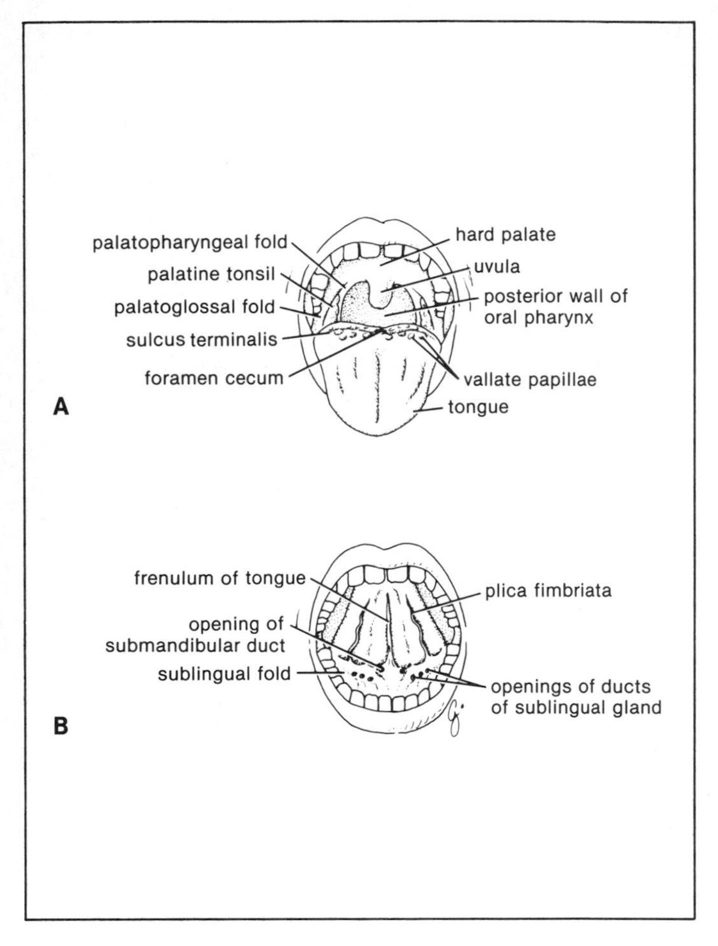

Figure 11–2. A. Cavity of the mouth. **B.** Undersurface of tongue showing openings of ducts of submandibular and sublingual salivary glands.

nerves, blood, and lymph vessels. Extensions of the pulp cavity pass into the roots of the tooth and are called **root canals.** The wall of the tooth is largely composed of a calcified connective tissue called **dentin.** This is protected in the crown by a covering of tough **enamel,** which is composed of calcium phosphate. The dentin in the root of the tooth is covered by a calcified connective tissue called **cementum.**

Tongue

The tongue is a mass of striated muscle covered with mucous membrane and forms part of the floor of the mouth and the oral pharynx. The tongue is divided into right and left halves by a **median fibrous septum.**

The muscles of the tongue are intrinsic and extrinsic. The **intrinsic muscles** are confined to the tongue and are not attached to bone. They consist of longitudinal, transverse, and vertical fibers. The function of the intrinsic muscles is to alter the shape of the tongue. The **extrinsic muscles** are attached to bones and the soft palate. They are the **genioglossus,** the **hyoglossus,** the **styloglossus,** and the **palatoglossus.** Contraction of these muscles changes the position of the tongue as in chewing and swallowing; they also move the tongue from side to side and move it in and out. The muscles of the tongue are supplied by the hypoglossal nerve. (The palatoglossus is an exception; it is a muscle of the soft palate and is supplied by the pharyngeal plexus.) The origin, insertion, nerve supply, and action of the tongue muscles are summarized in *Table 11-1.*

The mucous membrane of the upper surface of the tongue may be divided into anterior and posterior parts by a V-shaped sulcus, the **sulcus terminalis** (*Fig. 11-2*). The sulcus serves to divide the tongue into the anterior two-thirds, or oral part, and the posterior third, or pharyngeal part. Three types of papillae are present on the upper surface of the anterior two-thirds of the tongue: (1) the filiform papillae, (2) the fungiform papillae, and (3) the vallate papillae. The **filiform papillae** are very numerous and cover the anterior two-thirds of the tongue on its upper surface. They

TABLE 11–1. MUSCLES OF THE TONGUE

Name of Muscle	Origin	Insertion	Nerve Supply	Action
Intrinsic Muscles				
Longitudinal Transverse Vertical	Median septum and submucosa	Mucous membrane	Hyoglossal nerve	Alters shape of tongue
Extrinsic Muscles				
Genioglossus	Superior genial spine of mandible	Blends with other muscles of tongue and body of hyoid bone	Hypoglossal nerve	Protrudes apex through mouth
Hyoglossus	Body and greater cornu of hyoid bone	Blends with other muscles of tongue	Hypoglossal nerve	Depresses tongue
Styloglossus	Styloid process of temporal bone	Blends with other muscles of tongue	Hypoglossal nerve	Draws tongue upward and backward
Palatoglossus	(Muscle of soft palate, *see Table 11–2)*			

form small conical projections and are whitish in color due to the thickness of the cornified epithelium. The **fungiform papillae** are much less numerous than the filiform papillae and are scattered on the sides and apex of the tongue. They are mushroom-shaped and possess a vascular connective tissue core, which imparts a reddish tinge to the papillae. The **vallate papillae** are 10 to 12 in number and are situated in a row immediately in front of the sulcus terminalis. Each papilla protrudes slightly from the surface and is surrounded by a circular furrow in the walls of which lie the tastebuds. In the midline anteriorly, the underface of the tongue is connected to the floor of the mouth by a fold of mucous membrane called the **frenulum of the tongue.**

Blood Supply. Lingual artery, tonsillar branch of facial artery, ascending pharyngeal artery.

Lymphatic Drainage. Tip drains into submental lymph nodes, sides of anterior two-thirds drain into submandibular and deep cervical nodes. Posterior one-third drains into deep cervical nodes.

Sensory Innervation. Anterior two-thirds, lingual nerve (general sensation), chorda tympani (taste). Posterior third, glossopharyngeal for general sensation and taste.

Movements of the Tongue. *Protrusion.* Genioglossus muscles on both sides. *Retraction.* Styloglossus and hyoglossus muscles on both sides. *Depression.* Hyoglossus muscles on both sides. *Retraction and Elevation of Posterior Third.* Styloglossus and palatoglossus on both sides. **Shape changes** are produced by the intrinsic muscles.

Palate

The palate forms the roof of the mouth and the floor of the nasal cavity. It may be divided into two parts: (1) the hard palate in front and (2) the soft palate behind.

The **hard palate** is formed by the palatine processes of the maxillae and the horizontal plates of the palatine bones. The upper and lower surfaces of the hard palate are covered with mucous membrane. The **soft palate** is a mobile fold attached to the posterior border of the hard palate. It is covered on its upper and lower surfaces by mucous membrane and contains an aponeurosis of dense fibrous tissue, muscle fibers, lymphoid tissue, glands, vessels, and nerves. Its free posterior border has in the midline a conical projection called the **uvula.** The soft palate is continuous at the sides with the lateral wall of the pharynx.

From the undersurface of the soft palate a muscular fold extends downward and forward on each side that runs to the side of the base of the tongue. It is covered with mucous membrane and is called the **palatoglossal arch (anterior pillar of the fauces).** The muscle contained within the fold is the **palatoglossus muscle.** Another similar fold extends from the undersurface of the soft palate downward and laterally to join the pharyngeal wall. It is also covered with mucous membrane and is called the **palatopharyngeal arch (posterior pillar of the fauces).** The muscle contained within the fold is the palatopharyngeus muscle. The **palatine tonsils,** which are masses of lymphoid, are situated between the palatoglossal and palatopharyngeal arches.

The anterior pillar of the fauces on each side serves as a landmark and is the place where the mouth cavity becomes continuous with the oral part of the pharynx. At the posterior border of the soft palate the oral part of the pharynx becomes continuous with the nasal part of the pharynx.

The muscles of the soft palate are the tensor veli palatini, the levator veli palatini, the palatoglossus, the palatopharyngeus, and the musculus uvulae. The origins, insertions, nerve supply, and actions of these muscles are summarized in *Table 11–2.*

Blood Supply. Greater palatine branch of maxillary artery, ascending palatine branch of facial artery, ascending pharyngeal artery.

Lymphatic Drainage. Deep cervical lymph nodes.

TABLE 11–2. MUSCLES OF THE SOFT PALATE

Name of Muscle	Origin	Insertion	Nerve Supply	Action
Tensor veli palatini	Spine of sphenoid, auditory tube	Forms palatine aponeurosis with muscle of opposite side	Nerve to medial pterygoid from mandibular nerve	Tenses soft palate
Levator veli palatini	Petrous part of temporal bone, auditory tube	Palatine aponeurosis	Pharyngeal plexus	Raises soft palate
Palatoglossus	Palatine aponeurosis	Side of tongue	Pharyngeal plexus	Pulls root of tongue upward and backward. Narrows oropharyngeal isthmus
Palatopharyngeus	Palatine aponeurosis and border of hard palate	Posterior border of thyroid cartilage	Pharyngeal plexus	Elevates wall of pharynx. Pulls palatopharyngeal folds medially
Musculus uvulae	Posterior border of hard palate	Mucous membrane of uvula	Pharyngeal plexus	Elevates uvula

Sensory Innervation. Greater and lesser palatine nerves, nasopalatine and glossopharyngeal nerves.

Movements of the Soft Palate. The channel between the oral part of the pharynx and the nasal part of the pharynx can be closed by raising the soft palate. Closure occurs during swallowing to prevent food from entering the nasal part of the pharynx and the back of the nose; closure also occurs during speech. The soft palate is raised by the contraction of the **levator veli palatini** muscle on each side. At the same time, the upper fibers of the **superior constrictor** muscle contract and pull the posterior pharyngeal wall forward. The **palatopharyngeus** muscles on both sides also contract, so that the posterior pillars are pulled medially, like side curtains.

SALIVARY GLANDS

There are three main pairs of salivary glands called the **parotid, submandibular,** and **sublingual glands** and in addition there are many small **buccal glands** scattered throughout the mouth.

Parotid Gland

The parotid gland is the largest of the salivary glands and is composed almost entirely of serous acini. It lies in a deep hollow below the external auditory meatus, behind the ramus of the mandible and in front of the sternocleidomastoid muscle. Its medial border lies against the pharyngeal wall. A process of the gland, the **glenoid process** extends upward behind the temporomandibular joint. The anterior margin of the gland extends forward over the masseter muscle to form the **facial process.** The facial nerve divides the gland into **superficial and deep lobes.**

The **parotid duct** emerges from the anterior border of the gland and passes forward over the lateral surface of the masseter. It then turns sharply medially and pierces the buccinator muscle and the mucous membrane of the mouth. It enters the vestibule of the mouth upon a small papilla opposite the upper second molar tooth. The surface marking of the duct is that it lies a finger breadth below the zygomatic arch and it can be rolled on the anterior border of the tight masseter muscle.

Structures Within the Parotid Gland. From lateral to medial, these are (1) the facial nerve, (2) the retromandibular vein, and (3) the external carotid artery. Embedded within the gland are several lymph nodes.

Blood Supply. External carotid artery and its branches.

Lymphatic Drainage. Parotid lymph nodes and deep cervical lymph nodes.

Nerve Supply. Parasympathetic secretomotor nerves from glossopharyngeal nerve. The axons pass to the gland via the tympanic branch, the lesser petrosal nerve, and the otic ganglion. Postganglionic fibers pass to the gland via the auriculotemporal nerve. Postganglionic sympathetic fibers reach the gland as a nerve plexus around the external carotid artery.

Submandibular Gland

The submandibular gland is a large salivary gland and is composed of a mixture of serous and mucous acini, the former predominating. It lies beneath the lower border of the body of the mandible. It can be divided into superficial and deep parts, which are continuous with each other around the posterior border of the mylohyoid muscle.

The **superficial part of the gland** projects downward into the digastric triangle of the neck. The **deep part of the gland** extends forward below the mucous membrane of the mouth between the side of the tongue and the body of the mandible. Here it lies in the interval between the hyoglossus medially and the mylohyoid laterally.

The **submandibular duct** emerges from the anterior end of the deep part of the gland and runs forward beneath the mucous membrane of the mouth. It is crossed laterally by the lingual nerve and is related to the sublingual salivary gland. It opens into the mouth on a small papilla, which is situated at the side of the frenulum of the tongue.

Blood Supply. Facial and lingual arteries.

Lymphatic Drainage. Submandibular and deep cervical lymph nodes.

Nerve Supply. Parasympathetic secretomotor fibers from the facial nerve via the chorda tympani, pass to the submandibular ganglion. Postganglionic nerve fibers pass directly to the gland or along the duct. Postganglionic sympathetic fibers reach the gland along the facial and lingual arteries.

Sublingual Gland

The sublingual gland lies beneath the mucous membrane (sublingual fold) of the floor of the mouth close to the frenulum of the tongue. It contains both serous and mucous acini, the latter predominating.

The **sublingual ducts** (8 to 20 in number) open into the mouth on the summit of the sublingual fold, but a few may open into the submandibular duct.

Blood Supply. Facial and lingual arteries.

Lymphatic Drainage. Submandibular and deep cervical nodes.

Nerve Supply. Parasympathetic secretomotor nerves from the facial nerve via the chorda tympani, pass to the submandibular ganglion. Postganglionic fibers pass to the gland via the lingual nerve. Postganglionic sympathetic fibers reach the gland along the facial and lingual arteries.

PHARYNX

The pharynx is situated behind the nasal cavities, the mouth, and the larynx (*Fig. 11-1*) and may be divided into nasal, oral, and laryngeal parts. The pharynx is funnel-shaped, its upper, wider end lying under the skull, and its lower, narrow end becoming continuous with the esophagus opposite the sixth cervical vertebra. The pharynx has a musculomembranous wall, which is deficient anteriorly. Here, it is replaced by the posterior opening into the nose (posterior nares), the opening into the mouth, and the inlet of the larynx. The mucous membrane is continuous with that of the nasal cavities, the mouth, and the larynx. By means of the auditory tube, the mucous membrane is also continuous with that of the tympanic cavity.

The muscles in the wall of the pharynx consist of the superior, middle, and inferior constrictor muscles, whose fibers run in a more or less circular direction, and the stylopharyngeus and salpingopharyngeus muscles, whose fibers run in a more or less longitudinal direction. The origin, insertion, nerve supply, and action of these muscles are summarized in *Table 11-3*.

Interior of the Pharynx

The pharynx is divided into three parts: the nasal pharynx, the oral pharynx, and the laryngeal pharynx.

Nasal Pharynx. This lies above the soft palate and behind the nasal cavities. In the submucosa of the roof there is a collection of lymphoid tissue called the **pharyngeal tonsil.** The **pharyngeal isthmus** is the opening in the floor between the soft palate and the posterior pharyngeal wall. On the lateral wall is the opening of the **auditory tube,** the elevated edge of which is called the **tubal elevation.** The **pharyngeal recess** is a depression in the pharyngeal wall behind the tubal elevation. The **salpingopharyngeal fold** is a vertical fold of mucous membrane covering the salpingopharyngeus muscle.

TABLE 11–3. MUSCLES OF THE PHARYNX

Name of Muscle	Origin	Insertion	Nerve Supply	Action
Superior constrictor	Medial pterygoid plate, pterygoid hamulus, pterygomandibular ligament, mylohyoid line of mandible	Pharyngeal tubercle of occipital bone. Pharyngeal raphe midline posteriorly	Pharyngeal plexus	Aids soft palate in closing off nasal pharynx. Propels bolus downward
Middle constrictor	Lower part of stylohyoid ligament, lesser and greater cornu of hyoid bone	Pharyngeal raphe midline posteriorly	Pharyngeal plexus	Propels bolus downward
Inferior constrictor	Lamina of thyroid cartilage, cricoid cartilage	Pharyngeal raphe midline posteriorly	Pharyngeal plexus	Propels bolus downward
Cricopharyngeus	Lowest fibers of inferior constrictor			Sphincter at lower end of pharynx
Stylopharyngeus	Styloid process of temporal bone	Posterior border of thyroid cartilage	Glossopharyngeal nerve	Elevates larynx during swallowing
Salpingopharyngeus	Auditory tube	Blends with palatopharyngeus	Pharyngeal plexus	Elevates pharynx
Palatopharyngeus	(Muscles of palate, *see Table 11–2)*			

Oral Pharynx. This lies behind the oral cavity. The floor is formed by the posterior one-third of the tongue and the interval between the tongue and the epiglottis. In the midline is the **median glossoepiglottic fold** and on each side the **lateral glossoepiglottic folds.** The depression on each side of the median glossoepiglottic fold is called the **vallecula.**

On the lateral wall on each side are the palatoglossal and the palatopharyngeal arches or folds and the **palatine tonsils** between them (*Fig. 11–2*). The **palatoglossal arch** is a fold of mucous membrane covering the palatoglossus muscle. The interval between the two palatoglossal arches is called the **oropharyngeal isthmus** and marks the boundary between the mouth and the pharynx. The **palatopharyngeal arch** is a fold of mucous membrane covering the palatopharyngeus muscle. The **tonsillar sinus** is a recess between the palatoglossal and palatopharyngeal arches that is occupied by the palatine tonsil.

Laryngeal Pharynx. This lies behind the opening into the larynx. The lateral wall is formed by the thyroid cartilage and the thyrohyoid membrane. The **piriform fossa** is a depression in the mucous membrane situated on each side of the laryngeal inlet. It is a common site for the lodging of fish bones.

Sensory Nerve Supply of the Mucous Membrane of the Pharynx. The nasal part is supplied mainly by the maxillary nerve; the oral part is supplied by the glossopharyngeal nerve; the mucous membrane around the entrance into the larynx is supplied by the internal laryngeal branch of the superior laryngeal branch of the vagus nerve.

Blood Supply of the Pharynx. Ascending pharyngeal, tonsillar branches of facial and branches of maxillary and lingual arteries.

Lymphatic Drainage of Pharynx. Directly into the deep cervical lymph nodes or indirectly via the retropharyngeal or paratracheal nodes into the deep cervical nodes.

Palatine Tonsils

The palatine tonsils are two masses of lymphoid tissue located in the lateral walls of the oral part of the pharynx (*Fig. 11–2*). Each tonsil is covered by mucous membrane, and its free medial surface projects into the cavity of the pharynx. The surface is pitted by numerous small openings, which lead into the **tonsillar crypts.** The tonsil is covered on its lateral surface by a layer of fibrous tissue, called the capsule. The tonsil reaches its maximum size during early childhood, but after puberty it diminishes considerably in size. The lymphatic drainage of the tonsil is into the upper deep cervical lymph nodes, just below and behind the angle of the mandible.

Swallowing (Deglutition)

The masticated food is formed into a ball or bolus on the dorsum of the tongue and voluntarily pushed upward and backward against the undersurface of the hard palate. This is brought about by the contraction of the styloglossus muscles on both sides, which pull the root of the tongue upward and backward. The palatoglossus muscles now squeeze the bolus backward into the pharynx. From this point onward the process of swallowing becomes an involuntary act.

The nasal part of the pharynx is shut off from the oral part of the pharynx by the elevation of the soft palate, the pulling forward of the posterior wall of the pharynx, and the pulling medially of the palatopharyngeal arches. This prevents the passage of food and drink into the nasal cavities.

The larynx and the laryngeal part of the pharynx are now pulled upward by the contraction of the stylopharyngeus, salpingopharyngeus, thyrohyoid, and palatopharyngeus muscles. The main part of the larynx is thus elevated to the posterior surface of the epiglottis and the entrance into the larynx is closed.

The bolus moves downward over the epiglottis, the closed entrance into the larynx, and reaches the lower part of the pharynx as the result of the successive contraction of the superior, middle, and inferior constrictor muscles. Some of the food slides down the groove on either side of the entrance into the larynx, i.e., down through the **piriform fossae.** Finally, the lower part of the pharyngeal wall relaxes and the bolus enters the esophagus.

ESOPHAGUS

The esophagus is a muscular tube about 25 cm (10 inches) long, extending from the pharynx to the stomach (*Fig. 11-1*). It begins at the level of the cricoid cartilage in the neck and descends in the midline behind the trachea. In the thorax, it passes downward through the mediastinum and enters the abdominal cavity by piercing the diaphragm at the level of the tenth thoracic vertebra. The esophagus has a short course of about 2.5 cm (1 inch) before it enters the right side of the stomach.

Relations of the Esophagus in the Neck. *Anteriorly.* Trachea, recurrent laryngeal nerves. *Posteriorly.* Longus colli muscles, vertebral column. *Laterally.* Thyroid gland, carotid sheath and on the left side the thoracic duct.

Blood Supply of the Esophagus in the Neck. *Arteries.* Inferior thyroid arteries. *Veins.* Inferior thyroid veins.

Lymphatic Drainage of the Esophagus in the Neck. Deep cervical lymph nodes.

Nerve Supply of the Esophagus in the Neck. Recurrent laryngeal nerves, branches from sympathetic trunks.

Relations of the Esophagus in the Thorax. *Anteriorly.* Trachea, left recurrent laryngeal nerve, left bronchus, left atrium of heart. *Posteriorly.* Vertebral column, thoracic duct, azygos veins, right posterior intercostal arteries, descending thoracic aorta. *Laterally—Right Side.* Mediastinal pleura, azygos vein. *Left Side.* Aortic arch, left subclavian artery, thoracic duct, mediastinal pleura.

Blood Supply of the Esophagus in the Thorax. *Arteries.* Upper part from descending thoracic aorta, lower third left gastric artery. *Veins.* These drain into the azygos veins and from the lower third they drain into the left gastric vein, a tributary of the portal vein.

Lymphatic Drainage of the Esophagus in the Thorax. Upper part into superior and posterior mediastinal nodes and from lower third into nodes along the left gastric blood vessels and the celiac nodes in the abdomen.

Nerve Supply of the Esophagus in the Thorax. Vagal trunks (left vagus lies anterior, right vagus lies posterior), esophageal plexus, sympathetic trunks, greater splanchnic nerves.

Relations of the Esophagus in the Abdomen. The esophagus lies posterior to the left lobe of the liver and in front of the left crus of the diaphragm.

STOMACH

The stomach is a dilated portion of the alimentary canal and is situated in the upper part of the abdomen, extending from the left hypochondriac region into the epigastric and umbilical regions. The greater part of the stomach lies under cover of the lower ribs. It is roughly J-shaped and has two openings, the **cardiac** and **pyloric orifices,** two curvatures known as the **greater and lesser curvatures,** and two surfaces, an **anterior** and a **posterior** surface (*Fig. 11–1*).

The stomach is relatively fixed at both ends, but is very mobile in between. It tends to be high and transversely arranged in the short person and elongated vertically in the tall person (J-shaped stomach). Its shape undergoes considerable variation in the same person and depends on the volume of its contents, the position of the body, and the phase of respiration.

The stomach may be divided into the following parts: The **fundus** is dome-shaped and projects upward and to the left of the cardiac orifice. It is usually full of gas. The **body** extends from the level of the cardiac orifice to the level of the **incisura angularis,** a constant notch in the lower part of the lesser curvature. The **pyloric antrum** extends from the incisura angularis to the proximal limit of the pylorus. The **pylorus** is the most tubular part of the stomach. Its thick muscular wall forms the **pyloric sphincter.** The cavity of the pylorus is called the **pyloric canal.**

The **lesser curvature** forms the right border of the stomach and extends from the cardiac orifice to the pylorus. The lesser omentum extends from the lesser curvature to the liver. The **greater curvature** is much longer than the lesser curvature and extends from the left of the cardiac orifice, over the dome of the fundus, and then sweeps around and to the right to the inferior part of the pylorus. The gastrosplenic omentum (ligament) extends from the upper part of the greater curvature to the spleen, and the greater omentum extends from the lower part of the greater curvature to the transverse colon.

The **cardiac orifice** is where the esophagus enters the stomach. Although no anatomical sphincter can be demonstrated here, a physiologic mechanism exists that prevents regurgitations of stomach contents into the esophagus. The **pyloric orifice** is formed by the **pyloric canal,** which is about 2.5 cm (1 inch) long. The circular muscle coat of the stomach is much thicker here and forms the anatomic and physiologic **pyloric sphincter.**

Relations of the Stomach. *Anteriorly.* Anterior abdominal wall, left costal margin, diaphragm, left lobe of liver. *Posteriorly.* Lesser sac of peritoneum, diaphragm, spleen, left suprarenal gland, left kidney, splenic artery, pancreas, transverse colon.

Blood Supply. *Arteries.* Right and left gastric arteries supply lesser curvature. Right and left gastroepiploic arteries supply greater curvature. Short gastric arteries, from the splenic, supply the fundus (*Fig. 11–3*). *Veins.* These drain into the portal circulation. The right and left gastric veins drain into the portal

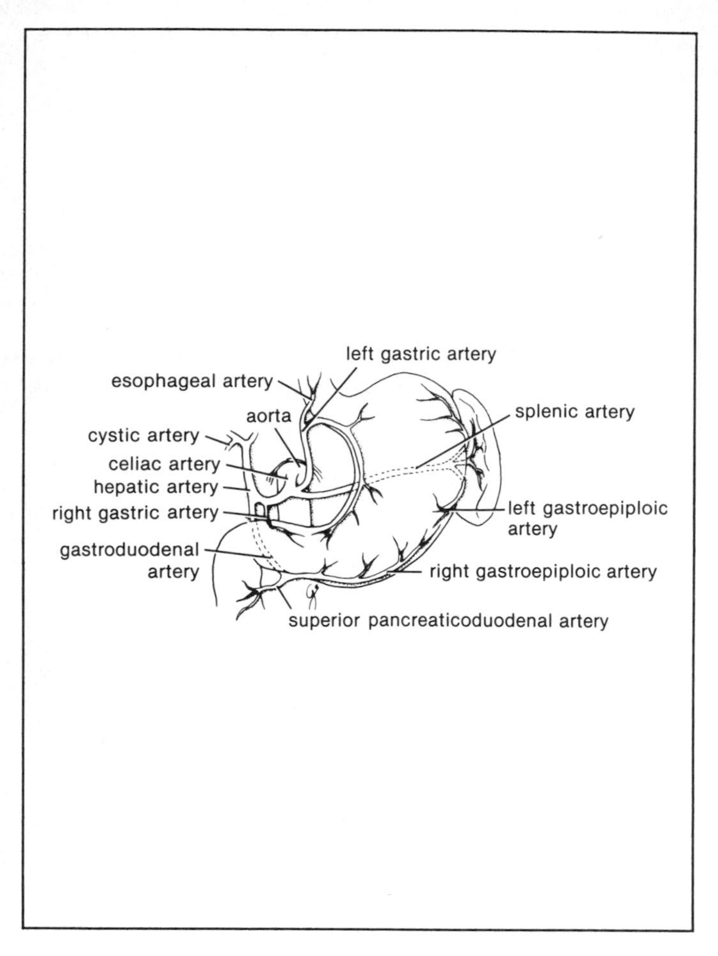

Figure 11–3. Arterial supply to the stomach.

vein, the short gastric and the left gastroepiploic veins drain into the splenic vein, the right gastroepiploic vein drains into the superior mesenteric vein.

Lymphatic Drainage. The lymph vessels follow the arteries into the left and right gastric nodes, the left and right gastroepiploic nodes and the short gastric nodes. All lymph from the stomach eventually passes to the celiac nodes.

Nerve Supply. Sympathetic from the celiac plexus and parasympathetic from the vagus nerves.

SMALL INTESTINE

The small intestine is the part of the alimenatry tract in which the greater part of digestion and food absorption takes place. The small intestine begins as a continuation of the pyloric end of the stomach (*Fig. 11–1*). It forms a much coiled tube that occupies the abdominal cavity and opens below into the large intestine at the ileocecal junction. It measures about 6.35 m (21 ft) in length and about 2.5 cm (1 inch) in diameter. The small intestine is divided into three parts: the duodenum, the jejunum, and the ileum.

Duodenum

The duodenum is a C-shaped tube about 25 cm (10 inches) long, which curves around the head of the pancreas (*Fig. 11–1*). It begins at the pyloric sphincter of the stomach and ends by becoming continuous with the jejunum. The duodenum is situated in the epigastric and umbilical regions. It has no mesentery. Although the first inch of the duodenum is covered on its anterior and posterior surfaces with peritoneum, and it has the lesser omentum attached to its upper border and the greater omentum attached to its lower border, the remainder of the duodenum is retroperitoneal.

The duodenum is divided into four parts:

The **first part** is 5 cm (2 inches) long and runs upward and backward on the transpyloric plane at the level of the first lumbar vertebra.

Relations of the First Part of the Duodenum. *Anteriorly.* Quadrate lobe of liver, gallbladder. *Posteriorly.* Common bile duct, portal vein, gastroduodenal artery, inferior vena cava. *Superiorly.* Entrance into lesser sac. *Inferiorly.* Head of the pancreas.

The **second part** of the duodenum is 8 cm (3 inches) long and runs vertically downward. Bile and main pancreatic ducts pierce the medial wall about half way down. They unite to form an ampulla that opens on the summit of a **major duodenal papilla.** The accessory pancreatic duct, if present, opens into the duodenum on a **minor duodenal papilla** about 1.9 cm (¾ inch) above the major duodenal papilla.

Relations of the Second Part of the Duodenum. *Anteriorly.* Fundus of gallbladder, right lobe of liver, transverse colon, small intestine. *Posteriorly.* Hilus of right kidney, right ureter. *Laterally.* Right colic flexure. *Medially.* Head of pancreas, bile duct, and pancreatic ducts.

The **third part** of the duodenum is 8 cm (3 inches) long and passes horizontally in front of the vertebral column.

Relations of the Third Part of the Duodenum. *Anteriorly.* Root of mesentery of small intestine, superior mesenteric vessels, coils of jejunum. *Posteriorly.* Right psoas muscle, inferior vena cava, aorta. *Superiorly.* Head of pancreas.

The **fourth part** of the duodenum is 5 cm (2 inches) long and runs upward and to the left and then forward as the **duodenojejunal junction.** The junction is held in position by the **ligament of Treitz** that ascends to the right crus of the diaphragm.

Relations of the Fourth Part of the Duodenum. *Anteriorly.* Root of mesentery of small intestine, coils of jejunum. *Posteriorly.* Left margin of aorta.

Blood Supply of the Duodenum. *Arteries.* The upper half is supplied by the superior pancreaticoduodenal artery, a branch of the gastroduodenal artery. The lower half is supplied by the inferior pancreaticoduodenal artery, a branch of the superior mesenteric artery. *Veins.* The superior pancreaticoduodenal vein joins the portal vein, the inferior vein joins the superior mesenteric vein.

Lymphatic Drainage of the Duodenum. The lymph vessels drain upward via pancreaticoduodenal nodes to gastroduodenal nodes and celiac nodes; and downward via pancreaticoduodenal nodes to superior mesenteric nodes.

Nerve Supply of the Duodenum. Sympathetic and vagus nerves via the celiac and superior mesenteric plexuses.

Jejunum and Ileum

The jejunum measures about 2.5 m (8 ft) long and the ileum about 3.6 m (12 ft) long (*Fig. 11–1*). Each has distinctive features, but there is a gradual change from one to the other. The jejunum begins at the duodenojejunal junction (*Fig. 11–4*), in the upper part of the abdominal cavity to the left of the midline. It is wider in diameter, thicker walled, and redder in color than the ileum. The coils of ileum occupy the lower right part of the abdominal cavity and tend to hang down into the pelvis. The ileum ends at the ileocecal junction. The coils of the jejunum and the ileum are suspended from the posterior abdominal wall by a fan-shaped fold of peritoneum known as the **mesentery of the small intestine.**

Blood Supply of the Jejunum and Ileum. *Arteries.* Branches of the superior mesenteric artery that anastomose with one another to form arcades. *Veins.* Drain into the superior mesenteric vein.

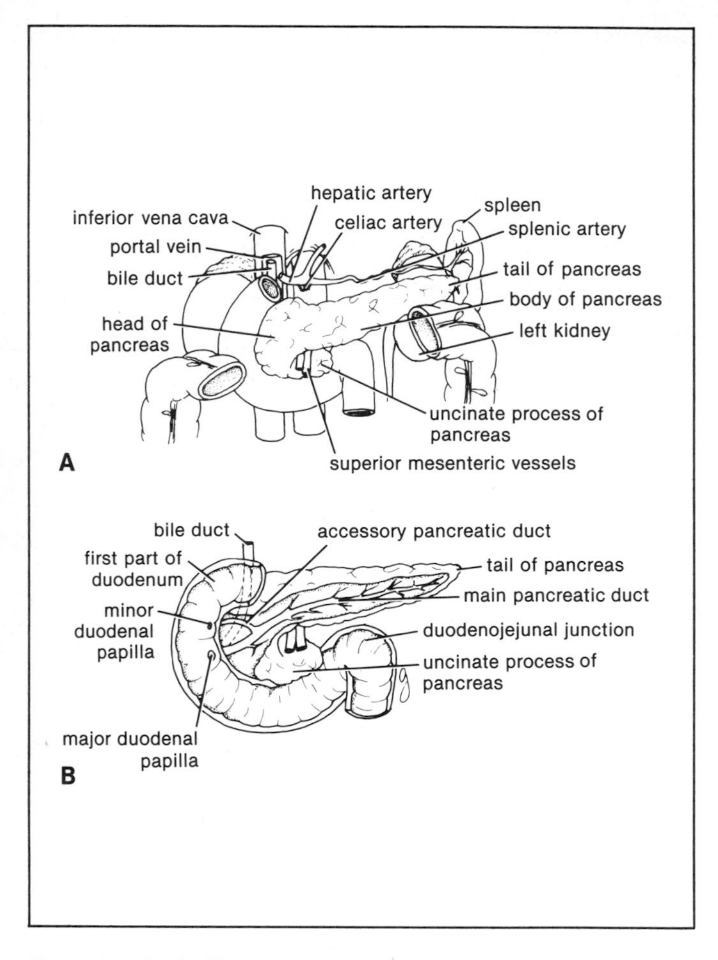

Figure 11–4. **A.** The pancreas and its immediate relations. **B.** The ducts of the pancreas and their openings into the duodenum.

Lymphatic Drainage of the Jejunum and Ileum. The lymph passes to the superior mesenteric nodes via a large number of intermediate mesenteric nodes.

Nerve Supply of the Jejunum and the Ileum. Sympathetic and vagus nerve fibers from the superior mesenteric plexus.

LARGE INTESTINE

The large intestine extends from the ileum to the anus (*Fig. 11–1*). It is divided into the cecum, the vermiform appendix, the ascending colon, the transverse colon, the descending colon, the sigmoid colon, the rectum, and the anal canal. The main functions of the large intestine include absorption of water, production of certain vitamins, storage of undigested food materials and formation of feces, and excretion of feces from the body.

Differences Between the Small and Large Intestines

External Differences

1. Small intestine is more mobile (exception is the duodenum), whereas the ascending and descending parts of the colon are fixed.
2. Diameter of full small intestine is smaller than that of full large intestine.
3. Small intestine has a mesentery (except duodenum), whereas large intestine is retroperitoneal (except transverse colon and sigmoid colon).
4. Longitudinal muscle of small intestine forms a continuous layer around the gut, whereas in the large intestine (with the exception of the appendix, rectum, and anal canal), the longitudinal muscle forms three bands, the **teniae coli.**
5. The small intestine has no fatty tags attached to its wall, whereas the large intestine has the **appendices epiploicae.**

6. The wall of the small intestine is smooth, whereas that of the large intestine is sacculated.

Internal Differences

1. The mucous membrane of the small intestine has permanent folds, called **plicae circulares,** which are absent from the large intestine.
2. The mucous membrane of the small intestine has **Peyer's patches,** whereas the large intestine has **solitary lymph follicles.**
3. The mucous membrane of the small intestine has villi, which are absent from the large intestine.

Cecum

The cecum is a blind-ended pouch and measures about 6 cm (2½ inches) long (*Fig. 11–1*). It lies within the right iliac fossa and is completely covered with peritoneum. At the junction of the cecum with the colon it is joined on the left side by the terminal part of the ileum. Attached to its posteromedial surface is the appendix.

Relations of the Cecum. *Anteriorly.* Anterior abdominal wall, greater omentum, small intestine. *Posteriorly.* Iliacus, psoas, femoral nerve and lateral cutaneous nerve of the thigh.

Blood Supply of the Cecum. *Arteries.* Anterior and posterior cecal arteries from ileocolic artery, a branch of the superior mesenteric artery. *Veins.* Drain into superior mesenteric vein.

Lymphatic Drainage of the Cecum. Mesenteric nodes and superior mesenteric nodes.

Nerve Supply of the Cecum. Sympathetic and vagus nerves via the superior mesenteric plexus.

Ileocecal (Ileocolic) Valve

The ileocecal valve is a rudimentary structure situated at the junction of the ileum and the cecum. It consists of two horizontal folds of mucous membrane that project around the orifice of the ileum. The valve plays little or no part in the prevention of reflux of colic contents into the ileum. The circular muscle of the lower end of the ileum (called the **ileocecal sphincter** by physiologists) serves as a sphincter and controls the flow of contents from the ileum into the colon. The smooth muscle tone is reflexly increased when the cecum is distended; the hormone gastrin produced by the stomach causes relaxation of the muscle tone.

Appendix

The vermiform appendix (*Fig. 11-1*) is an organ containing a large amount of lymphoid tissue. It varies in length from 8 to 13 cm (3 to 5 inches). The base is attached to the posteromedial surface of the cecum about 2.5 cm (1 inch) below the ileocecal junction. The remainder of the appendix is free. It has a complete peritoneal covering, which is attached to the mesentery of the small intestine by a short mesentery of its own, the **mesoappendix.** The appendix lies in the right iliac region and its tip or free end is subject to a considerable range of movement and may be found in the following positions: (1) coiled up behind the cecum; (2) hanging down into the pelvis against the right pelvic wall; and (3) in front or behind the terminal part of the ileum.

The surface marking of the base of the appendix is located on the anterior abdominal wall one-third of the way up the line joining the right anterior superior iliac spine to the umbilicus (**McBurney's point**). Inside the abdomen, the base of the appendix can be found by tracing the teniae coli of the cecum and following them to the appendix, where they converge to form a continuous muscle coat.

Blood Supply of the Appendix. *Arteries.* Appendicular artery, a branch of the posterior cecal artery. *Veins.* Appendicular vein drains into the posterior cecal vein.

Lymphatic Drainage of the Appendix. One or two nodes in mesoappendix then eventually into superior mesenteric lymph nodes.

Nerve Supply of the Appendix. Sympathetic and vagus nerves from superior mesenteric plexus.

Ascending Colon

The ascending colon is about 13 cm (5 inches) long and is located in the right iliac region (*Fig. 11–1*). It extends upward from the cecum to the inferior surface of the right lobe of the liver, where it turns sharply to the left, forming the **right colic flexure,** and becomes continuous with the transverse colon. The peritoneum covers the front and the sides of the ascending colon and binds it to the posterior abdominal wall. The ascending colon lies posteriorly on the iliacus, quadratus lumborum, and the lower pole of the right kidney.

Blood Supply of the Ascending Colon. *Arteries.* Ileocolic and right colic branches of superior mesenteric artery. *Veins.* Drain into the superior mensenteric vein.

Lymphatic Drainage of the Ascending Colon. Colic lymph nodes and superior mesenteric nodes.

Nerve Supply of the Ascending Colon. Sympathetic and vagus nerves from the superior mesenteric plexus.

Transverse Colon

The transverse colon is about 38 cm (15 inches) long and passes across the abdomen, occupying the umbilical and hypogastric regions (*Fig. 11–1*). It begins at the right colic flexure below the right lobe of the liver and hangs downward, suspended by the transverse mesocolon from the pancreas. It then ascends to the **left colic flexure** immediately inferior to the spleen. The left colic flexure is higher than the right colic flexure and is held up by the **phrenicocolic ligament.** The **transverse mesocolon** or mesentery of the transverse colon is attached to the superior border of the transverse colon, and the posterior layers of the greater omentum are attached to the inferior border.

Blood Supply of the Transverse Colon. *Arteries.* Proximal two-thirds is supplied by the middle colic artery, a branch of the superior mesenteric artery. Distal third is supplied by the left colic artery, a branch of the inferior mesenteric artery. *Veins.* Drain into the superior and inferior mesenteric veins.

Lymphatic Drainage of the Transverse Colon. Proximal two-thirds into colic nodes and then the superior mesenteric nodes; distal third into colic nodes and then the inferior mesenteric nodes.

Nerve Supply of the Transverse Colon. Proximal two-thirds is innervated by sympathetic and vagal nerves through the superior mesenteric plexus. The distal third is innervated by sympathetic and parasympathetic pelvic splanchnic nerves through the inferior mesenteric plexus.

Descending Colon

The descending colon is about 25 cm (10 inches) long and lies in the left iliac region (*Fig. 11-1*). It extends downward from the left colic flexure to the pelvic brim, where it becomes continuous with the sigmoid colon. The peritoneum covers the front and the sides and binds it to the posterior abdominal wall. The descending colon lies posteriorly on the left kidney, quadratus lumborum, and the iliacus muscle.

Blood Supply of the Descending Colon. *Arteries.* Left colic branch and sigmoid branches of inferior mesenteric artery. *Veins.* Inferior mesenteric vein.

Lymphatic Drainage of the Descending Colon. Colic nodes and inferior mesenteric nodes.

Nerve Supply of the Descending Colon. Sympathetic and parasympathetic pelvic splanchnic nerves through the inferior mesenteric plexus.

Sigmoid Colon (Pelvic Colon)

The sigmoid colon is about 25 to 38 cm (10 to 15 inches) long and begins as a continuation of the descending colon in front of the pelvic brim (*Fig. 11–1*). Below, it becomes continuous with the rectum in front of the third sacral vertebra. It hangs down into the pelvic cavity in the form of a loop.

The sigmoid colon is attached to the posterior pelvic wall by the fan-shaped **sigmoid mesocolon.** In the male the sigmoid colon is related anteriorly to the urinary bladder, and in the female the posterior surface of the uterus and the posterior fornix of the vagina.

Blood Supply of the Sigmoid Colon. *Arteries.* Sigmoid branch is the inferior mesenteric artery. *Veins.* Inferior mesenteric vein.

Lymphatic Drainage of the Sigmoid Colon. Colic nodes and inferior mesenteric nodes.

Nerve Supply of the Sigmoid Colon. Sympathetic and parasympathetic pelvic splanchnic nerves through the hypogastric plexuses.

Rectum

The rectum is about 13 cm (5 inches) long and begins in front of the third sacral vertebra as a continuation of the sigmoid colon (*Fig. 11–1*). It passes downward, following the curve of the sacrum and the coccyx, and ends in front of the tip of the coccyx by piercing the pelvic floor and becomes continuous with the anal canal. The lower part of the rectum lies immediately above the pelvic floor and is dilated to form the rectal **ampulla.** The peritoneum covers only the upper two-thirds of the rectum. The teniae coli of the sigmoid colon come together, so that the longitudinal muscle fibers form a broad band on the anterior and posterior surfaces of the rectum. The mucous membrane of the rectum, together with the circular muscle layer, form three semicircular folds, two are placed on the left rectal wall and one on the right wall. They are called the **transverse folds of the rectum.**

Relations of the Rectum. *Anteriorly in the Male.* Rectovesical pouch, sigmoid colon, coils of ileum, bladder, vas deferens, seminal vesicles, prostate. *Anteriorly in the Female.* Rectouterine pouch (pouch of Douglas), vagina. *Posteriorly.* Sacrum, coccyx, piriformis and coccygeus muscles, levatores ani muscles, sacral plexus, sympathetic trunks.

Blood Supply of the Rectum. *Arteries.* Superior rectal artery, a branch of the inferior mesenteric artery, is the chief artery and supplies the mucous membrane; middle rectal artery, a branch of the internal iliac artery, supplies the muscle wall; the inferior rectal artery, a branch of the internal pudendal artery, also supplies the muscle wall. *Veins.* The superior rectal vein drains into the inferior mesenteric vein and is a tributary of the portal circulation. The middle and inferior rectal veins drain into the internal iliac and internal pudendal veins, respectively. The anastomosis between the rectal veins is an important portal–systemic anastomosis.

Lymphatic Drainage of the Rectum. The lymph passes to **pararectal nodes** then upward to the **inferior mesenteric nodes.** Some lymph vessels pass to the **internal iliac nodes.**

Nerve Supply of the Rectum. Sympathetic and parasympathetic pelvic splanchnic nerves through the hypogastric plexuses.

Anal Canal

The anal canal is about 4 cm (1½ inches) long and passes downward and backward from the rectal ampulla to open on the surface at the **anus.** Except during defecation, its lateral walls are kept in apposition by the levatores ani muscles and the anal sphincters.

Relations of the Anal Canal. *Posteriorly.* Anococcygeal body, coccyx. *Anteriorly in the Male.* The perineal body, the urogenital diaphragm, the membranous part of the urethra, the bulb of the penis. *Anteriorly in the Female.* The perineal body, the urogenital diaphragm, the lower part of the vagina. *Laterally.* Fat-filled ischiorectal fossa.

The **mucous membrane** of the upper half of the anal canal shows vertical folds called **anal columns.** These are connected together at their lower ends by small semilunar folds called **anal valves.** The mucous membrane of the lower half of the canal is smooth and merges with the skin at the anus.

The **muscular coat,** as in other parts of the intestinal tract, is divided into an outer longitudinal and an inner circular layer of smooth muscle. The circular coat is thickened at the upper end of the anal canal to form the **involuntary internal sphincter.** Surrounding the internal sphincter of smooth muscle is a collar of striped muscle called the **voluntary external sphincter.** The external sphincter is divided into three parts: (1) subcutaneous, (2) superficial, and (3) deep. The attachment of these parts are given in *Table 3–14.*

The **puborectalis** fibers of the two levatores ani muscles form a sling, which is attached anteriorly to the pubic bones. The muscular sling pases backward around the junction of the rectum and the anal canal pulling them forward so that the rectum joins the anal canal at an acute angle.

At the junction of the rectum and the anal canal, the internal sphincter, the deep part of the external sphincter, and the puborectalis form a distinct ring called the **anorectal ring,** which can be palpated on rectal examination.

Blood Supply of the Anal Canal. *Arteries.* The superior rectal artery supplies the upper half and the inferior rectal artery the lower half. *Veins.* The upper half is drained by the superior rectal vein into the inferior mesenteric vein; the lower half is drained by the inferior rectal vein into the systemic circulation. The anastomosis between the rectal veins forms an important portal–systemic anastomosis.

Lymphatic Drainage of the Anal Canal. Lymph from the upper half ascends to the pararectal nodes and joins the inferior mesenteric nodes. Lymph from the lower half drains into the medial group of superficial inguinal nodes.

Nerve Supply of the Anal Canal. The mucous membrane of the upper half is sensitive to stretch and is innervated by fibers that ascend through the hypogastric plexuses. The lower half is sensitive to pain, temperature, touch, and pressure and is innervated by the inferior rectal nerves.

The internal anal sphincter is supplied by sympathetic fibers from the inferior hypogastric plexus. The voluntary external anal sphincter is supplied by the inferior rectal nerves.

Defecation

The act is preceded by a wave of peristalsis, which passes down the descending and sigmoid parts of the colon. The rectum becomes distended by the entrance of the feces, which gives rise to the desire to defecate. If the time and place are favorable, a coordinated reflex act occurs that results in the emptying of the descending colon, sigmoid colon, rectum, and anal canal. The intra-abdominal pressure is raised by the descent of the diaphragm, the closure of the glottis, and the contraction of the muscles of the anterior abdominal walls and the levatores ani muscles. The external pressure applied to the colon and the waves of peristalsis in the wall of the colon force the feces onward. The tonic contraction of the internal and external anal sphincters, including the puborectalis part of the levator ani muscles, is now voluntarily inhibited. The feces are evacuated through the anal canal. Depending on the laxity of the submucous coat, the mucous membrane of the lower part of the anal canal is extruded through the anus ahead of the fecal mass. At the end of the act, the mucosa is returned to the anal canal by the tone of the longitudinal smooth muscle fibers of the anal walls and the contraction and upward pull of the levatores ani muscle. The empty lumen of the anal canal is now closed by the tonic contraction of the anal sphincters.

LIVER

The liver is the largest organ in the body (*Fig. 11–1*). It is soft and pliable and occupies the upper part of the abdominal cavity just beneath the diaphragm. It is located in the right hypochondrium and the epigastrium. The greater part of the liver is situated under cover of the ribs and the costal cartilages and the diaphragm separates it from the pleura, the lungs, the pericardium, and the heart. The convex upper surface of the liver is molded to the undersurface of the domes of the diaphragm. The undersurface or visceral surface is molded to adjacent viscera and is therefore irregular in shape; it lies in contact with a small part of the esophagus, the stomach, the duodenum, the right colic flexure, the right kidney and suprarenal gland, and the gallbladder.

The liver may be divided into a large **right lobe** and a small **left lobe** by the attachment of the peritoneum of the falciform ligament (*Fig. 11–5*). The right lobe is further divided into a **quadrate lobe** and a **caudate lobe** by the presence of the gallbladder, the fissure for the ligamentum teres, the inferior vena cava, and the fissure for the ligamentum venosum. It has been shown that the quadrate and caudate lobes are a functional part of the left lobe of the liver. Thus, the right and left branches of the hepatic artery and portal vein, and the right and left hepatic ducts, are distributed to the right lobe and the left lobe (plus quadrate plus caudate lobes), respectively. There is apparently very little overlap between the two sides.

The liver is partially surrounded by peritoneum and is completely enclosed by a fibrous capsule. The structure of the liver is organized into a large number of basic units known as the **liver lobules.** The so-called **portal canals** are found in the connective septa between lobules. These canals include branches of the hepatic artery, the portal vein, and the bile ducts.

Porta Hepatis, Fissures, Grooves, and Fossae. The **porta hepatis** or hilus of the liver is found on the posteroinferior surface (*Fig. 11–5*). It lies between the caudate and the quadrate lobes

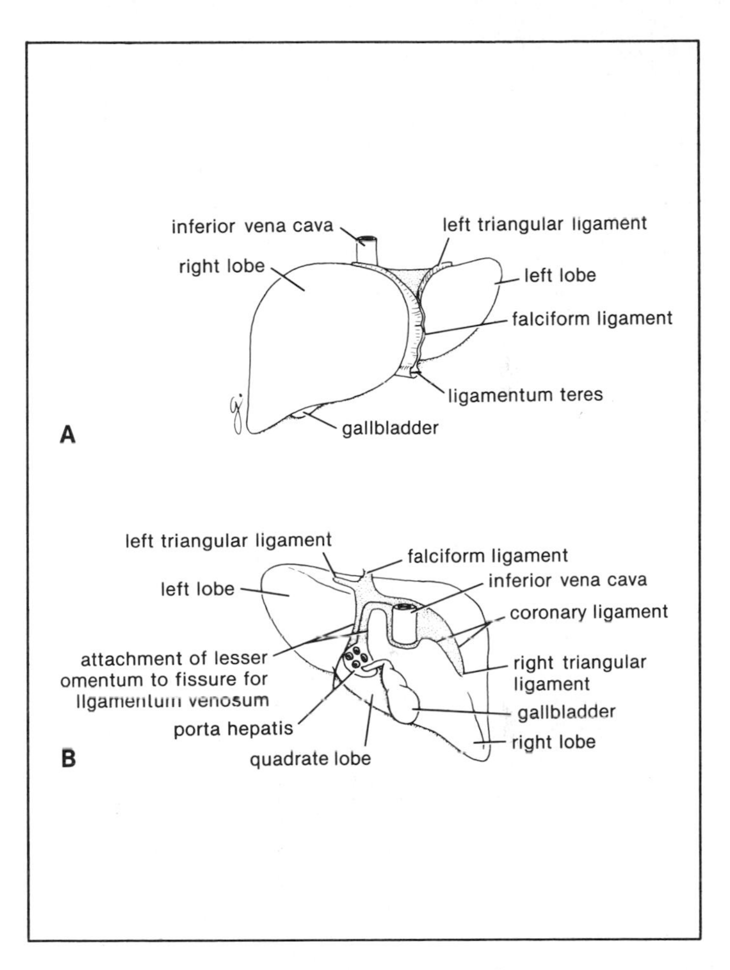

Figure 11–5. Liver. **A.** Anterior view. **B.** Posterior view.

and has attached to its margins the upper part of the free edge of the lesser omentum. Within the porta hepatis lie the right and left hepatic ducts, the right and left branches of the hepatic artery and portal vein, nerves, and lymphatics. The hepatic ducts are in front of the arteries and the branches of the portal vein are posterior.

The **fissure for the ligamentum teres** contains the ligamentum teres, which is the fibrous remains of the **umbilical vein.** It lies between the left lobe and the quadrate lobe.

The **fissure for the ligamentum venosum** contains the ligamentum venosum, which is the fibrous remains of the **ductus venosus.** It lies between the left lobe and the caudate lobe and has attached to its margins the upper part of the lesser omentum (*Fig. 11–5*).

The **groove for the inferior vena cava** is occasionally a tunnel. It lodges the inferior vena cava and it is here that the hepatic veins join the inferior vena cava. The groove lies between the right lobe and the caudate lobe (*Fig. 11–5*), and is separated from the porta hepatis by the caudate process.

The **fossa for the gallbladder** lies between the right lobe and the quadrate lobe (*Fig. 11–5*). There is no peritoneum between the gallbladder and the right lobe of the liver.

Peritoneal Ligaments of the Liver. The **falciform ligament** is a two-layered fold of peritoneum that is attached above to the diaphragm and below to the anterior abdominal wall down as far as the umbilicus (*Fig. 11–5*). It has a sickle-shaped free margin that contains the **ligamentum teres**, the remains of the umbilical vein. The falciform ligament passes to be attached to the anterior and then the superior surface of the liver. The right layer then forms the upper layer of the coronary ligament; the left layer forms the upper layer of the left triangular ligament.

The **coronary ligament** is formed by the reflexion of the peritoneum from the superior and posterior surfaces of the liver to the diaphragm (*Fig. 11–5*). The peritoneal layers forming the coronary ligament are widely separated, leaving a "bare area" of liver devoid of peritoneum.

The **right triangular ligament** is a V-shaped fold of peritoneum formed from the right extremity of the coronary ligament (*Fig. 11–5*). It connects the posterior surface of the right lobe of the liver to the diaphragm.

The **left triangular ligament** is formed by the reflexion of the peritoneum from the upper surface of the left lobe of the liver to the diaphragm (*Fig. 11–5*). Its left free edge lies anterior to the abdominal part of the esophagus.

The **lesser omentum** is described on *page 410*. Its upper end is attached to the margins of the porta hepatis and the fissure for the ligamentum venosum. It passes down to be attached to the lesser curvature of the stomach.

Blood Supply of the Liver. The **hepatic artery,** a branch of the celiac artery, divides into right and left terminal branches that enter the porta hepatis. The **portal vein** divides into right and left terminal branches that enter the porta hepatis behind the arteries. The **hepatic veins,** often three or more in number, emerge from the posterior surface of the liver and immediately drain into the inferior vena cava.

Lymphatic Drainage of the Liver. The lymph enters a number of lymph nodes in the porta hepatis and then drains to the celiac nodes. Some lymph vessels from the bare area pass up to the posterior mediastinal nodes.

Nerve Supply of the Liver. Sympathetic and parasympathetic (vagal) fibers from the celiac plexus. The left vagal trunk gives rise to a large hepatic branch, which travels directly to the liver.

GALLBLADDER AND BILE DUCTS

Gallbladder

The gallbladder is a pear-shaped sac lying on the undersurface (visceral surface) of the liver (*Fig. 11–6*). Its long axis is directed upward, backward, and to the left. It is divided into the fundus,

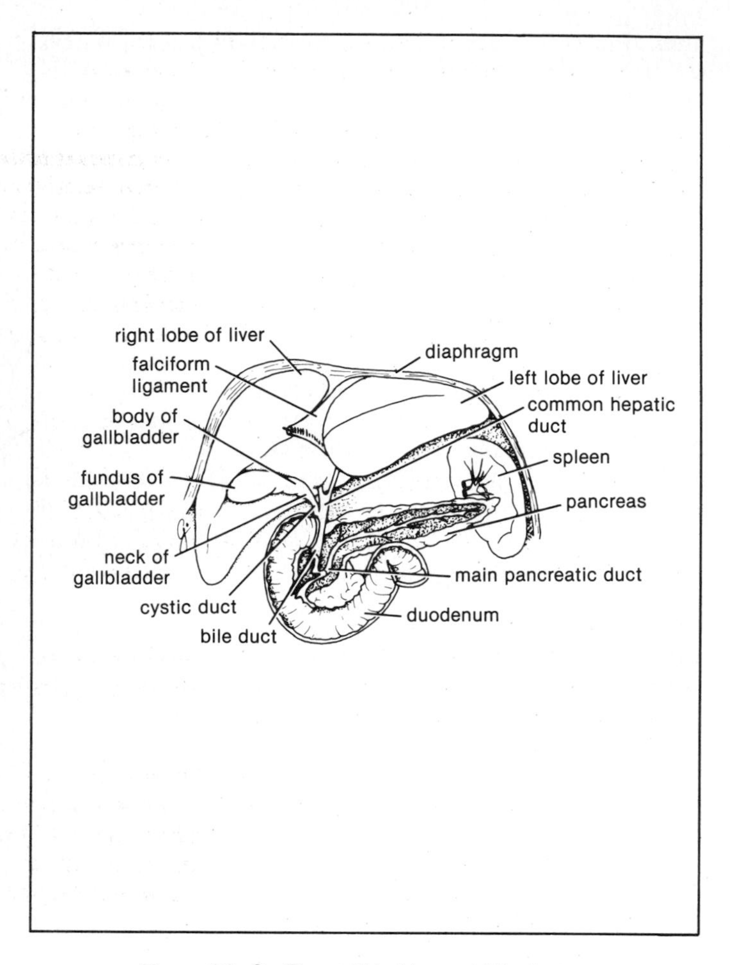

Figure 11–6. The gallbladder and bile ducts.

body, and neck. The **fundus** is rounded and usually projects below the inferior margin of the liver, where it is related to the anterior abdominal wall at the level of the tip of the right ninth costal cartilage. The **body** lies in direct contact with the visceral surface of the liver. The **neck** becomes continuous with the cystic duct.

The peritoneum completely covers the fundus of the gallbladder and binds the body and the neck to the visceral surface of the liver.

Relations of the Gallbladder. *Anteriorly.* Anterior abdominal wall, visceral surface of liver. *Posteriorly.* Transverse colon, first and second parts of duodenum.

Blood Supply of the Gallbladder. *Arteries.* Cystic artery, branch of right hepatic artery. *Veins.* Cystic vein drains into portal vein.

Lymphatic Drainage of the Gallbladder. Cystic lymph node located near neck of gallbladder then hepatic nodes and finally celiac nodes.

Nerve Supply of the Gallbladder. Sympathetic and parasympathetic vagal fibers from the celiac plexus.

Bile Ducts

Cystic Duct. The **cystic duct** is an S-shaped duct, about 4 cm (1½ inches) long (*Fig. 11–6*). It connects the neck of the gallbladder to the common hepatic duct to form the bile duct.

Hepatic Ducts. The **right** and **left hepatic ducts** emerge from the right and left lobes of the liver in the porta hepatis. Each hepatic duct has been formed by the union of small bile ducts within the liver known as **bile canaliculi.** After a short course the hepatic ducts unite to form the common hepatic duct (*Fig. 11–6*).

The **common hepatic duct** is about 4 cm (1½ inches) long and descends within the lesser omentum. It is joined on the right side by the cystic duct from the gallbladder to form the bile duct (*Fig. 11–6*).

Bile Duct. The bile duct (common bile duct) is about 8 cm (3 inches) long and lies in the right free edge of the lesser omentum having the portal vein behind and the hepatic artery on the left. It descends in front of the opening into the lesser sac and passes behind the first part of the duodenum (*Fig. 11–6*). The bile duct then lies in a groove on the posterior surface of the head of the pancreas. The bile duct ends below by piercing the medial wall of the duodenum about halfway down its length (*Figs. 11–4 and 11–6*). It is usually joined by the main pancreatic duct, and together they open into a small ampulla in the duodenal wall, called the **ampulla of Vater.** The ampulla itself opens into the lumen of the duodenum by means of a small papilla, the **major duodenal papilla** (*Fig. 11–4*). The terminal parts of both ducts and the ampulla are surrounded by circular smooth muscle fibers, known as the **sphincter of Oddi.** Occasionally, the bile and pancreatic ducts open separately into the duodenum.

PANCREAS

The pancreas is a soft, lobulated organ that lies on the posterior abdominal wall behind the stomach and behind the peritoneum (*Fig. 11–1*). It is an elongated structure that may be divided into a head, neck, body, and tail (*Fig. 11–4*). The **head** is disc-shaped and lies within the concavity of the C-shaped duodenum. The **uncinate process** is a projection that extends to the left behind the superior mesenteric vessels. The **neck** is narrow and connects the head to the body. It lies anterior to the beginning of the portal vein. The **body** passes upward and to the left across the midline. The **tail** extends forward to the hilus of the spleen in the lienorenal ligament.

Relations of the Pancreas. *Anteriorly.* Transverse colon, transverse mesocolon, lesser sac of peritoneum, stomach. *Poste-*

riorly. Bile duct, portal and splenic veins, inferior vena cava, aorta, left psoas muscle, left suprarenal gland, left kidney, hilus of spleen.

Pancreatic Ducts. The **main pancreatic duct** runs the length of the gland and opens into the second part of the duodenum with the bile duct on the major duodenal papilla (*Fig. 11-4*). Sometimes the main duct drains separately into the duodenum. The **accessory duct,** when present, drains the upper part of the head and then opens into the duodenum on the minor duodenal papilla.

Blood Supply of the Pancreas. *Arteries.* Splenic artery and superior and inferior pancreaticoduodenal arteries. *Veins.* The splenic vein joins the superior mesenteric vein to form the portal vein. The remaining veins drain into the portal vein.

Lymphatic Drainage of the Pancreas. Lymph nodes situated along the arteries and then into celiac and superior mesenteric nodes.

Nerve Supply of the Pancreas. Sympathetic and parasympathetic vagal nerve fibers from the celiac plexus.

PERITONEUM

The peritoneum is the serous membrane lining the abdominal and pelvic cavities and clothing the viscera (*Figs. 11-7 and 11-8*). The **parietal layer** lines the walls of the abdominal and pelvic cavities and the **visceral layer** covers the organs. The potential space between the parietal and visceral layers of peritoneum is called the **peritoneal cavity.** In the male this is a closed cavity, but in the female there is a communication with the exterior through the uterine tubes, the uterus, and the vagina. The peritoneal cavity may be divided into two parts, the greater sac and the lesser sac (*Fig. 11-7*). The **greater sac** is the main compartment of the per-

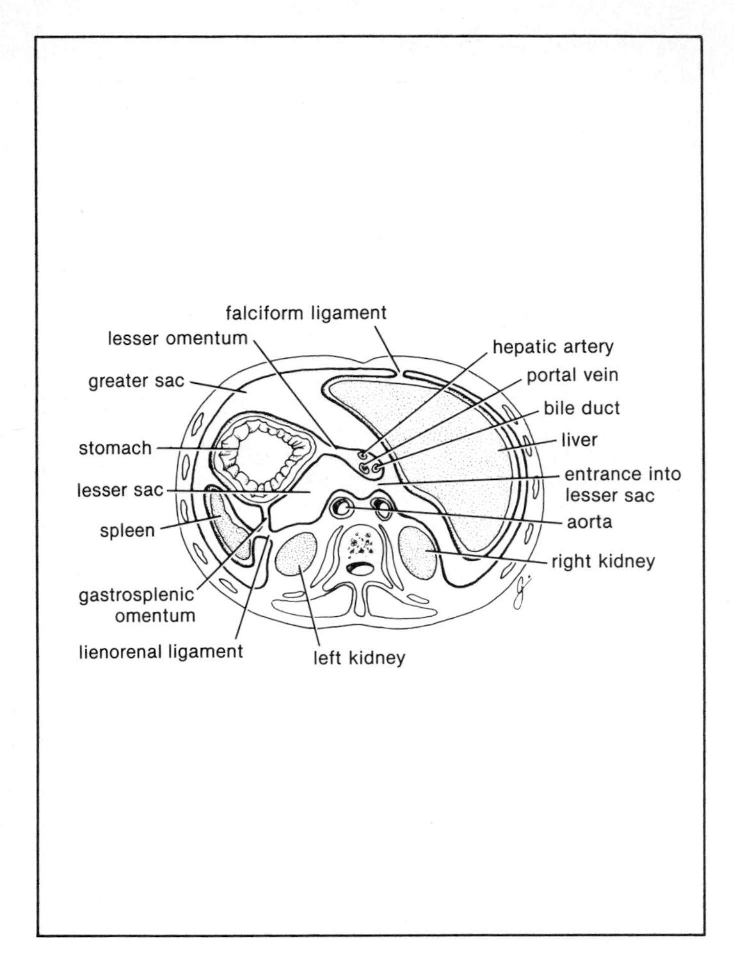

Figure 11–7. Transverse section of the abdomen at the level of the twelfth thoracic vertebra showing the arrangement of the peritoneum.

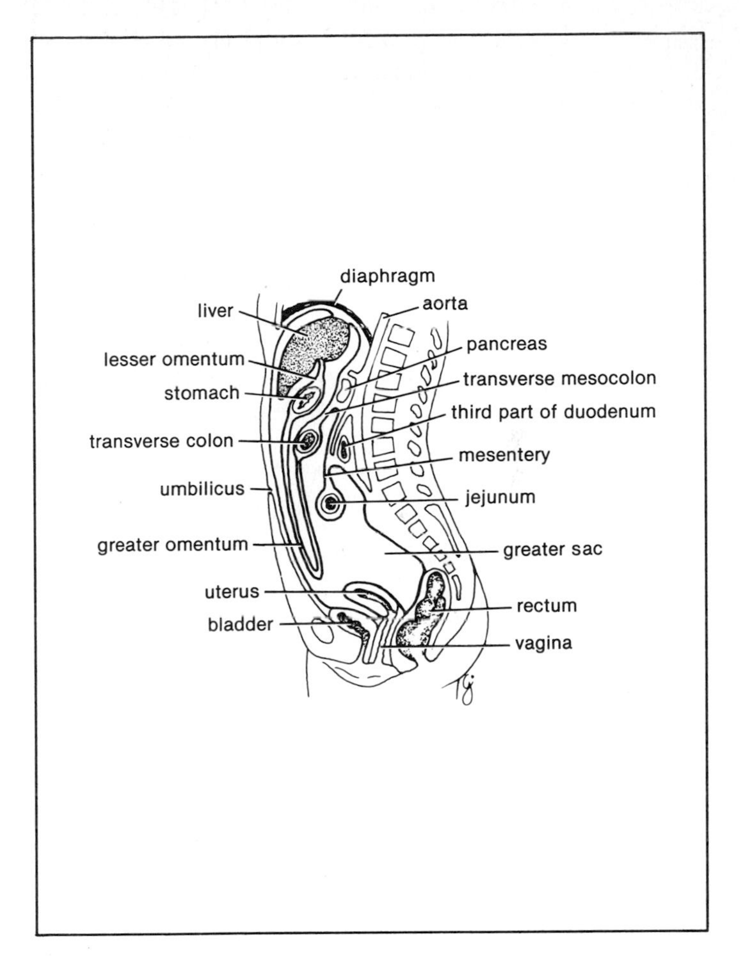

Figure 11–8. Sagittal section of the female abdomen, showing the arrangement of the peritoneum.

itoneal cavity and extends across the whole breadth of the abdomen, and from the diaphragm down into the pelvis. The **lesser sac** is the smaller compartment and lies behind the stomach. The greater and lesser sacs are in free communication with one another. The peritoneum secretes a small amount of serous fluid, which lubricates the surfaces of the peritoneum and facilitates free movement between the viscera.

Peritoneal Ligaments, Omenta, and Mesenteries

Peritoneal ligaments are two-layered folds of peritoneum that attach solid viscera to the abdominal walls. The liver, for example, is attached to the diaphragm by the **falciform ligament,** the **coronary ligament,** and the **right and left triangluar ligaments** (*Figs. 11-5, 11-7, and 11-8*).

Omenta are two-layered folds of peritoneum that connect the stomach to another viscus. The **greater omentum** connects the greater curvature of the stomach to the transverse colon (*Fig. 11-8*). It hangs down like an apron in front of the coils of the small intestine and is folded back on itself. The **lesser omentum** suspends the lesser curvature of the stomach to the fissure for the ligamentum venosum and the porta hepatis of the liver (*Fig. 11-8*). The **gastrosplenic omentum** (ligament) attaches the stomach to the hilus of the spleen.

Mesenteries are two-layered folds of peritoneum attaching parts of the intestines to the posterior abdominal wall, for example, the mesentery of the small intestine, the transverse mesocolon, and the sigmoid mesocolon (*Fig. 11-8*).

The peritoneal ligaments, omenta, and mesenteries permit blood, lymphatic vessels, and nerves to reach the various viscera.

Intraperitoneal and Retroperitoneal Relationships

The terms intraperitoneal and retroperitoneal are used to describe the relationship of various organs to their peritoneal covering. An organ is said to be intraperitoneal when it is almost totally covered with visceral peritoneum. The stomach, the jejunum, the ileum, and the spleen are good examples of intraperitoneal organs. Retroperitoneal organs are those that lie behind the peritoneum and are only partially coverd with visceral peritoneum. The pancreas and the ascending and descending parts of the colon are examples of retroperitoneal organs. It should be emphasized from the above definitions that no organ is actually within the peritoneal cavity. An "intraperitoneal" organ, such as the stomach, appears to be surrounded by the peritoneal cavity, but it is covered with visceral peritoneum and is attached to other organs by omenta.

Peritoneal Sacs and Fossae

Lesser Sac. The lesser sac is an extensive peritoneal pouch behind the stomach and lesser omentum (*Figs. 11–7 and 11–8*). It extends upward as far as the diaphragm and downward between the layers of the greater omentum. The left margin is formed by the spleen and the gastrosplenic omentum and lienorenal ligament. The right margin of the sac opens into the greater sac, i.e., the main part of the peritoneal cavity, through the **epiploic foramen.**

The boundaries of the epiploic foramen are: *Anteriorly.* Free border of lesser omentum, bile duct, hepatic artery, portal vein. *Posteriorly.* Inferior vena cava. *Superiorly.* Caudate process of caudate lobe of liver. *Inferiorly.* First part of duodenum.

Duodenal Fossae. Close to the duodenojejunal junction there may be four small pouches of peritoneum called the **superior duodenal fossa,** the **inferior duodenal fossa,** and the **paraduodenal fossae.**

Cecal Fossae. Folds of peritoneum in the vicinity of the cecum produce three peritoneal fossae called the **superior ileocecal,** the **inferior ileocecal,** and the **retrocecal fossae.**

Subphrenic Spaces. The complicated arrangement of the peritoneum in the region of the liver leaves spaces between the diaphragm and the liver that are called the **right and left anterior**

and posterior subphrenic spaces. These spaces are important clinically because they may provide sites for the accumulation of pus in patients with peritonitis.

Paracolic Gutters. The arrangement of the peritoneum covering the ascending colon and descending colon, the attachments of the transverse mesocolon and the mesentery of the small intestine to the posterior abdominal wall, result in the formation of four important paracolic gutters. These gutters lie on the lateral and medial sides of the ascending and descending colons, respectively, (*Fig. 11–7*) and provide channels for the movement of infected fluid in the peritoneal cavity in patients with peritonitis.

Nerve Supply of the Peritoneum

The **parietal peritoneum** is sensitive to pain, temperature, touch, and pressure. The parietal peritoneum lining the anterior and lateral abdominal walls is supplied by the lower six thoracic and first lumbar nerves, i.e., the same nerves that innervate the overlying muscles and skin. The parietal peritoneum in the pelvis is mainly supplied by the obturator nerve. The **visceral peritoneum** and the peritoneum that forms the mesenteries is sensitive to stretch and is innervated by the autonomic nerves that supply the viscera or are travelling in the mesenteries.

To understand the arrangement of the peritoneum it is helpful to trace the peritoneum around the abdominal cavity, first, in a horizontal and then in a vertical direction (*Figs. 11–7 and 11–8*).

Horizontal Disposition of the Peritoneum at the Level of the Twelfth Thoracic Vertebra

The parietal peritoneum lining the anterior abdominal wall forms a sickle-shaped fold called the **falciform ligament.** This connects the anterior surface of the liver to the anterior abdominal wall above the umbilicus and to the diaphragm. It lies slightly to the right of the midline. In the free border of the ligament, where the two layers of peritoneum are continuous with each other, lies the **ligamentum teres** (*Fig. 11–7*).

If the parietal peritoneum is followed around the abdominal wall on the left side, it reaches the lateral margin of the left kidney (*Fig. 11–7*). Here, it becomes continuous with the visceral peritoneum covering the lateral margin and part of the anterior surface of the left kidney. The peritoneum then leaves the kidney and passes to the hilus of the spleen as the posterior layer of the **lienorenal ligament.** The visceral peritoneum covers the spleen and, on reaching the hilus again, is reflected onto the greater curvature of the stomach as the anterior layer of the **gastrosplenic omentum** (ligament). The visceral peritoneum covers the anterior surface of the stomach and leaves the lesser curvature to form the anterior layer of the **lesser omentum.** On the right, the lesser omentum has a free border, and here the peritoneum folds around the **bile duct,** the **hepatic artery,** and the **portal vein.** The free border of the lesser omentum forms the anterior margin of the opening into the lesser sac.

The peritoneum forms the posterior layer of the **lesser omentum** and becomes continuous with the visceral layer of peritoneum covering the posterior wall of the stomach. Here, the peritoneum forms the anterior wall of the lesser sac. At the greater curvature of the stomach, the peritoneum leaves the stomach, forming the posterior layer of the **gastrosplenic omentum** (ligament) and reaches the hilus of the spleen. Here, it is reflected backward to the posterior abdominal wall, forming the anterior layer of the **lienorenal ligament.** The peritoneum now covers the anterior surface of the pancreas, the aorta, and the inferior vena cava, forming the posterior wall of the lesser sac. The peritoneum passes onto the anterior surface of the right kidney and sweeps around the lateral abdominal wall to reach the anterior abdominal wall. Note that the peritoneum forms a continuous layer around the abdomen (*Fig. 11–7*).

Vertical Disposition of the Peritoneum

The parietal peritoneum lining the anterior abdominal wall may be traced upward to the left of the falciform ligament to reach the undersurface of the diaphragm (*Fig. 11–8*). Here, it is reflected onto the upper surface of the liver as the anterior layer of the **left triangular ligament.** The visceral peritoneum then covers the anterior and inferior surfaces of the liver until it reaches the **porta hepatis.** Here, the peritoneum passes to the lesser curvature of the stomach as the anterior layer of the **lesser omentum.** The peritoneum then covers the anterior surface of the stomach, and leaves the greater curvature forming the anterior layer of the **greater omentum.**

The greater omentum hangs down as a fold in front of the coils of intestine and contains within it the lower part of the lesser sac. Having reached the lowest limit of the greater omentum, the peritoneum folds upward and forms the posterior layer of the greater omentum. On reaching the inferior border of the transverse colon, the peritoneum covers its posterior surface and then leaves the colon to form the posterior layer of the **transverse mesocolon.** The peritoneum then passes to the anterior border of the pancreas and runs downward anteriorly to the third part of the duodenum.

The peritoneum now leaves the posterior abdominal wall as the anterior layer of the **mesentery of the small intestine.** The visceral peritoneum covers the jejunum and then forms the posterior layer of the mesentery. On returning to the posterior abdominal wall, the peritoneum runs downward into the pelvis and covers the anterior surface of the upper part of the rectum. From here, it is reflected onto the posterior surface of the upper part of the vagina, forming the important **rectouterine pouch (pouch of Douglas).** In the male the peritoneum is reflected onto the upper part of the posterior surface of the bladder and the seminal vesicles, forming the **rectovesical pouch.**

The peritoneum passes over the upper surface of the uterus in the female and is reflected from its anterior surface onto the surface of the bladder. In both sexes the peritoneum passes from the bladder onto the anterior abdominal wall. Once again note that the peritoneum forms a continuous layer around the abdomen (*Fig. 11-8*).

12

Urinary System

The urinary system consists of two kidneys situated on the posterior abdominal wall, two ureters, which run down on the posterior abdominal wall and enter the pelvis, one urinary bladder located within the pelvis, and one urethra, which passes through the perineum (*Fig. 12–1*). The urethra in the male not only conducts urine to the surface but is an excretory duct for the reproductive system, conveying the semen to the exterior. The function of the urinary system is to excrete from the body most of the waste products of cellular metabolism and to play a major role in controlling the water and electrolyte balance within the body.

KIDNEYS

The kidneys are paired reddish-brown organs that are bean shaped. Each adult kidney measures about 11.25 cm (4 inches) long, 5 to 7.5 cm (2 to 3 inches) wide, and 2.5 cm (1 inch) thick. The kidneys lie behind the peritoneum high up on the posterior abdominal wall on either side of the vertebral column (*Fig. 12–1*). They are located above the waistline on the transpyloric plane and are largely under cover of the costal margin being partially protected by the eleventh and twelfth pairs of ribs. The right kid-

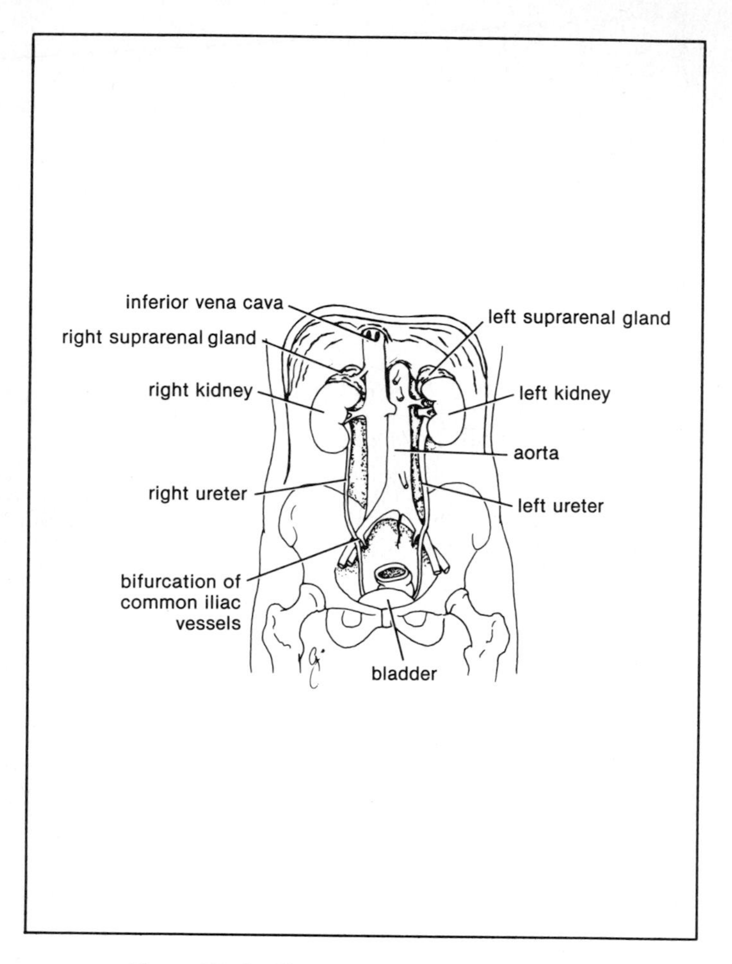

Figure 12–1. The parts of the urinary system.

ney lies slightly lower than the left kidney, due to the large size of the right lobe of the liver. With contraction of the diaphragm during respiration, both kidneys move downward in a vertical direction by as much as 2.5 cm (1 inch).

The kidneys are surrounded by a **fibrous capsule**, which is closely applied to its outer surface (*Fig. 12–2*). Outside the fibrous capsule is a covering of fat known as **perirenal fat (perinephric fat)**. The **renal fascia** surrounds the perirenal fat and encloses the kidneys and the suprarenal glands. The renal fascia is a condensation of areolar tissue. Both the perirenal fat and the renal fascia support the kidneys and hold them in position on the posterior abdominal wall.

On the medial concave border of each kidney is a vertical slit, which is bounded by thick lips of renal substance and is called the **hilus** (*Fig. 12–2*). Extending inward from the hilus is a large cavity, the **renal sinus**. The hilus transmits the renal pelvis, which is the upper expanded end of the ureter, the renal artery, the renal vein, lymph vessels, and sympathetic nerve fibers.

Relations of the Kidneys

Right Kidney. Anteriorly. Right lobe of liver, second part of duodenum, right colic flexure, coils of small intestine, right suprarenal gland. *Posteriorly.* Diaphragm, twelfth rib, psoas, quadratus lumborum, and transversus abdominis muscles, subcostal (T12), iliohypogastric, ilioinguinal nerves (L1).

Left Kidney. Anteriorly. Spleen, stomach, pancreas, left colic flexure, coils of jejunum, left suprarenal gland. *Posteriorly.* Diaphragm, eleventh and twelfth ribs, psoas, quadratus lumborum, transversus abdominis muscles, subcostal (T12), iliohypogastric, ilioinguinal nerves (L1) (*Fig. 12–3*).

Renal Structure

A coronal section through the kidney (*Fig. 12–2*) shows that it is made up of a dark reddish brown outer **cortex** and a lighter col-

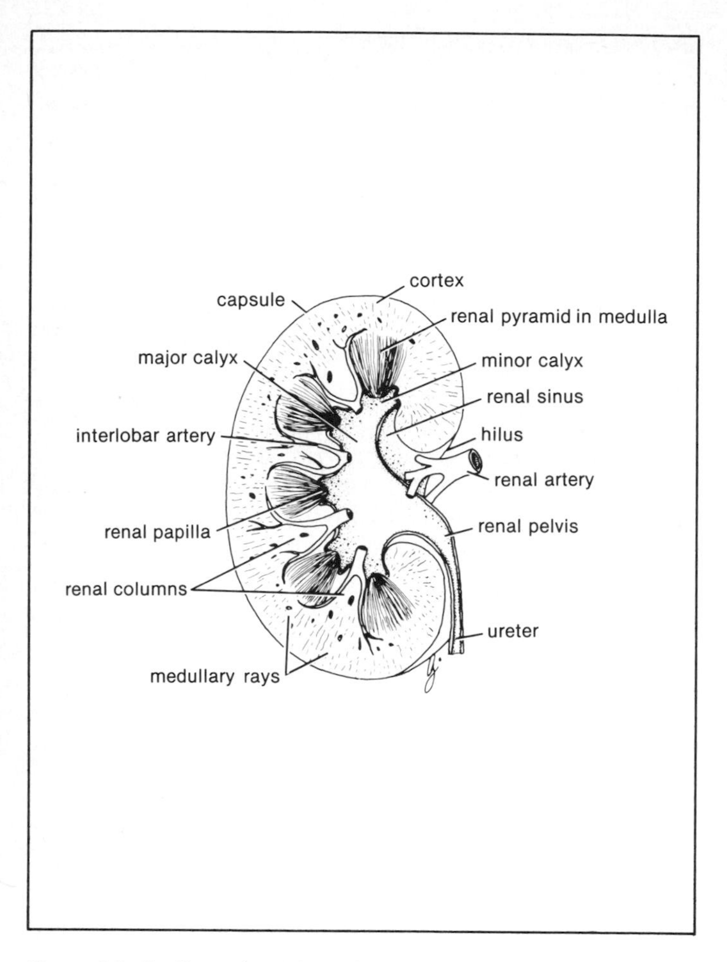

Figure 12–2. Coronal section of the kidney showing cortex, medulla, pyramids, renal papillae, and calyces.

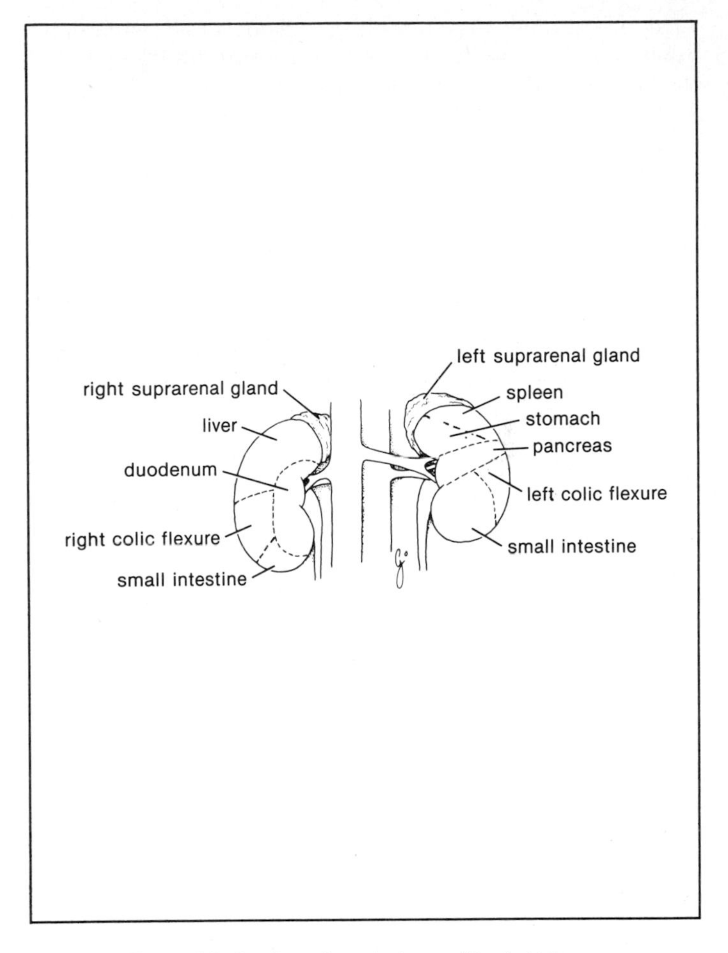

Figure 12–3. Anterior relations of both kidneys.

ored inner **medulla**. The medulla is composed of about a dozen **renal pyramids**, each having its base oriented toward the cortex and its apex, the **renal papilla**, projecting medially. The cortex extends into the medulla between adjacent pyramids as the **renal columns**. Extending from the bases of the renal pyramids into the cortex are striations known as **medullary rays**.

A **renal lobe** may be defined as a renal pyramid with the cortical tissue overlying its base and lying along its sides. A **renal lobule** is a medullary ray and its neighboring cortical tissue. Renal lobes and lobules are not separated from one another by connective tissue septa as is seen in many other organs in the body.

Within the renal sinus, the upper expanded end of the ureter, the **renal pelvis,** divides into two or three **major calyces**, each of which divides into two or three **minor calyces** (*Fig. 12–2*). Each minor calyx is indented by the apex of the renal pyramid, the **renal papilla**. Within the renal sinus the tributaries of the renal vein lie anteriorly, whereas posteriorly is the renal pelvis, with branches of the renal artery in front and behind.

Blood Supply of the Kidneys. *Arteries.* Renal artery a branch of the aorta. *Veins.* Renal vein drains into the inferior vena cava.

Lymphatic Drainage of the Kidneys. Lateral aortic lymph nodes around the origin of the renal artery.

Nerve Supply of the Kidneys. Renal sympathetic plexus.

URETERS

The two ureters are muscular tubes that extend from the kidneys to the posterior surface of the urinary bladder (*Fig. 12–1*). Each ureter measures about 25 cm (10 inches) long and less than 1.25 cm (½ inch) in diameter. At its upper end the ureter is expanded to form a funnel called the renal pelvis. The renal pelvis lies within the hilus of the kidney where it receives the major calyces.

The ureter emerges from the hilus of the kidney and runs vertically downward behind the parietal peritoneum on the psoas muscle. It enters the pelvis by crossing the bifurcation of the common iliac artery in front of the sacroiliac joint (*Fig. 12–1*).

Relations of the Ureters in the Abdomen

Right Ureter. Anteriorly. Peritoneum, duodenum, terminal part of ileum, right colic and ileocolic vessels, right testicular or ovarian vessels, root of mesentery of small intestine. *Posteriorly.* Psoas, which separates it from the transverse processes of the lumbar vertebrae, bifurcation of right common iliac artery.

Left Ureter. Anteriorly. Peritoneum left colic vessels, left testicular or ovarian vessels, sigmoid colon. *Posteriorly.* Psoas, which separates it from the transverse processes of the lumbar vertebrae, bifurcation of left common iliac artery.

The inferior mesenteric vein lies along the medial side of the left ureter.

Relations of the Ureters in the Pelvis. Each ureter runs down the lateral wall of the pelvis to the region of the ischial spine and turns forward to enter the lateral angle of the bladder. The ureters pierce the bladder wall obliquely and this provides a valvelike action, that prevents a reverse flow of urine toward the kidneys as the bladder fills.

In the male the ureter is crossed near its termination by the vas deferens. In the female the ureter leaves the region of the ischial spine by turning forward and medially beneath the base of the broad ligament; here it is crossed by the uterine artery.

Ureteric Constrictions. The ureter possesses three constrictions: (1) where the renal pelvis joins the ureter, (2) where it is kinked as it crosses the pelvic brim, and (3) where it pierces the bladder wall.

Blood Supply of the Ureters. *Arteries.* (1) Upper end, the renal artery, (2) middle portion, the testicular or ovarian artery, (3) inferior end, superior vesical artery. *Veins.* Veins that correspond to the arteries.

Lymphatic Drainage of the Ureters. Lateral aortic and iliac nodes.

Nerve Supply of the Ureters. Renal, testicular (or ovarian) and hypogastric plexuses.

URINARY BLADDER

The urinary bladder is located immediately behind the pubic bones within the pelvis (*Fig. 12–1*). Its function is to receive urine from the kidneys via the ureters and store it. The bladder has a maximum capacity of about 500 ml.

The empty bladder is pyramidal in shape, having an apex, a base, and a superior and two inferolateral surfaces (*Fig. 12–4*); it also has a neck. When the bladder fills, it loses its pyramidal shape and becomes ovoid; the posterior surface and neck are largely unchanged in position, but the superior surface rises into the abdomen. In the young child the empty bladder projects upward into the abdomen; later when the pelvis enlarges, the bladder sinks to become a pelvic organ.

The **apex** of the bladder points anteriorly and is connected to the umbilicus by the **median umbilical ligament** (remains of urachus). The **base** of the bladder faces posteriorly and is triangular in shape. The ureters enter the superolateral angles and the urethra leaves the inferior angle. The **superior surface** of the bladder is covered with peritoneum, which is reflected laterally on to the lateral pelvic walls. As the bladder fills, the superior surface bulges upward into the abdominal cavity peeling the peritoneum off from the lower part of the anterior abdominal wall. The **neck** of the bladder points inferiorly.

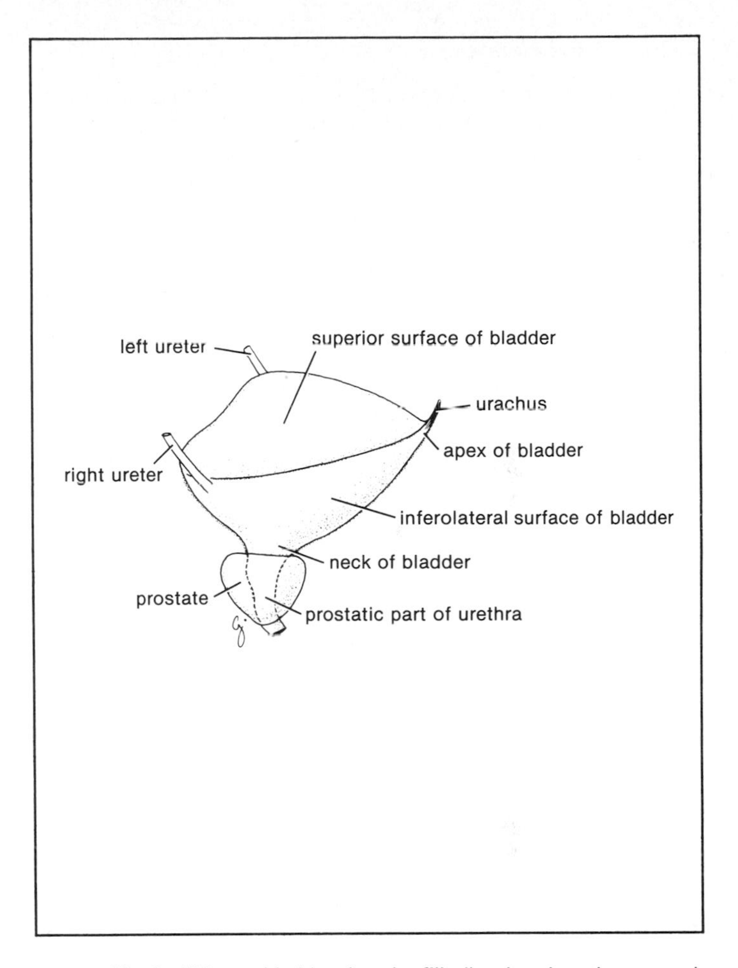

Figure 12–4. Urinary bladder (partly filled), showing the general shape and surfaces. Note the position of the apex anteriorly and the neck inferiorly.

Interior of the Bladder

The area of mucous membrane covering the internal surface of the base of the bladder is called the **trigone** (*Fig. 12–5*). Here the mucous membrane is always smooth, because it is firmly adherent to the underlying muscular coat. The trigone shows at its lateral angles the small, slitlike openings of the ureters and below the crescentic opening of the urethra. The trigone is limited above by a muscular ridge, which runs from one ureteric orifice to the other, and is known as the **interureteric crest or ridge**. In the male the median lobe of the prostate bulges upward into the bladder slightly, behind the urethral orifice, to form a swelling, the **uvula vesicae**.

Muscle Coat of the Bladder

The muscle coat, the **detrusor muscle**, consists of three layers of smooth muscle fibers, an outer longitudinal layer, a middle circular layer, and an inner longitudinal layer. At the neck of the bladder just above the exit of the urethra, the middle layer of circular muscle fibers forms the **sphincter vesicae**.

Relations of the Bladder. *Anteriorly.* Symphysis pubis. Retropubic pad of fat. *Superiorly in the Male.* Coils of ileum, sigmoid colon. *In the Female.* Uterovesical pouch, coils of ileum, body of uterus. *Posteriorly in the Male.* Vasa deferentia, seminal vesicles (*Fig. 12–5*), rectovesical pouch, coils of ileum, sigmoid colon. *In the Female.* Vagina and rectum. *Laterally in Both Sexes.* Obturator internus and levator ani muscles. *Inferiorly in the Male.* Prostate. *In the Female.* Urogenital diaphragm.

Ligaments of the Bladder. Anterior, lateral and posterior ligaments are formed of pelvic fascia.

Blood Supply of the Bladder. *Arteries.* Superior and inferior vesical arteries, branches of the internal iliac artery. *Veins.* Vesical veins drain into the internal iliac veins.

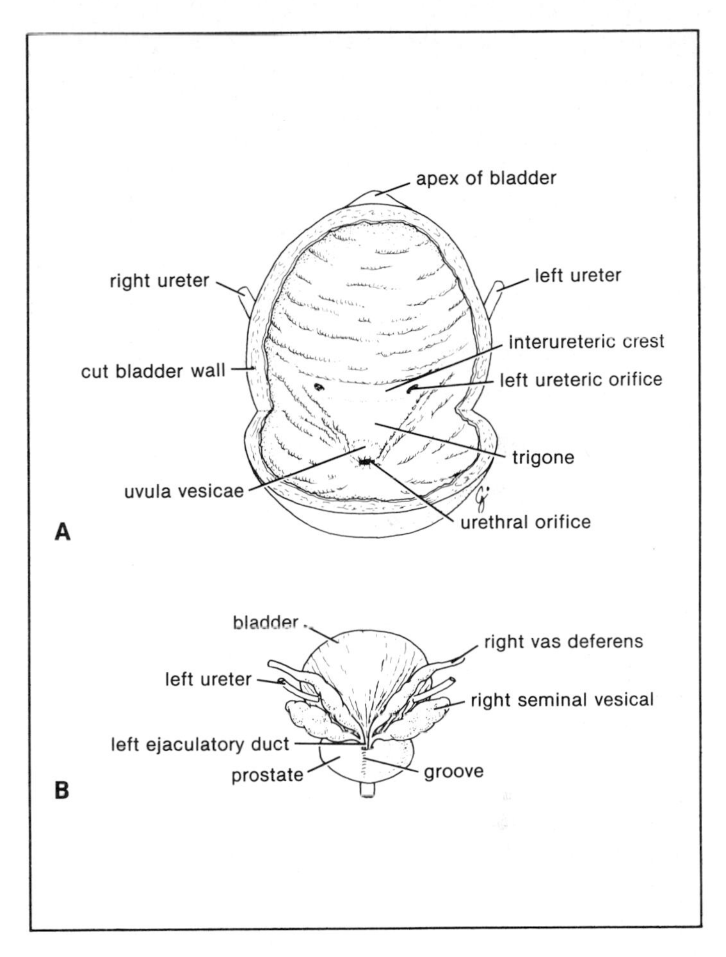

Figure 12–5. A. Interior of urinary bladder in the male, as seen from in front. **B.** Bladder, seminal vesicles, and prostate as seen from behind.

Lymphatic Drainage of the Bladder. Internal and external iliac nodes.

Nerve Supply of the Bladder. Sympathetic and parasympathetic fibers from the inferior hypogastric plexuses.

Micturition

The maximum capacity of the adult bladder is about 500 ml. When the volume of urine reaches about 300 ml, stretch receptors in the wall of the bladder transmit impulses to the central nervous system and the individual has a conscious desire to micturate.

Micturition Reflex. Micturition is a reflex action, which in the toilet-trained individual, is controlled by higher centers in the brain. The reflex is initiated by the stretching of the bladder muscle as the organ fills with urine. The afferent impulses enter the second, third, and fourth sacral segments of the spinal cord. Efferent impulses leave the cord from the same segments and pass via the parasympathetic preganglionic nerve fibers in the hypogastric plexuses to the bladder wall, where they synapse with postganglionic neurons. By means of this nervous pathway, the detrusor muscle of the bladder wall contracts, and the sphincter vesicae relaxes. Efferent impulses also pass to the urethral sphincter via the pudendal nerve and this undergoes relaxation. Once urine enters the urethra, additional afferent impulses pass to the spinal cord from the urethra and reinforce the reflex action. Micturition can be assisted by contraction of the abdominal muscles that raises the pressure within the abdominal and pelvic cavities and exerts pressure on the bladder from the outside.

In young children, micturition is a simple reflex act and takes place whenever the bladder becomes distended. In the adult, this simple stretch reflex is inhibited by the activity of the cerebral cortex until the time and place for micturition are favorable. The inhibitory fibers pass downward with the pyramidal tracts to the second, third, and fourth sacral segments of the cord. The contraction of the sphincter urethrae, which closes the urethra, is under voluntary control. Voluntary control of micturition is normally developed during the second or third year of life.

URETHRA

The urethra is a small tube leading from the neck of the bladder to the exterior. The opening of the urethra on the surface is called the **urinary meatus**.

The **male urethra** is about 20 cm (8 inches) long and extends from the neck of the bladder to the external urinary meatus on the glans penis (*see Fig. 13-1*). It is divided into three parts: (1) prostatic, (2) membranous, and (3) penile. The **prostatic urethra** runs through the prostate from the base to the apex. It is the widest and most dilatable portion of the entire urethra. On the posterior wall is a longitudinal ridge called the **urethral crest**. On each side of this ridge is a groove called the **prostatic sinus** into which opens the prostatic glands. On the summit of the urethral crest is a depression, the **prostatic utricle**, on the edges of which open the two ejaculatory ducts.

The **membranous urethra** lies within the urogenital diaphragm and is surrounded by the sphincter urethrae. It extends from the apex of the prostate to the bulb of the penis. It is the shortest and least dilatable part of the urethra.

The **penile urethra** is surrounded by the erectile tissue of the bulb and the corpus spongiosum of the penis. The external meatus is the narrowest part of the entire urethra. The part of the urethra that lies within the glans penis is dilated to form the **fossa terminalis** (navicular fossa). The bulbourethral glands open into the penile urethra below the perineal membrane.

The **female urethra** is about 3.8 cm (1½ inches) long. It extends from the neck of the bladder to the vestibule of the perineum, where it opens about 2.5 cm (1 inch) below the clitoris. It traverses the sphincter urethrae and lies immediately in front of the vagina.

Sphincter Urethrae Muscle

The sphincter urethrae muscle lies within the urogenital diaphragm. It arises from the pubic arch and passes medially to surround the membranous part of the urethra. The muscle is innervated by the perineal branch of the pudendal nerve. The sphincter urethrae is under voluntary control and micturition is stopped by the muscle compressing the membranous urethra.

13

Male Reproductive System

The male reproductive system consists of a pair of gonads, the testes, their excretory ducts and the accessory glands, and the penis (*Fig. 13–1*). The excretory ducts on each side are the epididymis, the vas deferens, and the ejaculatory duct. The accessory glands are a pair of seminal vesicles, a pair of bulbourethral glands, and the prostate gland.

The external genital organs consist of the penis and the scrotum.

SCROTUM

The scrotum is a pouch of loose skin and superficial fascia that supports the testes and the epididymides (*Fig. 13–1*). Internally the scrotum is divided into right and left halves by a septum. The **skin** of the scrotum is thin, rugose, and pigmented. A raised ridge in the midline indicates the line of fusion of the two lateral labioscrotal swellings in the embryo. The **superficial fascia** is continuous with the superficial fascia of the anterior abdominal wall. The **fatty layer of Camper (Camper's fascia)** is, however, replaced in the scrotum by smooth muscle, the **dartos muscle**, which is inervated by sympathetic nerve fibers. The membranous layer

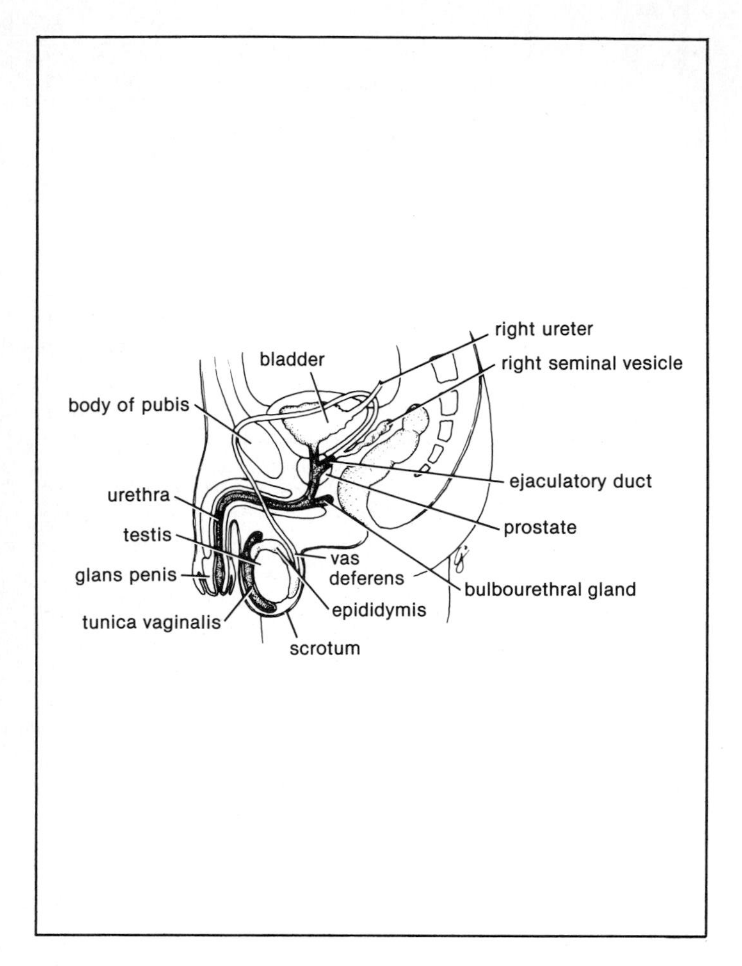

Figure 13–1. Male reproductive system as seen in sagittal section.

of the superficial fascia, the **fascia of Scarpa** of the anterior abdominal wall, becomes the **Colles' fascia**. It is attached posteriorly to the perineal body and the posterior edge of the perineal membrane. Laterally, it is attached to the ischiopubic rami. The superficial fascia contributes to the septum of the scrotum.

The **spermatic fasciae** are situated deep to the superficial fascia. The **external spermatic fascia** is derived from the external oblique aponeurosis (*see p. 105*). The **cremasteric fascia** is derived from the internal oblique muscle (*see p. 105*). The **internal spermatic fascia** is derived from the fascia transversalis. The cremaster muscle is supplied by the genital branch of the genitofemoral nerve (L1 and L2).

The **tunica vaginalis** is located deep to the spermatic fasciae. It covers the anterior, medial, and lateral surfaces of each testis. It is a closed sac that is invaginated from behind by the testis.

The scrotum is suspended outside the body cavities and its internal temperature is about 3°F below that of the body temperature. This is of great importance in providing a suitable environment for the production of spermatozoa in the testes, because normal development of spermatozoa (**spermatogenesis**) cannot take place at body temperature. Should the temperature of the scrotum fall, the **dartos muscle** in the scrotal wall contracts causing the testes to move closer to the body. At the same time the **cremaster muscle** in the spermatic cord and scrotal wall reflexly contracts elevating the testes toward the pelvis. A rise in temperature within the scrotum causes the dartos and cremaster muscles to relax so that the testes move away from the body and cool.

TESTES

The testes are paired ovoid organs measuring about 5 cm (2 inches) long and are slightly flattened from side to side (*Fig. 13–2*). They are responsible for the production of the male germ cells, the **spermatozoa**, and the male sex hormone, **testosterone**. Each testis is a mobile organ lying within one-half of the scrotum, the

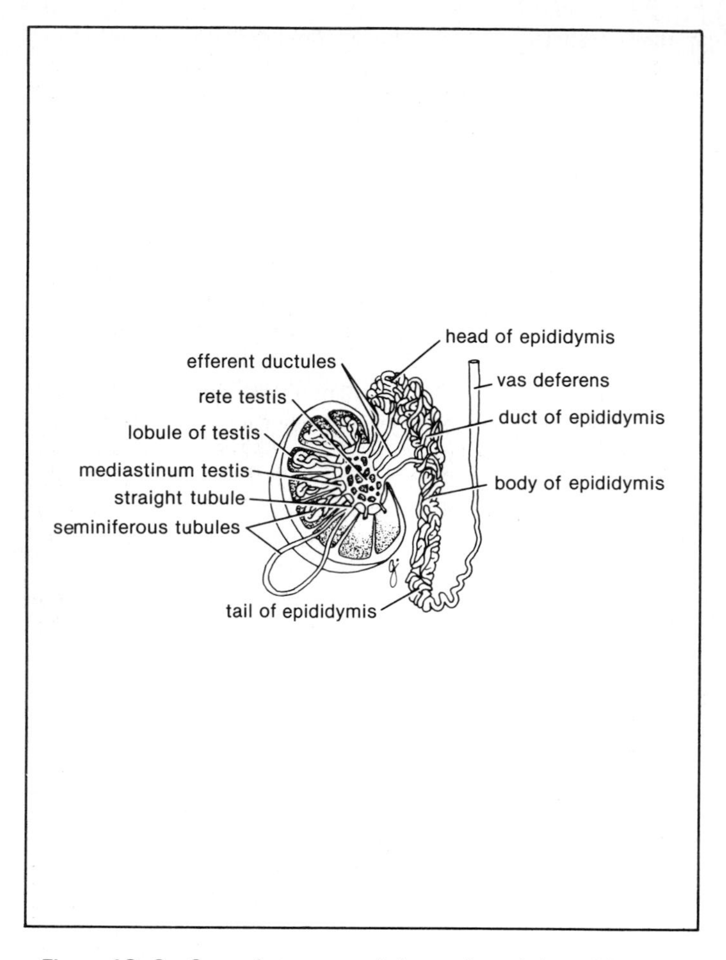

Figure 13–2. General structure of the testis and the epididymis.

left testis usually lies at a lower level than the right. In early fetal life, the testes are situated in the abdominal cavity, near the kidneys. As the fetus matures, the testes descend and just before birth pass through the inguinal canal to enter the scrotum. The descent of the testes from the abdominal cavity into the scrotum is important, because it has been found that spermatogenesis will only take place normally if the testes are at a temperature lower than that of the abdominal cavity (see above).

Each testis has a fibrous capsule, the **tunica albuginea** (*Fig. 13–2*), which is thickened posteriorly to form the **mediastinum testis**. Extending from the inner surface of the capsule to the mediastinum is a series of fibrous septa that divide the interior of the organ into about 250 lobules. Lying within each lobule are one to three coiled **seminiferous tubules**. Each tubule is in the form of a loop, the ends of which are continuous with a **straight tubule**. The straight tubules open into a network of channels within the mediastinum testis called the **rete testis**. Situated within each lobule between the seminiferous tubules are delicate connective tissue and groups of rounded **interstitial cells (Leydig cells)** that produce the male sex hormone **testosterone**. The rete testis is drained by efferent ductules into the long much-coiled duct, the **epididymis**, which is situated on the posterior surface of the testis (*Fig. 13–2*).

Blood Supply of the Testes. The **testicular artery** is a branch of the abdominal aorta. It descends on the posterior abdominal wall, passes through the inguinal canal, and supplies the testis and the epididymis. The **testicular vein** emerges from the testis and the epididymis as an extensive venous network, the **pampiniform plexus**. The plexus becomes reduced in size as it ascends through the inguinal canal and becomes a single vein at the deep inguinal ring. The right testicular vein drains into the inferior vena cava and the left vein joins the left renal vein.

Lymphatic Drainage of the Testes. Paraaortic lymph nodes on the side of the aorta at the level of the first lumbar vertebra.

Nerve Supply of the Testes. Sympathetic nerve fibers from renal or aortic sympathetic plexuses.

EPIDIDYMIS

The epididymis is a comma-shaped firm structure located along the posterior border of each testis in the scrotum (*Fig. 13–2*). It consists of a highly coiled tube held together by connective tissue. The larger superior portion is called the **head**, below this lies the **body**, and the narrow inferior portion is called the **tail**. Laterally, there is a groove between the testis and the epididymis, which is lined with the inner visceral layer of the tunica vaginalis, called the **sinus of the epididymis.**

The efferent ductules of the testis drain into the duct of the epididymis (*Fig. 13–2*). The duct of the epididymis measures about 6 m (20 ft) long and is embedded in connective tissue. The tube emerges from the tail of the epididymis to become the vas deferens, which enters the spermatic cord.

The function of the epididymis is to reabsorb fluid from the lumen and thus concentrate the spermatozoa. It also serves as a storage area for the spermatozoa and allows time for the sperm to undergo further maturation and become fully motile. The smooth muscle in the wall expels the sperm from the epididymis during ejaculation.

Blood Supply, Lymphatic Drainage, and Nerve Supply of the Epididymis. Same as for the testis.

VAS DEFERENS

The vas deferens is a single tube that measures about 45 cm (18 inches) long. It emerges from the lower end or tail of the epididymis and passes up through the inguinal canal into the abdomen (*Fig. 13–1*). It then descends into the pelvis and crosses the ureter

Lymphatic Drainage of the Prostate. Internal iliac nodes.

Nerve Supply of the Prostate. Inferior hypogastric plexuses.

BULBOURETHRAL GLANDS (COWPER'S GLANDS)

The bulbourethral glands are two small glands located on either side of the membranous urethra in the deep perineal pouch (*Fig. 13-1*). The ducts of the glands open into the penile urethra below the urogenital diaphragm. The secretion is mucuslike and is poured into the penile urethra as the result of erotic stimulation. The secretion lubricates the vagina during sexual intercourse.

SPERMATIC CORD

The spermatic cord is a collection of blood and lymphatic vessels and autonomic nerves that accompany the vas deferens through the inguinal canal. The inguinal canal is a slitlike passage on each side of the body that passes through the lower part of the anterior abdominal wall just above the medial half of the inguinal ligament. The blood vessels are the testicular artery and vein, which supply the testis and the epididymis. The lymphatic vessels drain the lymph from the testis and the epididymis up into the lymph nodes in the abdominal cavity (paraaortic lymph nodes). In addition there are the following: (1) the cremasteric artery, a branch of the inferior epigastric artery, which supplies the cremasteric fascia; (2) the artery to the vas deferens, a branch of the inferior vesical artery; and (3) the genital branch of the genitofemoral nerve, which supplies the cremaster muscle.

Coverings of the Spermatic Cord

The spermatic cord is covered by (1) the internal spermatic fascia (from the fascia transversalis), (2) the cremasteric fascia (from the internal oblique), and (3) the external spermatic fascia (from the external oblique).

PENIS

The penis has two functions: (1) it is the organ of copulation and serves to introduce spermatozoa into the vagina and (2) it is part of the urinary system and assists in conveying urine to the exterior. The penis has a cylindrical **body** that hangs free and a fixed **root** (*Fig. 13–3*). The **body of the penis** has an expanded distal end called the **glans penis**. The **prepuce** or **foreskin** is a hoodlike fold of skin that covers the glans.

The interior of the body of the penis is composed of three cylinders of erectile tissue enclosed in a tubular sheath of fascia. The erectile tissue is made up of two dorsally placed **corpora cavernosa** and a single **corpus spongiosum** applied to their ventral surface (*Fig. 13–3*). At its distal extremity, the corpus spongiosum expands to form the glans penis, which covers the distal ends of the corpora cavernosa. Running through the center of the corpus spongiosum is the penile part of the urethra that opens onto the surface of the glans penis at the **external urethral meatus**.

The **root of the penis** is located in the perineum and is made up of three masses of erectile tissue, which are called the **bulb of the penis** and the **right and left crura of the penis**. The bulb is situated in the midline. It is traversed by the urethra and is covered on its outer surface by the **bulbospongiosus muscles**. Each crus is attached to the side of the pubic arch and is covered on its outer surface by the **ischiocavernosus muscle**. The bulb of the penis is continuous anteriorly with the corpus spongiosum and the two crura are continuous anteriorly with the corpora cavernosa in the body of the penis.

The erectile tissue of the penis consists of a spongelike network of connective tissue and smooth muscle lined with endothelium and filled with blood. The blood enters the erectile tissue from branches of the internal iliac artery and is drained by veins that enter the internal pudendal vein. During sexual excitement the

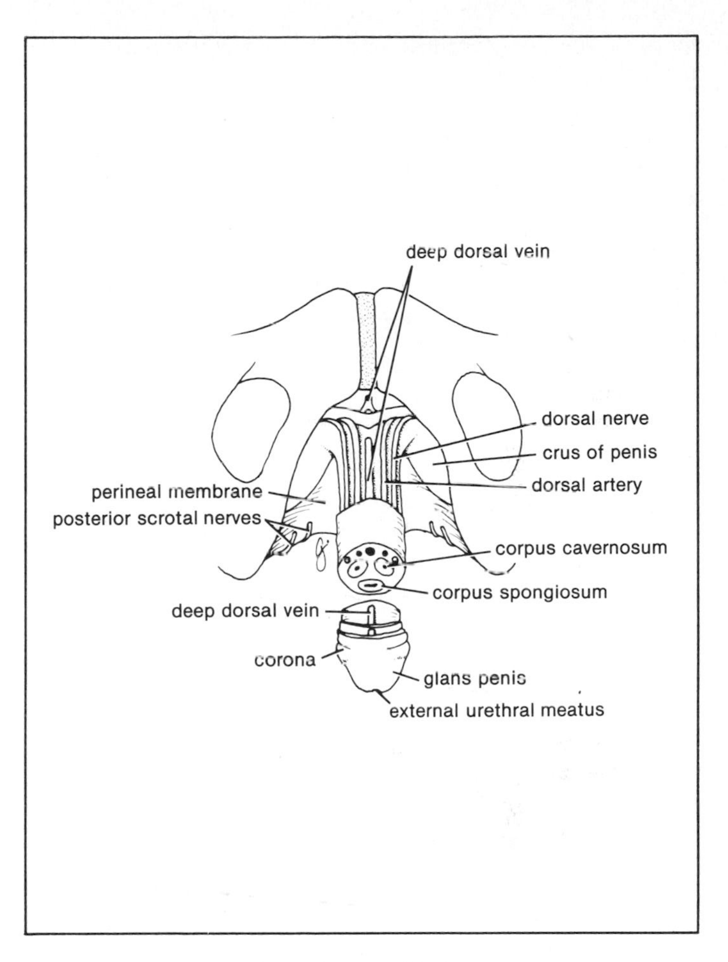

Figure 13–3. Root and body of penis.

blood flow into the erectile tissue increases resulting in the erection of the penis.

Blood Supply of the Penis. *Arteries.* Deep arteries of the penis and branches of the dorsal arteries of the penis. *Veins.* Deep dorsal vein.

Lymphatic Drainage of the Penis. Glans penis drains into the deep inguinal and external iliac nodes. The skin of the remainder of organ drains into the superficial inguinal nodes. The erectile tissue drains into the internal iliac nodes.

Nerve Supply of the Penis. Pudendal nerve.

Erection of the Penis

Efferent nervous impulses pass down the spinal cord to the parasympathetic outflow in the second, third, and fourth sacral segments. The parasympathetic preganglionic fibers enter the inferior hypogastric plexuses and synapse on the postganglionic neurons. The postganglionic fibers join the internal pudendal arteries and are distributed along their branches, which enter the erectile tissue at the root of the penis. Vasodilation of the arteries now occurs, producing a great increase in blood flow through the blood spaces of the erectile tissue. The corpora cavernosa and corpus spongiosum become engorged with blood and expand, compressing their draining veins against the surrounding fascia. By this means, the outflow of blood from the erectile tissue is retarded so that the internal pressure is further accentuated and maintained. The penis thus increases in length and diameter, becomes firm, and assumes the erect position.

Ejaculation

A nervous discharge occurs along the sympathetic nerve fibers to the smooth muscle of the duct of the epididymis and the vas deferens on each side, the seminal vesicles, and the prostate. The smooth muscle contracts, and the spermatozoa, together with the secretions of the seminal vesicles and the prostate, are discharged into the prostatic urethra. The fluid now joins the secretions of the bulbourethral glands and is then ejaculated from the penile urethra as a result of the rhythmic contractions of the bulbospongiosus muscles, which compress the urethra.

14

Female Reproductive System

The female reproductive system consists of a pair of ovaries, a pair of uterine tubes, a uterus, a vagina, and the external genital organs (*Fig. 14–1*). The mammary glands or breasts may also be considered a part of the female reproductive system.

OVARIES

The ovaries are the organs responsible for the production of the female germ cells, the **ova**, and the female sex hormones, **estrogen** and **progesterone**. Each ovary is an almond-shaped organ measuring 4 × 2 cm (1½ by ¾ inches) and is located in the upper part of the pelvic cavity. It is attached to the back of the broad ligament by the **mesovarium**. Usually the ovary lies with its long axis vertical (*Fig. 14–2*), but it shares in any movement of the broad ligament and uterus. The ovary is suspended from the lateral wall of the pelvis by that part of the broad ligament that extends between the mesovarium and the lateral pelvic wall; it is known as the **suspensory ligament of the ovary** and contains the ovarian vessels and nerves. The ovarian vessels and nerves enter the ovary at the **hilum**. The ovaries are surrounded by a thin fibrous capsule, the **tunica albuginea**. The ovary has an outer **cor-**

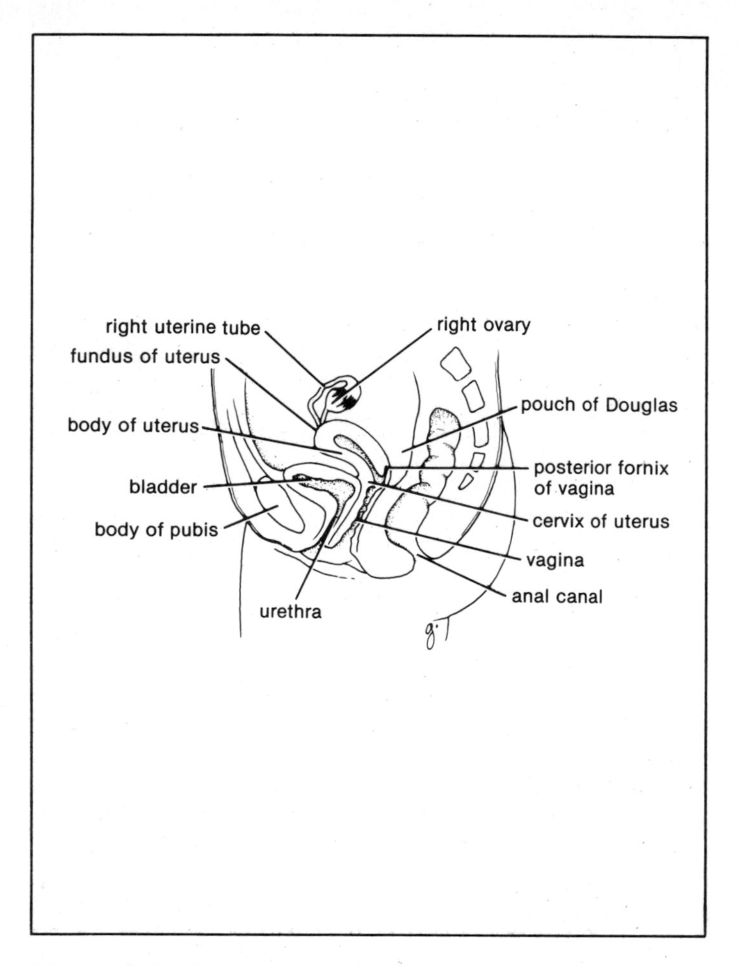

Figure 14–1. Female reproductive system as seen in sagittal section.

tex and an inner **medulla**, but the division between the two is ill defined. Embedded in the connective tissue of the cortex are the **ovarian follicles** in different stages of development and degeneration. The medulla consists of very vascular connective tissue.

The ovary usually lies against the lateral wall of the pelvis in a depression called the **ovarian fossa**, bounded by the external and internal iliac arteries.

Relations of the Ovaries. *Anteriorly.* Broad ligament. *Posteriorly.* Ureter and internal iliac vessels. *Laterally.* Obturator nerve; obturator internus muscle. *Medially.* Rectouterine pouch of Douglas.

During pregnancy, the enlarging uterus pulls on the broad ligament and the round ligament of the ovary, and raises the ovary up into the abdominal cavity.

Ligaments of the Ovaries

Suspensory Ligament. This is the lateral part of the broad ligament connecting the mesovarium to the lateral pelvic wall (*Fig. 14–2*). It contains the blood and the lymphatic vessels and nerves supplying the ovary.

Round Ligament of the Ovaries. This is the remains of the upper part of the gubernaculum and extends from the medial margin of the ovary to the lateral wall of the uterus (*Fig. 14–2*). (The round ligament of the uterus is the remains of the lower part of the gubernaculum.)

Blood Supply of the Ovaries. The **ovarian artery**, a branch of the abdominal aorta. The **ovarian vein** drains into the inferior vena cava on the right side and into the left renal vein on the left.

Lymphatic Drainage of the Ovaries. Paraaortic nodes at the level of the first lumbar vertebra.

Nerve Supply of the Ovaries. Aortic plexus; the branches accompany the ovarian artery.

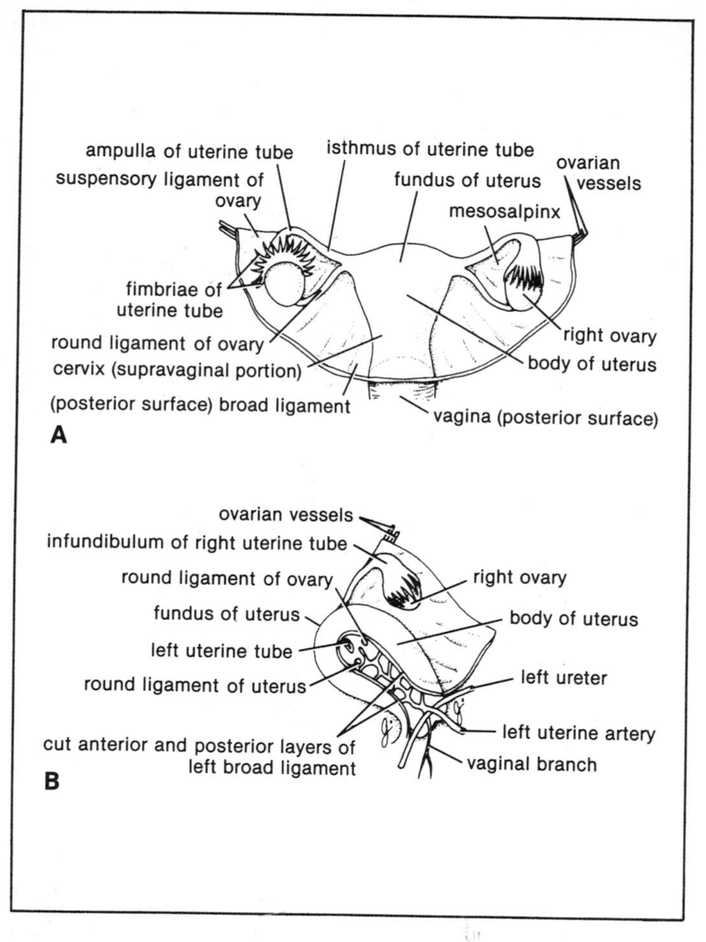

Figure 14–2. **A.** Posterior surface of the uterus and the broad ligaments, showing the position of the ovaries. **B.** Lateral view of uterus, showing attachment of broad ligament and the relationship between the left uterine artery and the left ureter.

UTERINE TUBES

The uterine tubes are two in number (*Fig. 14-1*). Each tube is about 10 cm (4 inches) long and lies in the upper border of the broad ligament. It connects the peritoneal cavity in the region of the ovary with the cavity of the uterus. It may be divided into four parts: the infundibulum, the ampulla, the isthmus, and the interstitial part (*Fig. 14-2*). The **infundibulum** is the funnel-shaped lateral extremity that projects beyond the broad ligament and overlies the ovary. The free edge of the funnel is broken up into a number of fingerlike processes, known as **fimbriae**, which are draped over the ovary (*Fig. 14-2*). The **ampulla** is the widest part of the tube. The **isthmus** is the narrowest part of the tube and lies just lateral to the uterus. The **intramural part** is the segment that pierces the uterine wall.

The uterine tube receives the ovum from the ovary, and provides a site where fertilization can occur. The secretions of the lining cells provide nourishment for the fertilized ovum and the actions of the cilia and the peristalsis of its walls transports the fertilized ovum to the cavity of the uterus. In addition the tube also provides a conduit along which the spermatozoa travel to reach the ovum.

Blood Supply of the Uterine Tubes. From the uterine and ovarian arteries. The veins correspond to the arteries. The uterine veins drain into the internal iliac veins and the ovarian veins ascend to the inferior vena cava on the right and the left renal vein on the left.

Lymphatic Drainage of the Uterine Tubes. The lymph vessels follow the arteries and drain into the internal iliac nodes and the paraaortic nodes.

Nerve Supply of the Uterine Tubes. Sympathetic and parasympathetic fibers from the superior and inferior hypogastric plexuses.

UTERUS

The uterus is located in the pelvis between the bladder and the rectum (*Fig. 14–1*). It is a hollow, pear-shaped organ with thick muscular walls and serves for the reception, retention, and nutrition of the fertilized ovum. In the young adult, who never had a child, it measures 8 cm (3 inches) long, 5 cm (2 inches) wide, and 2.5 cm (1 inch) thick. The uterus is divided up into the fundus, the body, and the cervix (*Fig. 14–2*).

The **fundus** is the part of the uterus that lies above the entrance of the uterine tubes. The **body** is the part of the uterus that lies below the entrance of the uterine tubes. It narrows below, where it becomes continuous with the **cervix**. The cervix pierces the anterior wall of the vagina and is divided into the **supravaginal** and **vaginal parts of the cervix**. The **cavity** of the uterine body is triangular in coronal section, but it is merely a cleft in the sagittal plane. The cavity of the cervix, the **cervical canal**, is spindle-shaped and communicates with the cavity of the body through the **internal os**, and with that of the vagina, through the **external os**.

In the majority of women, the long axis of the uterus is bent forward on the long axis of the vagina, forming an angle of 90 degrees (**antiversion of the uterus**). The long axis of the body of the uterus is bent forward on the long axis of the cervix, forming an angle of about 170 degrees (**anteflexion of the uterus**). In the erect positon, with the bladder empty, the uterus lies in an almost horizontal plane.

Relations of the Uterus. *Anteriorly.* Uterovesical pouch and superior surface of the bladder. The vaginal cervix is related to the anterior fornix of the vagina. *Posteriorly.* Rectouterine pouch of Douglas, coils of ileum, sigmoid colon. *Laterally.* Broad ligament, uterine artery and vein, uterine tube, round ligaments of ovary and uterus. The supravaginal cervix is related to the ureter (*Fig. 14–2*) and the vaginal cervix is related to the lateral fornix of the vagina.

Supports of the Uterus

The uterus is supported mainly by the tone of the levatores ani muscles and by three important ligaments.

Levatores Ani Muscles and Perineal Body. The levatores ani muscles form a broad muscular sheet stretching across the pelvic cavity, and together with the pelvic fascia on their upper surface, they effectively support the pelvic viscera. Some of the fibers of levator ani are inserted into a fibromuscular structure called the **perineal body**. This body lies in the perineum between the vagina and the anal canal. It is slung up to the pelvic walls by the levatores ani, and thus supports the vagina and indirectly supports the uterus.

Transverse Cervical, Pubocervical, and Sacrocervical Ligaments. These three ligaments are condensations of pelvic fascia on the upper surface of the levatores ani muscles. They are attached to the cervix and the vault of the vagina, and play an important part in supporting the uterus and keeping the cervix in its correct position. The **transverse cervical ligaments** pass to the cervix and the upper end of the vagina from the lateral walls of the pelvis. The **pubocervical ligaments** pass to the cervix from the posterior surface of the pubis. They are positioned on either side of the neck of the bladder, to which they give some support (**pubovesical ligaments**). The **sacrocervical ligaments** pass to the cervix and the upper end of the vagina from the lower end of the sacrum.

The **broad ligaments** are two-layered folds of peritoneum that extend across the pelvic cavity from the lateral margins of the uterus to the lateral pelvic walls (*Fig. 14-2)*. Superiorly, the two layers are continuous and form the upper free edge. Inferiorly, at the base of the ligament, the layers separate to cover the pelvic floor. The ovary is attached to the posterior layer by the **mesovarium**. The part of the broad ligament that lies lateral to the attachment of the mesovarium is sometimes referred to as the **suspensory ligament of the ovary**.

Each broad ligament contains the following:

1. The uterine tube in its upper free border.
2. The round ligament of the ovary and the round ligament of the uterus.
3. The uterine and ovarian blood vessels, lymphatics, and nerves.

Clinically the broad ligaments are considered to play a very minor role in supporting the uterus.

The **round ligament of the uterus** extends between the superolateral angle of the uterus, through the inguinal canal, in the anterior abdominal wall, to the subcutaneous tissue of the labium majus. It helps to keep the uterus anteverted (tilted forward) and anteflexed (bent forward), but it is considerably stretched during pregnancy.

Blood Supply of the Uterus. **Uterine artery** from the internal iliac artery and also the **ovarian artery**. The **veins** correspond to the arteries.

Lymphatic Drainage of the Uterus. From the fundus they follow the ovarian artery to the paraaortic nodes at the level of the first lumbar vertebra. From the body and cervix they drain into the internal and external iliac nodes. A few vessels pass through the inguinal canal to the superficial inguinal nodes.

Nerve Supply of the Uterus. Sympathetic and parasympathetic nerve fibers from the inferior hypogastric plexus.

VAGINA

The vagina is the female organ for sexual intercourse. It serves as the excretory duct for the menstrual flow from the uterus, and forms part of the birth canal. The vagina is a muscular tube lined with mucous membrane that extends upward and backward from the vulva to the uterus; it measures about 8 cm (3 inches) long. The vaginal orifice possesses a thin mucosal fold called the **hymen** that is perforated at its center. The vagina has anterior and posterior walls, which are normally in appositon. The area of the vaginal lumen that surrounds the cervix of the uterus is divided into four regions or **fornices**: anterior, posterior, right lateral, and left lateral. The upper half of the vagina lies within the pelvis between the bladder anteriorly and the rectum posteriorly; the lower half lies within the perineum between the urethra anteriorly and the anal canal posteriorly (*Fig. 14–1*).

Relations of the Vagina. *Anteriorly.* Bladder above, urethra below. *Posteriorly.* Rectouterine pouch of Douglas, ampulla of rectum, perineal body. *Laterally.* Ureter (upper part of vagina), levator ani, urogenital diaphragm, bulb of vestibule.

Supports of the Vagina. The upper part of the vagina is supported by the levatores ani muscles and the transverse cervical, pubocervical, and sacrocervical ligaments. The middle part of the vagina is supported by the urogenital diaphragm. The lower part of the vagina, especially the posterior wall, is supported by the perineal body.

Blood Supply of the Vagina. **Vaginal artery**, a branch of the internal iliac artery, **vaginal branch of uterine artery**. **Vaginal veins** drain into the internal iliac veins.

Lymphatic Drainage of the Vagina. The upper third drains into the internal and external iliac nodes, the middle third into the internal iliac nodes, and the lower third into the superficial inguinal nodes.

Nerve Supply of the Vagina. Inferior hypogastric plexuses.

VULVA

The vulva is the name applied to the female external genitalia and includes the mons pubis, the labia majora, the labia minora, the clitoris, and the greater vestibular glands.

Mons pubis is a rounded, hair-bearing elevation of skin found in front of the pubis. **Labia majora** are prominent folds of skin extending posteriorly from the mons pubis to unite in the midline. They contain adipose tissue and are covered with hair on their outer surfaces. The labia majora are equivalent to the scrotum in the male. **Labia minora** are two smaller folds of soft skin devoid of hair that lie between the labia majora. Their posterior ends are united to form a sharp fold, the **fourchette**. Anteriorly, they split to enclose the clitoris, forming an anterior **prepuce** and a posterior **frenulum**.

Vestibule

This is a smooth triangular area bounded laterally by the labia minora, with the clitoris at its apex and the fourchette at its base. It is perforated by the urethra (the orifice lies immediately behind the clitoris) and the vagina.

Vaginal Orifice. This is protected by a thin mucosal fold called the **hymen**, which is perforated at its center to allow the menstrual flow to escape from the vagina. After childbirth only a few tags of the hymen remain.

Orifices of the Ducts of the Greater Vestibular Glands. These are small orifices, one on each side, in the groove between the hymen and the posterior part of the labium minus.

Clitoris. The clitoris, which corresponds to the penis in the male, is situated at the apex of the vestibule anteriorly. It has a structure similar to the penis, and the **glans** of the clitoris is partly hidden by the **prepuce**. The **root of the clitoris** is made up of three masses of erectile tissue, which are called the bulb of the vestibule and the right and left crura of the clitoris. The **bulb of the vestibule** corresponds to the bulb of the penis, but because of the presence of the vagina, it is divided into two halves. It is attached to the undersurface of the urogenital diaphragm and is covered by the **bulbospongiousus muscles**. Anteriorly, the two halves unite to form the glans clitoris. The **crura of the clitoris** correspond to the crura of the penis. They are covered by the **ischiocavernosus muscles**.

Paraurethral Glands

The paraurethral glands, which correspond to the prostate in the male, open into the vestibule by small ducts on either side of the urethral orifice.

Greater Vestibular Glands

The greater vestibular glands are a pair of small mucus-secreting glands that lie under cover of the posterior parts of the bulb of the vestibule and the labia majora. Each drains its secretion into the vestibule by a small duct, which opens into the groove between the hymen and the posterior part of the labium minus.

MAMMARY GLANDS

The mammary glands are specialized accessory glands of the skin that are capable of secreting milk (*Fig. 14–3*). They are present in both sexes. In the male and the immature female they are similar in structure. The **nipples** are small and surrounded by a colored area of skin called the **areola**. The breast tissue consists of little more than a system of ducts embedded in connective tissue and they do not extend beyond the margin of the aerola.

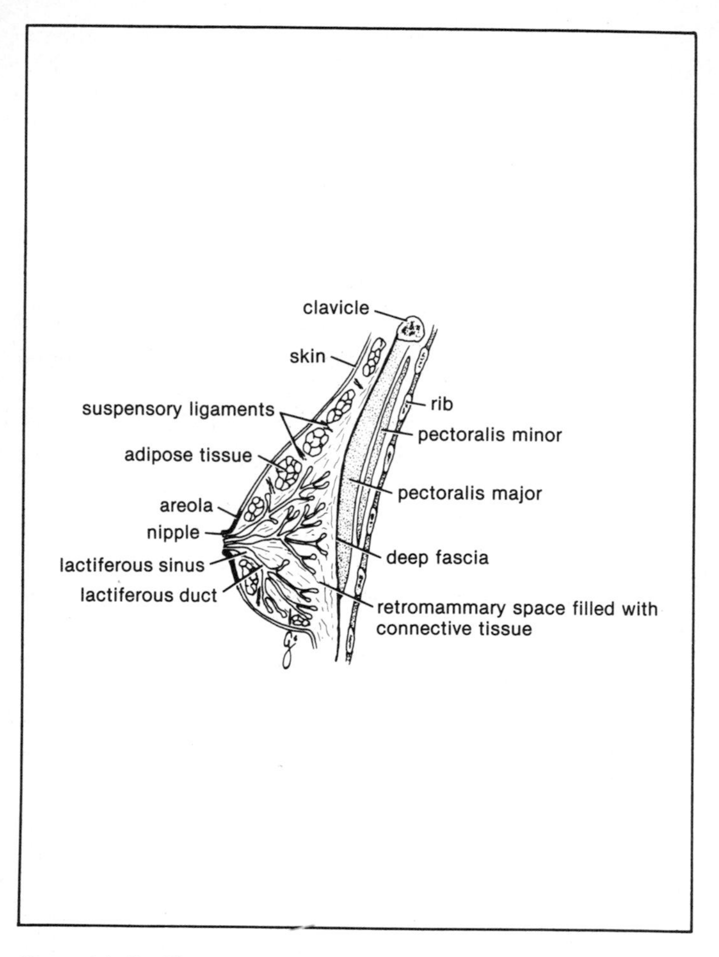

Figure 14-3. Mature mammary gland seen in vertical section. Shows the general structure.

At puberty, in the female under the influence of the ovarian hormones, the mammary glands gradually enlarge and assume their hemispherical shape (*Fig. 14–3*). The ducts elongate, but the increased size of the glands is mainly due to the deposition of adipose tissue. The base of the breast extends from the second to the sixth rib and from the lateral margin of the sternum to the midaxillary line. The greater part of the gland lies in the superficial fascia. A small part, called the **axillary tail** extends upward and laterally, pierces the deep fascia at the lower border of the pectoralis major muscle and enters the axilla. The mammary glands are separated from the deep fascia covering the underlying muscle by an area of loose areolar tissue known as the **retromammary space**.

In young women, the breasts tend to protrude forward from a circular base; in older women they tend to be pendulous. Each gland consists of 15 to 20 **lobes** that radiate out from the nipple. There is no capsule. Each lobe is separated from its neighbor by connective tissue septa containing adipose tissue that, in the upper part of the gland, are well developed and extend from the skin to the deep fascia and serve as **suspensory ligaments**. The main lactiferous duct from each lobe opens separately on the summit of the nipple and possesses a dilated **ampulla or lactiferous sinus** just prior to its termination (*Fig. 14–3*).

The **nipple** is a conical projection from the anterior surface of the gland and is traversed by 15 to 20 lactiferous ducts that open by small orifices on its tip. The skin of the nipple is pink or brown in color. There are numerous circular and longitudinally-arranged smooth muscle fibers in the connective tissue of the nipple that cause the nipple to become erect when it is mechanically stimulated. The areola is an area of pigmented skin that surrounds the base of the nipple. There are numerous sebaceous and sweat glands in the areola.

Blood Supply of the Mammary Glands. *Arteries.* **Lateral thoracic and thoracoacromial arteries**, branches of the axillary artery. **Perforating branches** of the internal thoracic and intercostal arteries. *Veins.* These correspond to the arteries.

Lymphatic Drainage of the Mammary Glands. Lateral part of the gland drains into the anterior axillary or pectoral nodes; medial part of the gland drains into the internal thoracic nodes; a few lymph vessels drain posteriorly into the posterior intercostal nodes whereas some communicate with the lymphatic vessels of the opposite breast and with those of the anterior abdominal wall.

15

Endocrine System

The endocrine system and the autonomic nervous system work closely together to regulate the metabolic activities of the different organs and tissues of the body so as to maintain homeostasis. The hypothalamus should be regarded as the higher nervous center for the control and integration of the activities of these two systems.

The endocrine system is made up of several glands: the pituitary gland (hypophysis cerebri), the pineal gland (body), the thyroid gland, the parathyroid glands, the suprarenal glands (adrenals), the islets of Langerhans of the pancreas, the gonads, and when present the placenta. In addition, there are groups of cells that form a small but nevertheless important part of the system and are not included in this chapter; they include the gastroenteroendocrine cells, certain thymic cells, certain kidney cells, and endocrine cells of the lungs.

The endocrine glands may be distinguished from the exocrine glands in that they do not possess ducts but pour their secretions (hormones) directly into the bloodstream.

PITUITARY GLAND (HYPOPHYSIS CEREBRI)

The pituitary gland is a small oval structure measuring about 1.25 cm (½ inch) in diameter (*Fig. 15–1*). It is attached to the tuber

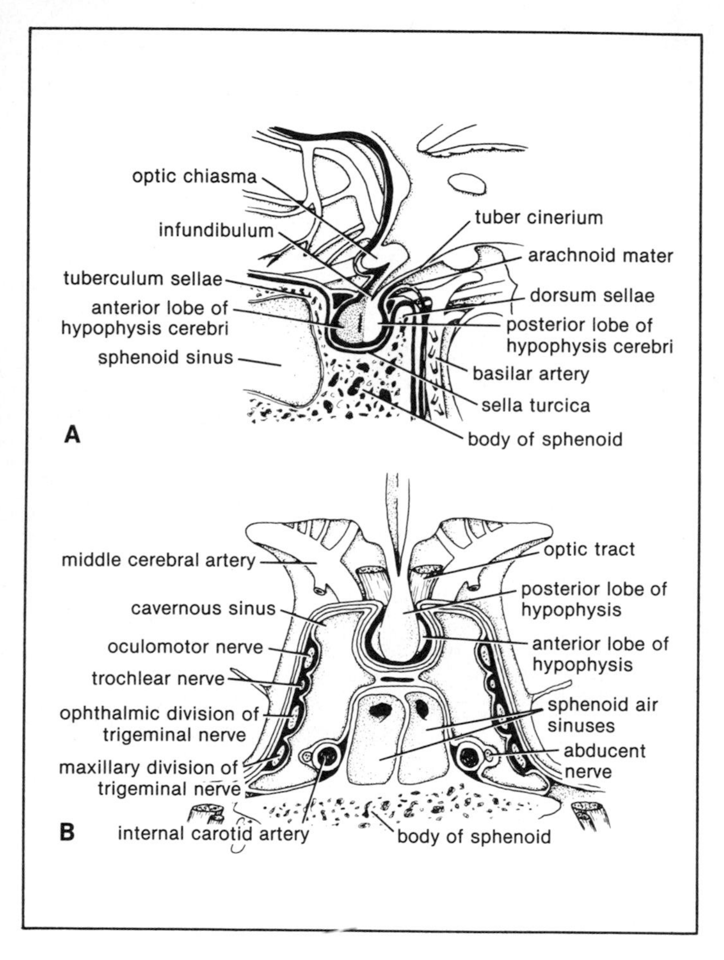

Figure 15–1. **A.** Sagittal section of the middle cranial fossa showing the hypophysis cerebri. **B.** Coronal section through the body of the sphenoid, showing the hypophysis cerebri and the cavernous sinuses.

cinereum on the undersurface of the brain by a stalklike structure, the **infundibulum**. The pituitary gland is well protected, being situated in the hypophyseal fossa, which is the deepest part of the sella turcica of the sphenoid bone.

The pituitary gland may be divided into an **anterior lobe or adenohypophysis** and **posterior lobe or neurohypophysis**. The anterior lobe may be subdivided into the **pars anterior** and the **pars intermedia** by a cleft that is a remnant of an embryonic pouch. A projection from the pars anterior, the **pars tuberalis** extends up along the anterior and lateral surfaces of the pituitary stalk.

Relations of the Pituitary Gland. *Anteriorly.* Sphenoid sinus. *Posteriorly.* Dorsum sellae, basilar artery, pons. *Superiorly.* Diaphragma sellae, optic chiasma. *Inferiorly.* Body of sphenoid, sphenoid sinus. *Laterally.* Cavernous sinus and its contents (internal carotid artery, abducent, oculomotor, and trochlear nerves, and ophthalmic and maxillary divisions of the trigeminal nerve) (*Fig. 15-1*).

Blood Supply of the Pituitary Gland. *Arteries.* Superior and inferior hypophyseal arteries, branches of internal carotid artery. *Veins.* These drain into the intercavernous sinuses. Note the importance of the **hypophyseal portal system** that extends from the median eminence to the anterior lobe of the pituitary and carries releasing hormones and release-inhibiting hormones.

Hypothalamohypophyseal Tract

This extends from the supraoptic and paraventricular nuclei of the hypothalamus into the posterior lobe of the pituitary. The hormones vasopressin and oxytocin are released at the axon terminals in the posterior lobe of the pituitary.

PINEAL GLAND

The pineal gland (or body) is a small cone-shaped body that projects posteriorly from the posterior end of the roof of the third ventricle of the brain. A small recess of the ventricle, called the **pineal recess**, extends into the base of the stalk. The pineal gland commonly becomes calcified in middle age and it may be visualized in radiographs. The pineal consists essentially of groups of cells **pinealocytes,** supported by glial cells. The gland has a rich blood supply and is innervated by postganglionic sympathetic nerve fibers. Although the function of the pineal gland is not fully understood, it has been shown to contain **melatonin, serotonin,** and **norepinephrine**.

THYROID GLAND

The thyroid gland is situated in the neck and is bound down to the larynx and the trachea by deep fascia. It consists of **right and left lobes** connected by a **narrow isthmus** (*Fig. 15–2*). The gland has a fibrous capsule. Each lobe of the gland is pear-shaped, with its apex being directed upward along the lateral side of the thyroid cartilage; its base lies below alongside the trachea. The isthmus extends across the midline in front of the second, third, and fourth tracheal rings. A **pyramidal lobe** is often present, and it projects upward from the isthmus. A muscular band, the **levator glandulae thyroideae** often connects the pyramidal lobe to the hyoid bone.

The thyroid gland is a very vascular organ, and is surrounded by a sheath formed of the pretracheal layer of deep fascia. The sheath attaches the gland to the larynx and the trachea.

Relations of the Lobes of the Thyroid Gland. *Anterolaterally.* Sternothyroid, superior belly of omohyoid, sternohyoid, anterior border of sternocleidomastoid. *Posterolaterally.* Common carotid artery, internal jugular vein, vagus nerve, deep cervical lymph nodes. *Medially.* Larynx, trachea, pharynx, esophagus, external laryngeal nerve, recurrent laryngeal nerve. *Posteriorly.* Superior and inferior parathyroid glands.

Relations of the Thyroid Isthmus. *Anteriorly.* Sternothyroids, sternohyoids, anterior jugular veins, fascia and skin. *Posteriorly.* Second, third, fourth tracheal rings.

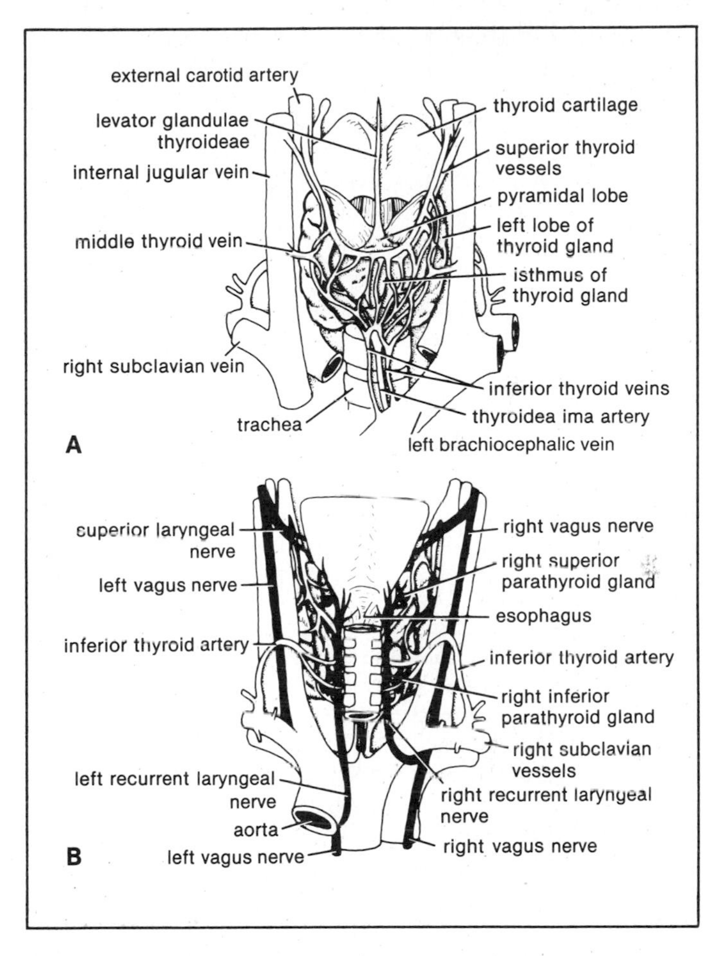

Figure 15–2. **A**. Thyroid gland, anterior view. **B**. Thyroid gland, posterior view; note the relationship between the recurrent laryngeal nerves and the inferior thyroid arteries.

Blood Supply of the Thyroid Gland. *Arteries.* Superior thyroid (related to external laryngeal nerve) from external carotid, inferior thyroid (related to recurrent laryngeal nerve) from thyrocervical trunk, thyroidea ima (if present) from brachiocephalic or aortic arch. *Veins.* Superior thyroid vein drains into the internal jugular vein, middle thyroid vein drains into the internal jugular vein, inferior thyroid vein drains into left brachiocephalic vein.

Lymphatic Drainage of the Thyroid Gland. Deep cervical and paratracheal lymph nodes.

Nerve Supply of the Thyroid Gland. Superior, middle, and inferior cervical sympathetic ganglia.

PARATHYROID GLANDS

The parathyroid glands are four small ovoid structures that are closely related to the posterior border of the thyroid gland in the neck (*Fig. 15–2*). They lie within the fascial capsule of the thyroid gland. The two **superior parathyroid glands** are the most constant in position and lie at the level of the middle of the posterior border of the thyroid gland. The two **inferior parathyroid glands** usually lie close to the inferior poles of the thyroid gland or sometimes they are found some distance from the thyroid gland outside the fascial capsule and may be located in the superior mediastinum in the thorax.

Blood Supply of the Parathyroid Glands. *Arteries.* Superior and inferior thyroid arteries. *Veins.* Superior, middle and inferior thyroid veins.

Lymphatic Drainage of the Parathyroid Glands. Deep cervical and paratracheal lymph nodes.

Nerve Supply of the Parathyroid Glands. Superior or middle cervical sympathetic ganglia.

SUPRARENAL GLANDS

The suprarenal (adrenal) glands are two in number and are situated close to the upper poles of the kidneys on the posterior abdominal wall (*see Fig. 12–1*). They are retroperitoneal and surrounded by perirenal fat; they are separated from the kidneys by the renal fascia. Each gland has a yellow-colored cortex and a dark brown medulla.

The right gland is pyramidal in shape and caps the right kidney. The left gland is crescentic in shape and extends along the medial border of the left kidney.

Relations of the Suprarenal Glands

Right Gland. *Anteriorly.* Right lobe of liver, inferior vena cava. *Posteriorly.* Diaphragm.

Left Gland. *Anteriorly.* Stomach, lesser sac of peritoneum, pancreas. *Posteriorly.* Diaphragm.

Blood Supply of the Suprarenal Glands. *Arteries.* Suprarenal branch of inferior phrenic, suprarenal branch of aorta, suprarenal branch of renal artery. *Veins.* A single vein on each side. Right suprarenal vein drains into inferior vena cava, left suprarenal vein drains into left renal vein.

Lymphatic Drainage of the Suprarenal Glands. Paraaortic nodes.

Nerve Supply of the Suprarenal Glands. Numerous preganglionic sympathetic nerves from splanchnic nerves. The majority of the fibers end on the cells in the suprarenal medulla.

ISLETS OF LANGERHANS OF PANCREAS

The pancreas is a soft, lobulated organ that lies on the posterior abdominal wall behind the peritoneum (*see p. 406*). The pancreas is both an exocrine and an endocrine gland. The greater part of the gland produces the exocrine secretion that passes into the duodenum. The endocrine part of the gland is formed of clusters of cells called the **islets of Langerhans**, which are scattered among the exocrine acini.

INTERSTITIAL CELLS OF TESTES (LEYDIG CELLS)

The testes are paired ovoid organs situated within the scrotum (*see p. 431*). Each testis is both an exocrine and an endocrine gland. The greater part of each gland is made up of seminiferous tubules whose function is to produce spermatozoa. The spermatozoa constitute the exocrine secretion that passes via ducts into the urethra.

The endocrine part of each testis consists of groups of rounded interstitial cells (Leydig cells) embedded in loose connective tissue between the seminiferous tubules.

OVARIES

The mature ovaries are paired ovoid organs situated within the pelvis (*see p. 443*). Each ovary has an outer cortex and an inner medulla, but the division between the two is ill defined. Embedded in the connective tissue of the cortex are the **ovarian follicles** in different stages of development.

The ovarian hormones are produced by the theca interna cells, which are the cells found in the stroma of the ovary immediately outside the graafian follicle, and by the cells of the corpus lutem.

Index

Note: All arteries, muscles, nerves, tendons, and veins are listed under main headings of Artery(ies); Muscle(s); Nerve(s); Tendon(s); Vein(s). Italic *f* indicates figure; *t* indicates table; *n* indicates footnote.